ABOUT TELEVISION

Books by Martin Mayer

MARTIN MAYER

ABOUT **T**ELEVISION

HARPER & ROW, PUBLISHERS

NEW YORK
SAN FRANCISCO
EVANSTON
LONDON

1817

Portions of this book have appeared in somewhat different form in *Audience, Change, Commentary, Esquire, Fortune, Harper's Magazine,* and *TV Guide.*

The lines from T. S. Eliot's "Burnt Norton," in *Four Quartets,* copyright 1943 by T. S. Eliot, on page 404, are reprinted by permission of Harcourt Brace Jovanovich and Faber and Faber Ltd.

FIRST EDITION

STANDARD BOOK NUMBER: 06-012879-8

LIBRARY OF CONGRESS CATALOG CARD NUMBER: 70-181633

For Tom and Jim, before they leave home

Contents

Preface

1

I started work on *About Television* knowing—as I had not known at the start of any previous book—what I was going to put on the opening pages. The symbolism of the great tower on the flat tableland was overwhelmingly right—the conquest of reality by economics and technology, the statement of what urban society has done to rural peace, the Ripleyesque oddity of the sight. I went to North Dakota on a weekend between two weeks of travel for other purposes. It was the weekend of Bobby Kennedy's funeral. In the farm states, some of the television stations were cutting into the long ride of the funeral train to show local commercials. . . .

Then everything languished for a while: this was a hard subject to get a grip on. As Stephen Hearst of the BBC wrote recently in the *Times Literary Supplement,* "When you turn to the literature or the journalism of a medium which simultaneously engages the hearts and minds of so many of us, you find the first so frail an infant as to force us to postpone literary 'confirmation' for years to come, and the second, with a few notable exceptions, a trivial pursuit." Let us begin, said the preacher; but he didn't say where.

What is surprising about the triviality of the literature of television is the mismatch with the obvious importance of the subject. Here is a social machine that has affected the daily lives of ordinary people more profoundly than anything since the mass production of automobiles in the 1920s. The reach is unimaginable—hundreds of millions of human beings watching at the same time as Neil Armstrong sets a pressurized boot on the moon, or Pele puts a cleated shoe behind a soccer ball in Mexico City. The technology is a miracle of the age, the economic impact is considerable, the social and political influence is believed to be all but dominant. And the decision-making machinery of television is interesting as a thing in itself, elaborate, variously efficient, somehow linked to public policy—but nobody knows how.

These are, of course, hard cases, sure prey to easy answers and bad law. Television has an extra dimension, because it is an extension of perception that is itself perceived, and in that context we are inevitably victims of what Maxwell Smart might call the old blind-men-and-the-elephant trick. After *Today* devoted a whole program to Barbara Walters' interview with Richard Nixon, the telephone lines to the show filled up with compliments both from Presidential assistants delighted that the medium had caught the warm humanity of their boss and from political antagonists overjoyed that television had finally shown the world what a cold fish this calculating politician really is. Sometimes the camera does hold one point of view, arriving with the police at the 1967 riots or with the demonstrators at the 1968 conventions, but usually viewers can write their own scenarios, even when the program is theatre and tries to say something. "Archie Bunker," said a friend of ours in the country recently, "tells it like it is"; and Alf Garnett—the English original from whom Archie has been derived—was the scourge of the Labour Party in 1970.

We are happiest with simplicity and/or conspiracy. The light bulb goes on because I flick the switch; violent, vapid, vulgar, *vicious* television shows are scheduled because They want to manipulate Us. We are perplexed and sometimes angered by a power that so rarely takes the side of the good, the true, and the beautiful (at least in one's own country). "Media barons," growls an FCC Commissioner, regretting Runnymede. But the power of mass communications is the power to introduce; what happens after the introduction is determined by forces much stronger and longer-lasting than a necessarily evanescent broadcast. In the end it is the audience, not the broadcaster, that *uses* television. The interplay of user and provider is the institutional frame for all those pictures trans-

mitted to all those homes; to see them outside the institutional
frame is to lose any sense of their significance.

That is what went wrong with the consideration of television as
a "medium," a word which still carries, if only subconsciously, the
connotation of an old lady in gypsy costume through whom the
Ouija board speaks. What is magic here is technology and talent,
what is mysterious is the human time scale, which is only super-
ficially controlled by clocks or even by the diurnal round. Television
has changed the way we spend our time. While increasing the
strength of our belief that we see what others see, it has (like the
automobile) locked us up into individual units divided from our
fellows during activities that were once communal. By confusing
the remote and the familiar, television has changed the range and
nature of the objects toward which emotion is directed.

These years may be the watershed time for television. Serious
people have stopped scorning it, and some have even outgrown
their fascination with the alleged good it could do or the alleged
harm it does. Political discussion has not yet caught up with the
change of attitudes—federal commissioners, lawyers, judges, government officials, Congressmen, foundation executives and media
freaks are still behaving as though the institutional structure had a
social purpose separate from the production of programs for people
to watch, as though the end result were rapping on the Ouija board
rather than the experience of artifice and artifact. But the consumers
of television, less dazzled now, are I think ready to look at it
straight-on.

There is a history here. Just as the typewriter keyboard was fixed
when the machine was new and slow-moving (and the first necessity
was to separate the angles of approach of letters likely to be struck
one right after the other), the organization of television broadcasting was essentially set by the earlier needs of radio broadcasting.
There is an industry here, too, capital and labor to be rewarded
through an astonishing marketplace where nothing is real but the
money or can be counted on to stay the same from day to day.
Great piles of computer print-out document the fickleness of taste;

the game is fast and talent for it is scarce; losers keep losing, but winners don't necessarily keep winning.

There is experience elsewhere to be compared against our own. If you want to look at these big subjects, you have to find some places to stand that are not in the middle of what you are trying to examine. Here and there throughout this book, then, are comparisons of American practice with what I found in Europe, especially in Britain and France, but also in Germany, Austria and Italy. In general, the aim has been to specify similarity and difference rather than to find better and worse; but sometimes one cannot avoid better and worse.

And then, as the saying goes, there is the future to consider. Coaxial cable is unrolling across the country, delivering programs a new way; shady investment bankers are offering letter stock in TV cassette operations; America is experimenting with a full-size "public" television service financed from tax revenues, while across the Atlantic and in Canada governments are trying to push more of the cost of television onto advertisers. The nice thing about the future is that it is soft—by definition, anything is possible. But lots of things are very unlikely, including all that green believed to be in the adjacent pasture just beyond the fence of time. In the last sections, then, *About Television* offers information—and, oh, just a soupçon of opinion—about probabilities for the future.

2

As I found when I went into the files, I have been nosing around television for a long time, inspired by magazine editors. In 1956 Jack Fischer of *Harper's* tossed me into the still young, pullulating world of television production with an assignment for a two-part piece on the making of programs; three years later, he asked me to view and criticize, ten thousand words' worth, everything that was supposed to be worth seeing in the 1959–60 season. Among the others who were prescient or supportive of this book in later years were Ralph Ginzburg and Harold Hayes on *Esquire*, who wanted (respectively) information about Milton Berle and Lou

Cowan; Charles Ramond of the Advertising Research Foundation, who wanted a pamphlet for laymen on broadcast ratings; Robert Kotlowitz on *Show*, who commissioned profiles of the networks for the first year of that ill-fated venture; Roger Youman and Merrill Panitt of *TV Guide*, who asked for articles on a large variety of subjects over a number of years; Louis Banks and Robert Lubar of *Fortune*, who meshed their plans with mine for a piece on the selling of network time to advertisers (and let me call it "How Will Television Feel After It Gives Up Smoking?" despite a house rule against titles that ask questions); Otto Friedrich of the *Saturday Evening Post*, who sent me to California to look at STV; and Arthur Singer and Stephen White of the Sloan Foundation, who commissioned a report on the possibilities of cable television for the arts, for the use of their Commission on Cable Communications.

Thanks to them all, and also to three ladies at Harper & Row—Gene Young, who pushed this project through the contract stage and stayed with it until her boss allowed others to offer greater opportunities; Marguerite Munson Glynn, who came out of retirement to work on this manuscript as she has on most of my books over the years; and Ann Harris, whose calmly affectionate perception and brute hard work greatly helped the grinding down and polishing of what came to her as a rough diamond indeed.

I have had help from hundreds of people at networks and broadcasting companies, program producers, research services, advertising agencies and government bureaus, and I can scarcely thank them all, though I am grateful to all. With one or two minor exceptions, I was able to see everyone I asked to see; the level of cooperation has been remarkably high. What is missing from these pages that should be here can be blamed on me alone—either I wasn't smart enough to know I needed it, or I wrongly sacrificed it myself in the savage cutting that was the last stage of the work on this book.

I would be remiss, however, if I did not acknowledge especially generous assistance I have received at one time or another over the last sixteen years from a number of people who work in this field. At random, then, extra thanks to David Adams, Pat Weaver, Robert Kintner, Reuven Frank and Herb Schlosser of NBC; Richard Jencks, Lou Cowan, Fred Friendly, Richard Salant, Bill Leonard, Perry

Wolff, Jack Cowden and Charles Steinberg of CBS; Ellis Moore, Garrett Blowers, Mari Yanofsky, Ell Henry, Av Westin, Bill Brademan, Roone Arledge and James Hagerty of ABC; Donald McGannon, Bob Schmidt and Martin Umansky of the station ownership world; Roy Danish of the TIO; Kenneth Cox, Lee Loevinger, and William Ray of the FCC; Joe Dine, Fritz Jacobi, Jay Levine, Bob Myhrum, John Macy and Richard Moore from public television; Peter Langhoff, Warren Cordell, Henry Rahmel, Jay Eliasberg, Julius Barnathan, Thomas Coffin and Bill Simmons from the world of research; Sandy McKee, Murray Chercover, Bud Garrett and Gordon Keeble in Canada; Peter Saynor, Barney Keelan, David Attenborough and John Rothwell in Britain; Franz-Josef Wild in Munich; Gerd Bacher in Vienna; Pierre Schaeffer, Mme. Claude Mercier and Mme. Jacqueline Baudrier in Paris—but, really, the list could go on too long, and still be nowhere near complete. The pleasure of this work is in the education, and one is always dependent on others for education. As they used to say in the news business before everybody got so preternaturally solemn, you meet such interesting people.

For those who have the fortitude to read further in this subject, I can recommend five books, in order of publication: Richard Hoggart's *The Uses of Literacy* (which is not supposed to be about television but is), Wilbur Schramm's *Television in the Lives of Our Children* (with Jack Lyle and Edwin B. Parker), Gary A. Steiner's *The People Look at Television*, William B. Stephenson's *The Play Theory of Mass Communications* and William Belson's *The Impact of Television*. The annual handbooks of BBC and ITA in England, and of NHK in Japan (available in English translation), are also worth perusing. The indispensable vade mecum of American television is *Broadcasting*, a trade weekly of unusually high quality.

As always, I am grateful for the tolerance of my wife and our boys, who had to put up with a great deal during the last months of the work on this book, and for the patience of all those who wrote me letters that didn't get opened (let alone answered) in the last months of 1971.

—Martin Mayer

New York
February 1972

CHAPTER 1

Towers and Other Landmarks

I believe television is going to be the test of the modern world, and that in this new opportunity to see beyond the range of our vision we shall discover either a new and unbearable disturbance of the general peace or a saving radiance in the sky. We shall stand or fall by television—of that I am quite sure.
> —E. B. WHITE, 1938

It is a most intriguing fact in the intellectual history of social research that the choice was made to study the mass media as agents of persuasion rather than agents of entertainment.
> —ELIHU KATZ and DAVID FOULKES,
> social scientists, 1962

And under all this vast illusion of the cosmopolitan planet, with its empires and its Reuter's agency, the real life of man goes on concerned with this tree or that temple, with this harvest or that drinking song, totally uncomprehended, totally untouched. And it watches from its splendid parochialism, possibly with a smile of amusement, motor-car civilisation going its triumphant way, outstripping time, consuming space, seeing all and seeing nothing, roaring on at last to the capture of the solar system, only to find the sun cockney and the stars suburban.
> —G. K. CHESTERTON, 1905

1

On a farm in Blanchard, Traill County, North Dakota, stands a tricornered latticework of steel beams painted red and white, rising straight up for two-fifths of a mile—the tallest man-made structure in the world, the transmission tower of station KTHI-TV, Channel 11, Grand Forks–Fargo. The station used to be identified by other call letters, but changed its name in 1963, while the tower was under construction, to permit the use of a new promotion symbol, a long-legged girl labeled "Katy High." Management likes to say that if you put the Eiffel Tower on top of the Great Pyramid at Gizeh, and the Washington Monument on top of *them*, the tip of the monument would still be below the KTHI antenna. "If a twenty-second commercial started at the same moment a baseball was dropped from the top of the KTHI tower, it would have ended nearly four seconds before the ball hit the ground." At some seasons, the ball might hit a Brown Swiss cow, because such cows—property of a Farmer Brown, H. Kenneth Brown, who owns the land—graze around the two-story cinder-block building at the foot of the tower.

At the top, the visual horizon is sixty miles away, and full power input of 304,000 watts will generate usable video signals on high-quality rooftop home antennae 105 miles off. "It's so flat here," said engineer David Chumley, "we have no shadow situation at all— you get good coverage over the whole area." This area is larger than the states of Massachusetts, New Jersey, Delaware, Connecticut and Rhode Island put together. "But the tower doesn't look like much, you know," said William P. Dix, formerly the station's general manager, a veteran of broadcasting wars in New York, "because there's nothing you can compare it to." That's almost true, but not quite: only a few thousand yards away stands a very slightly smaller stick, phallus for a rival station that also aims to reach Fargo and Moorhead, Grand Forks and Valley City, Thief River Falls and Devils Lake, and all the farms and crossroad towns around them.

Even in 1963 the cost of raising such a tower was more than three-quarters of a million dollars. Much more than a steel frame is involved. The cable carrying the power to the antenna must be protected against moisture by a blanket of nitrogen under pressure, and against freezing by a circulating antifreeze solution pumped all the way to the top. Forty percent of the power will be lost in the climb, not because electricity cares about up or down but simply because the cable is so long. Structurally, the tower is sustained by guy wires as thick as a man's thumb, twenty-seven of them, angling down from various heights to coffins of reinforced concrete sunk twenty-five feet under the surface of Farmer Brown's quarter section. Tower and guy wires add up to almost six hundred tons of steel.

Though carefully lit on fourteen levels, beyond FAA specifications, the tower is a hazard to tourist aviation, light planes with pilots who can't resist coming in for a closer look: they can see the tower all right, but not the guy wires. "Channel 6 went down," Dix recalled with true competitive relish, "on the first day of a ratings week, when a crop duster hit a guy wire. When the plane fell, it cut our power wire, and we were off for one minute before our generator came on. The hose broke, and there was antifreeze all over the floor."

At Farmer Brown's, a dirt road with deep mud holes led to the building at the base of the tower. The ground floor was a garage, with a small electrical shop and a huge Allis Chalmers emergency generator, plus tanks for the fuel to feed it, for the antifreeze, for the nitrogen. On the second floor was an ordinary windowless control room, temperature-controlled through noisy air-conditioning ducts, featuring the usual monitor screens, control panels, banks of gun-metal gray equipment studded with dials and meters. Beside the stairs was a little room with a window, offering a cot, a stove and a refrigerator. The engineer on duty was Larry Johnson, a matter-of-fact young veteran who had previously run an Army closed-circuit system in El Paso and lived 150 miles from Blanchard. His weekly tour was from 12:30 Wednesday afternoon to 9 o'clock Friday night—56½ hours a week, a schedule that permitted the facility to be staffed constantly for the price of three men. Nevertheless, a tower like KTHI isn't cheap to operate. The electricity alone must run $30,000 a year, and even in low-wage North Dakota the three engineers cost at least as much again. And there are special problems, too. "We have a leak in the nitrogen hose, up near the top," Johnson said, looking at the ceiling. "A GE man came out here to work on it, but he got hysterics when he was only four hundred feet up."

Most of the time, what comes out of this tower is ABC network programs, though KTHI has produced local news shows, *Romper*

Room for kiddies and a *Top Ten Dance Party* for bigger kiddies, at its studios in both Fargo and Grand Forks. The station also carried several hours a day of movies and syndicated shows (mostly former network programs, now peddled on a station-to-station basis). On Sunday afternoon, Dix used to go on camera himself, in a program called *Let Me Speak to the Manager,* answering questions in letters written by the audience. "It's our sixth-highest daytime rating," Dix reported, not without awe. One result of this program was a late-night movie on weekends, making KTHI the only station in the area to be on after midnight. "A lady wrote in asking why we signed off early," Dix recalled, "and I said on the air, 'If you want a late show, you can have it—write in.' A couple of days later we got a petition from the Air Force base, with 150 signatures, and the man who sent it wrote, 'I got these in one hour—if you need more, let me know.' So I went to our advertisers, and said, 'Do you want to *own* an audience?' We charge $175 for half the movie, which gives the advertiser four spots." If it's a local advertiser, KTHI will make the commercials for him, at a gross cost normally under $50 per reusable minute.

An investment like the tower, however, is not made for the sake of local storekeepers—indeed, by extending the station's range far beyond any one market area, the tower gives local advertisers an unusually high fraction of waste circulation. Fuqua Industries, the conglomerate that then owned KTHI, built the tallest structure in the world because in the early 1960s the advertising agencies worked on a rule of thumb that made 100,000 homes the minimum size for a major market that would be bought routinely in national advertising campaigns on television. And the only way you could get 100,000 homes in the reception area of a single transmitter in North Dakota was by hoisting it up more than two thousand feet. Two very tall towers stand in Traill County, involving a total investment of $1.5 million, though one could serve both channels (as the one atop New York's Empire State Building serves all that city's television stations), because broadcasters, the people, the Federal Communications Commission and the Department of Justice all believe that competition is part of the American way. That's why the world's two tallest structures are out in the back of beyond in North Dakota.

2

It was a long way, and not just geographically, from Farmer Brown's west quarter in Traill County to the Ground Floor, the restaurant in Black Rock, the extraordinary Eero Saarinen pillar of dark stone and dark glass that houses the offices of the Columbia Broadcasting System on Sixth Avenue in Manhattan. The Ground Floor was a world of artifice very pure and very simple. Its stunning brass, brown and black decor was squared off in every detail: the many chandeliers from the high ceiling were bright glass boxes, the chairs were a cubist's dream of straight edges. The restaurant started in 1965, as the building was being finished, as a toy for William S. Paley, who bought the infant Columbia Broadcasting System in 1928 and is now chairman of its board and a great gourmet. He had been growing restive because (though his decisions continued to be final when he made them) he couldn't really run the broadcasting company any more. It was decided that rather than lease the space in the new corporate headquarters to outside restaurateurs, CBS would go into the restaurant business itself. A maître d'hôtel and an assistant maître d'hôtel were hired for starters, and were asked to pick a name for the place *tout de suite* and get it to Mr. Paley. A groggy weekend later, they brought in a list of forty assorted French restaurant names not already in use in New York, and Paley brushed them all aside. "No, no," he said. "I'm meeting my friends for lunch in my building on the ground floor, and . . ." He looked around with wild surmise, and the restaurant had a name.

For more than a year after the restaurant opened, Paley himself, white-haired, gracious, the squire of landed acres, came down from the executive floor every day just before noon and wandered around the kitchen, tasting the soup and occasional other goodies. (Then he would return, of course, to his own executive dining room and his personal chef.) A CBS executive one day asked me to lunch with him and a friend for whom he thought I might be able to do a favor, and after we met at his office he trotted us off, somewhat to my surprise, to another elaborate feeding trough called the Four Seasons. Asked why he had neglected what was still an excellent restaurant in his own building, he said, "You don't dare eat there—it's too embarrassing. Paley comes down around two o'clock and comes over and asks you what you had for lunch, and then suggests you come up to his office and discuss the main course with him." Alas, this operation did not make money despite high prices, and CBS does not retain money-losing subsidiaries. The facilities were franchised out, the quality declined, and eventually the place became part of the Restaurant Associates chain as the Ground Floor Café, with leatherette chairs in the off colors that make you feel you've already been here a while and should let someone else have your table.

In better restaurants every weekday lunch, directors and writers and agents and stars and producers of movies, a few at a time most of the year but many at once in February, can be found picking over the bones of some unlucky exotic bird and putting on various moods of confidence or despair. This is their marketplace. A block north on Sixth Avenue, on what became in the 1960s one of New York's most dense promenades of office buildings, is the headquarters of the American Broadcasting Company, in a much less interesting but even newer dark brown skyscraper; two blocks south, in the building once called Radio City, the headquarters of the National Broadcasting Company. These are the networks; among them, simply to fill the three hours of "prime time" between eight and eleven every night, they need 3,285 hours of program each year. Even after all repeated programs and coverage of actual events are deducted, they need annually, just for the evenings, more than twice as much film as Hollywood ever produced for theatrical distribution in its biggest single year. The critics complain with varying degrees of savagery that most of it isn't very good.

About 85 percent of the time Americans spend watching television will be devoted to programs produced to the specifications of the networks and fed by them around the country, through facilities leased from AT&T, to the stations and their transmitting towers. In the winter months, during the three prime-time hours, more than three-fifths of all the homes in America will be watching television. They don't have to watch network programs—in 1971 three-quarters of American homes could receive at least five different television stations, only three of which would be hooked into, "affiliated with," a commercial network. But in fact more than 90 percent of those watching television—in winter, well over 50 percent of all American homes—will watch network programs during the prime-time hours. The decisions made in the three big office buildings along Sixth Avenue determine what most Americans pay attention to on most evenings of the year.

The responsibility is too great for anyone—or any three—to bear. It almost never happens that anyone in the three buildings thinks about what he might want to do with the attention of so many people; he feels instinctively, and almost certainly correctly, that if he tried such self-assertion he would lose their attention to someone more simpatico in one of the other buildings. Instead, the network program man tries to figure out what all those people would like to see, and to judge which of the film-makers soliciting his business in the restaurants and bars can provide the shows most satisfying to most people. Guessing lies at the heart of this work. Later, when decisions must be made about which of a number of "pilot" films is most likely to result in a series people will wish to watch every week, the network vice presidents will have audience research information to help them; but when they commission the pilots, they steer by dead reckoning.

Their starting point, of course, is the fact that communication is not for the human animal a merely utilitarian activity. "At its best," the maverick sociologist William Stephenson writes, "mass communication allows people to become absorbed in *subjective play*"; it gives "communications-pleasure," which "brings no material gain and serves no 'work' function, but it does induce certain elements of self-enchantment." Direct conversation, Stephenson adds, relying on the maverick psychiatrist Thomas Szasz, is the highest form of such pleasure: "Two people meet and converse [and] say afterwards how much they enjoyed it." Television gives this pleasure without effort: "The television set in the house is almost like another member of the family," says Perry Wolff, a serious, literary-minded man with thinning black hair and a nervous manner, who came to CBS after sitting at the feet of Gertrude Stein in Paris and has been producing public-affairs shows for that network for almost two decades. "In many ways it's a *closer* member of the family—it comforts us; it even gives us options."

The phenomenon is not strictly—even mostly—American. There are 25 million television receivers in the Soviet Union, and nearly as many in Japan; 17 million each in West Germany and in Britain, nearly 4 million in Poland, 3 million in Czechoslovakia, 2 million in Hungary and in Yugoslavia, almost a million in Egypt. When the Argentine government suspended television broadcasting in Buenos Aires to help tide the city over a chronic power shortage, the citizenry successfully petitioned to have the street lamps dimmed instead. Where people are dependent on cable for their television service (as is true in many parts of Canada), a break in the wire can put the telephone switchboards out of business as householders call the cable company in fury. If psychologists and economists could construct an index that measured satisfactions, they would find that for most of mankind in the industrialized countries the quantity of pleasure derived from television is greater than the quantity ascribable to any other facet of their lives. Perhaps this should not be so, but it is; and anyone who approaches this phenomenon with the notion that most people don't like what television offers them will never begin to understand the subject.

For the program designers in the Sixth Avenue office buildings, the measurement of pleasure is quantitative—the "share of audience" drawn by each network in each time period (it is axiomatic that the total size of the audience watching television in the evening will not be influenced in the least by the programs offered, though it can be influenced by extraneous factors—Alfred Hitchcock's movie *The Birds* once held the record for most audience watching a movie, because it was aired on the night of the biggest snowstorm the Eastern half of the country had seen in ten years). What makes the share so overwhelmingly important is its influence on what advertisers will do and thus on another set of numbers which appear annually on "the bottom line"—the profit-and-loss statement of the broadcasting company that owns the network.

3

It must be said that neither commercial sponsorship nor network domination was ever a necessary condition of American broadcasting, and neither was contemplated in the laws to license broadcasting stations. The Federal Radio Commission, which issued the licenses in the years before the Federal Communications Commission, noted in 1928 that "Such benefit as is derived by advertisers must be incidental and entirely secondary to the interest of the public." In 1929 the National Association of Broadcasters published a code insisting that "commercial announcements, as the term is generally understood, shall not be broadcast between seven and eleven P.M."

Even in the 1970s a noncommercial broadcasting system in the United States is not inconceivable, though it is awfully unlikely. Of the $2.8 billion in net revenues received by the television networks and stations in 1970, something just over $1.5 billion was spent for programs and their transmission. The audience for television is so enormous that a tax of $25 per television household per year—50¢ a week—would support the institution in the style to which it has become accustomed. Communications experts who work for charitable foundations seem to regard a fee of $60 a year as a moderate charge even against the income of poor people for the extra benefits that might or might not come from the introduction of cable television. On all colorations and intensities of the political spectrum, however, leaders seem agreed that we have better uses for $1.5 billion of tax money than the support of televised services that are now delivered free of direct charge.

Anyway, most people don't much mind commercials. In 1960 only 7 percent of Gary Steiner's large viewing sample mentioned commercials at all when asked, "How do you feel about television in general?"; even of those who said they were "negative" or "extremely negative" about television, only 14 percent mentioned commercials as a source of their displeasure. While two-fifths said they would prefer television without commercials, only a quarter said they would be willing to pay "a small amount yearly if I could have television without commercials"—and three-quarters agreed with the statement that "Commercials are a fair price to pay for the entertainment you get." More recently, in Canada, a third of all those queried for a government survey called *Mass Media* said they thought commercials were more interesting than programs.

In most European countries where television started as a tax-supported service, commercials are now being broadcast, though advertisers are not permitted to interrupt programs with their messages. As the costs of television programming have grown, governments have decided they would rather let the private sector pay them. In France, where the Gaullists broke with the intellectual community in ramming through approval for up to eight minutes a night of commercial messages on l'ORTF, the 8 o'clock commercial break has become one of the most popular time periods on television, especially with children, whose love for predictability and repetition and regularity is perfectly matched by television advertising. (Incidentally, the French government retained control over which products could be advertised on television, and among those still forbidden in 1971 was the automobile, presumably for fear that television advertising would help Volkswagen take business from state-owned Renault.) In Italy, the *"Carosèllo"* ("Carrousel"), the grouping of commercials which starts off each evening's entertainment, draws a bigger audience than any of the programs.

American broadcasting did not know commercials in its very earliest years. The first radio broadcasts, in 1920, were paid for by equipment manufacturers, Westinghouse and RCA, which could enter the costs of transmitters and programs as promotion expense for the sale of receivers. The first advertising on the air came in 1922, when the Queensborough Corporation, a real-estate firm, bought ten minutes on WEAF in New York, then operated by American Telephone & Telegraph. Though not many New Yorkers had radios, inquiries about the new apartments in Queens jumped dramatically, and presently department stores were inquiring about the availability of radio time for advertising purposes.

The equipment manufacturers—especially RCA under David

Sarnoff—resisted the idea of selling time, but they were even more violently opposed to the notion of a government-operated broadcasting system supported by tax revenues. In a statement that reads very strangely half a century after its utterance, Sarnoff told a Congressional committee in 1924 that "The air belongs to the people. Its main highways should be maintained for the main travel. To collect a tax from the radio audience would be a reversion to the days of toll roads and bridges; to the days when schools were not public or free and when public libraries were unknown." But broadcasting clearly could not be supported by the promotion budgets of the set-makers, and even if it could have been, there were strong arguments of public policy against turning over such a resource to so small a handful of men—especially in the 1920s, when RCA itself was controlled by a consortium of Westinghouse, GE and AT&T. If the system was not to be supported by taxes, advertising would have to pay the bills.

But "direct advertising," using radio to deliver sales pitches, was considered improper. Instead, advertisers sought to "sponsor" the programs people turned on their sets to hear. The sponsorship announcement in the first days of commercial radio was as chaste as the announcement now offered by noncommercial television that Mobil or Xerox put up the money that made the program possible. Gilbert Seldes reported "the total advertising, direct or indirect, [spoken on] a very popular and successful program" in 1925: "Tuesday evening means The Ever-Ready Hour, for it is on this day and at this time each week that the National Carbon Company, makers of Ever-Ready flashlights and radio batteries, engages the facilities of these fourteen radio stations to present its artists in original radio creations."

Commercially, the aim was to exploit a "gratitude factor"—people would buy the product to thank the sponsor for the show. Even in today's cynical world, incidentally, this can work—in 1970 Talman Federal Savings & Loan in Chicago reported the biggest gain in deposits of any bank in that city, and credited its success mostly to its sixty hours a week of classical music sponsorship on WFMT.

It will be noted that we are now en route to networking: Ever-

Ready had bought time on fourteen stations, and paid the telephone company to link them together, because no one station could deliver an audience large enough to justify the expenditure necessary to produce a star-studded musical program. For some reason, this development came as a surprise: the government had licensed stations to localities, and had assumed that each would remain a local operation, like a newspaper. "A broadcasting station," the Federal Radio Commission said as late as 1928, "may be regarded as a sort of mouth on the air for the community it serves." But the 1920s were the time when movies were beginning to rip apart the vaudeville business, and the superiority of national to local entertainment was beginning to seem obvious. Radio was destined to be a national medium like magazines and movies, not a local medium like newspapers. Moreover, there were public values to be gained: as early as 1923 a chain of six stations, connected by ordinary telephone wire, carried Calvin Coolidge's State of the Union Message live from the Capitol. And it was obviously uneconomic to assemble a special network for every advertiser or every event.

In 1926, as part of the deal that put AT&T out of the broadcasting business, RCA established the first permanent network, the National Broadcasting Company, connecting a flagship originating station in New York (WEAF, acquired from the Bell System) to "affiliates" all over the country. RCA guaranteed AT&T a minimum of a million dollars in line rentals every year for ten years. The service was launched (as far out as Kansas City) with a concert commanding the services of, among others, Walter Damrosch and Mary Garden. The coast-to-coast network came into being in the fall of 1927, just in time to broadcast an account of the Harvard-Yale game.

The broadcasting network was and is a program supplier to local stations, and in any rational economic organization of the system the local stations would pay for the service. The idea that networks should pay stations to broadcast their programs is on its face as ludicrous as the notion that wire services ought to pay newspapers for printing their copy. In its earliest years, in fact, NBC did charge each local station a flat fee of $90 an hour for those programs which went over the wire without sponsorship. But the local stations had been selling their time to advertiser-made *ad hoc* networks, and they regarded the permanent network as an advertiser—a customer, not a supplier. They objected bitterly to paying for programs, and in 1928 some of the biggest shifted their affiliation to CBS when that newly organized network offered sustaining (that is, unsponsored) programs for free in return for an option on any and all of a station's time for which the network could in fact make the sale. NBC in self-protection soon adopted the CBS plan, which gave network time salesmen the advantage of a guaranteed national coverage for a sponsor's show.

The notion that the network "bought" time on the station became so ingrained in broadcasting practice that the Federal Communications Commission established "sale of station time to networks" as a standard item of report required from all licensees. The stations themselves are, of course, independently owned; in the television band, the FCC allows a single company (which may be a network or just a corporate group like Metromedia or Westinghouse) to control up to five channels in different cities. Under today's FCC rules, network "affiliation contracts" require the network to offer the programs, but permit the stations to refuse them.

In theory, option time created a radio broadcasting system centrally controlled to a degree unprecedented in American industry. In every city, at a nod from New York, stations would have to carry the same program at the same time. In fact, the networks wound up with remarkably little control over radio programming. Just as stations regarded networks as the functional equivalent of advertisers that bought time, the advertisers considered networks as technical conveniences, offering interconnections at bargain prices and saving the cost and nuisance of arranging purchases from many different stations. Mostly, the advertisers kept for themselves (with help from talent agencies and production departments in advertising agencies) the prerogative of producing the programs they would sponsor.

What the advertiser bought from a radio network was a time period, usually half an hour, on a list of stations across the country. The cost was somewhat less than separate purchases on all the stations, and included the interconnections and the use of network studio facilities. (All programs were broadcast live: prerecorded material was deeply frowned on, even after records became technically better, and even locally. "The public in large cities," said the FRC sternly, "can easily purchase and use phonograph records of the ordinary commercial type." The first recorded material ever carried by the NBC Radio network was Herbert A. Morrison's famous description—"Oh, the humanity . . . Oh, this is terrible!"— of the explosion of the *Hindenburg*.) The network paid its affiliates about a third of their normal time rate for the period. The local stations in return received programming much more popular than anything they could do themselves, plus the right to sell "spot commercials"—from Bulova Watch Time to Plymouths—in "adjacencies" or "station breaks."

Most of the deep thinkers in advertising had always believed in regularity and repetition, and in radio the gratitude factor added another reason to seek frequency rather than reach; so advertisers took the same period every week, all year long (sometimes omitting the summer, when broadcast audiences were, and are, smaller). Having secured his time, the advertiser would go buy a program to put on it. In retrospect, radio programs seem remarkably cheap. "People forget," says Joe Iaricci, a vice president in NBC-TV network sales, who goes back to the radio days, "but advertisers weren't the sponsors of shows; *products* were. Pepsodent, not Bristol-Myers, sponsored Bob Hope; Jello, not General Foods, sponsored Jack Benny; Raleigh, not Brown & Williamson, sponsored Red Skelton."

Even when the program was officially produced by the network, which did happen, it was produced for the advertiser, who as buyer exercised ultimate control. The man from the sponsor who wouldn't let a comedian tell a joke about booze or girls became one of the archetypal figures of fun in the 1930s, much beloved of Fred Allen. And he still lives in television, in vestigial form: "One of the major buys on *The Courtship of Eddie's Father*," says Henry Miller, program administrator on the West Coast for ABC-TV, "is a cereal company, and sometimes the show has a breakfast scene. Of course, a cereal client doesn't want them eating bacon and eggs." The classic case was a *Playhouse 90* episode dealing with the Nuremberg Trials, sponsored in part by the natural-gas industry. "In going through the script," an agency man told a House committee, "we noticed gas referred to in half a dozen places that had to do with the death chambers. This was just an oversight on somebody's part."

By the early 1940s, with agency-produced soap operas dominating their daytime schedules and agency-produced comedy shows and spook dramas drawing the mass audience at night, the networks were trivial factors in determining what was put on the radio for the American people. Contrasting broadcast and print advertising, Neil Borden's classic *Advertising: Text and Cases* suggested in 1951 that "In constructing radio or television programs, the advertiser is, in a sense, his own editor, building his own audience appeal." The FCC, later to become very upset about network domination of program sources, praised in 1946 "the 'package' program, selected, written, casted and produced by the network or station itself, and sold to the advertiser as a ready-built package, with the time specified by the station or network. In order to get a particular period of time the advertiser must take the package program which occupies that period. This practice, still far from general, appears to be a step in the direction of returning control of programs to those licensed to operate in the public interest."

Citing this comment in an appearance before a Senate committee in 1956, president Frank Stanton of CBS added, "We have not gone as far as the Commission urged us to go in 1946; we do not tie in time to program." The occasion for his appearance, however, was a Congressional effort to diminish network control of programming, which program producers said was keeping them from finding markets for their product and owners of unaffiliated stations said was making it impossible for them to get hold of good material. And within a few years Stanton's network, and the others, would tie time to program irrevocably, leading the FCC to abolish all option contracts and to cry out that "the public interest requires limitation on network control."

Obviously, something very important had happened between

Neil Borden's 1940s assumption that advertisers could construct their own programs for their own purposes and the FCC's desperate scramble in the 1960s to find ways to assure that television programming would not become entirely an expression of network purposes. Equally obviously, the something very important was the replacement of radio by television.

4

Oddly enough, television had been around for a long time. The first patents for a device to send pictures by wire were issued in Germany in 1884, and by 1930 the Englishman J. L. Baird was selling television sets to the public for about $130 each and broadcasting video signals on BBC transmitting equipment after the close of the radio broadcast day. This was a mechanical system that involved a disc with pinholes and a light behind it; by synchronizing the very fast spinning of the disc with varying intensity of the light, a picture —well, a silhouette—could be constructed. Similar systems were being tried out in the United States, where Secretary of Commerce Herbert Hoover appeared on a television transmission over wires from Washington to New York in 1927, and General Electric telecast a play in Schenectady in 1928.

As early as 1923, however, Vladimir K. Zworykin, a Russian refugee working in America, had won a patent on a practical electronic television camera, though he never got the thing working right until he joined RCA in 1929. (Gilbert Seldes once speculated that if Zworykin had gone to work in Hollywood rather than in New York the movie industry rather than the radio broadcasters might have controlled television.) And in Britain engineers working for Electrical and Musical Industries developed an improved camera based on Zworykin's. The BBC in 1936 ran a formal test of this "Emitron" as against the Baird system, alternating weeks of broadcasting in each. The EMI system was an easy winner for picture quality, and was officially adopted. In 1937 the BBC televised the coronation of King George VI, and in 1938 regular television broadcasts began, with a page of programs announcing the television schedule amidst the radio schedules in *Radio Times*. American television first went to the public in 1939, when RCA put a few hundred sets in the stores at $625 each for rich New Yorkers presold by the enthusiastic promotion for the new device that dominated the RCA Exhibition Hall at the newly opened World's Fair. But the license RCA held from the Federal Communications Commission was for "experimental" broadcast only, and in fact the system RCA used in 1939 was not what finally became the standard broadcasting system in America.

Technically, television required a great deal of decision-making before service could begin; radio was simple by comparison. Sound waves translate directly into electrical terms in the broadcasting studio, and translate equally directly back into speech or music in the radio at home. But picture information is much more complicated, and there were various ways to code it and to reconstruct it from the code at the opposite ends of the broadcasting system.

Very quickly and too simply: a television camera contains many dots of an element which is sensitive to light and generates a tiny electrical charge when light hits it. An electron beam in the camera scans these dots, moving very rapidly from left to right in a pattern of parallel horizontal lines, and discharges the light-sensitive element. At the receiving end, in the television set, a "cathode ray" scans dots of a fluorescing element on the inside of the television screen, making an identical pattern of lines, causing each dot on the screen to light up or stay dark according to the charges carried on the electron beam in the camera in the studio. Between the studio and the television set, of course, are interposed all the paraphernalia of broadcast, generators, transmission towers, home antennae and much else.

Obviously this system is going to work only if the camera in the studio and the set in the home can be synchronized— the number of lines in the horizontal pattern, and the length of time required to complete each "frame" of these lines (like the individual picture "frames" in a strip of movie film) must be the same in both camera and receiver. The simplest way to synchronize the length of time needed to complete the frame is to use the frequency of the alternating current that drives both camera and receiver and all the other electrical appliances in the city. In Britain and on the European continent, the electrical current alternates at 50 cycles per second. The British and European systems make a separate frame for every two cycles of AC, or 25 frames per second. In the United States the current alternates at 60 cycles per second. American television also generates a separate frame for each two cycles of AC, so the American television standard is 30 frames per second.

The choice of the number of lines to be used in a television picture was an arbitrary decision. British television has been operating at 409 lines; the Europeans operate at 625 lines (and the British will slowly convert to 625 over the next decade); and American television operates at 525 lines (RCA's false start in 1939 had been at 441 lines). Every so often somebody publishes a magazine article about a glorious future in which satellites will send the same television picture to homes all over the world, but it can't happen. No American set, demanding a 525-line frame 30 times a second, can make any picture at all from a European transmission of a 625-line frame 25 times a second—or vice versa. In fact, American and European broadcasting companies can't use each other's videotapes. As late as mid-1971 there was only one machine in the whole world, in London, that could convert television signals from one system to signals usable in all other systems. Except for news, all television pictures that cross the Atlantic are sent on film. This is a great nuisance, but nothing can be done about it: each country is locked into its own system by the immense public investment in television sets.

Another technicality relating to frame speeds: professional Hollywood 35-mm. movies are made to play at a rate of 24 frames per second. In Europe this presents no serious problem: the movie projector is simply speeded up to show the television camera 25 frames per second, which is what the television camera needs; the pitch of the sound track rises very slightly, and the movie gets finished just a little quicker—a 1-hour, 12-minute movie plays in 1 hour, 7-½ minutes. In the United States, however, the gap between the 24-frame standard of the movies and the 30-frame standard of television transmission was too great to be bridged by speeding up the film.

Originally, nobody in television worried about this problem—it was just inconceivable to the pioneers of the new medium that their glorious invention, with its unique capacity to disseminate live action, live drama, live performances, would ever be employed simply as a carrier of film. Eventually, however, the engineers perfected the "telecine chain," which uses the fact that the 30-frame-per-second television picture is actually two 60-frame-per-second pictures with each frame repeated once. The aperture of the movie projector used in the telecine chain is geared to show one frame of the film for 2/60th of a second and the next frame for 3/60th of a second. The two frames together use up 1/12th of a second, and 24 frames a second pass the lens of the movie camera, making 30 frames per second on the television tube. The human eye is nowhere near fast enough to catch the deception. Indeed, the telecine system is so acceptable that movies and series made for television only—even commercials with no other imaginable use—are produced on film running 24 frames per second. The use of these same telecine chains to televise old silent movies on public television, however—as Pauline Kael complained bitterly in The New Yorker in fall 1971—produced a 50 percent speed-up of the action, because many silents were made originally to play at 16 frames per second.

One last technicality: the "channel." All broadcasting stations are distinguished from each other by the "frequency" of the "carrier signal" (i.e., the speed of alternation in the current the transmitter pumps into the air: the "information" needed to convert electrical impulse into sound or picture is carried as a modulation of that basic frequency). Because sound is relatively simple to broadcast, a narrow "band" of frequencies is all a radio station needs to deliver a sound without interfering with broadcasts from other stations. AM radio broadcasting operates at frequencies from 55,000 to 165,000 cycles per second, and stations with carrier frequencies only 4,500 cycles apart will not interfere with each other.

But the quantity of information needed to reconstruct a telecast picture requires a band width of 4.5 *million* cycles per second, and the carrier frequency must be much faster than that. In 1945 the FCC approved for use in America some thirteen carrier frequencies, each with the exclusive right to 6 million cycles ("megacycles") of band width, falling in the range from 44 to 216 million cycles, which the engineers called Very High Frequency (VHF). These channels were to be shared with nonbroadcast radio-telephone use (police and firecalls, etc.). The Commission recognized that over the long run these thirteen channels might not provide enough stations, and set aside for possible future exploration an Ultra High Frequency (UHF) range from 480 to 920 megacycles.

Radio waves at relatively low frequencies, like the AM broadcast band, meander about and bounce off the layers of the atmosphere and in general wander many hundreds of miles from their transmitters, especially at night; the FCC had learned that it could not authorize too many broadcasters to use the same frequency, even in fairly widely separated cities. But the VHF waves used for television go in straight lines, zipping out into space as the earth curves inward—that's why KTHI in North Dakota has to go so far up to garner a larger coverage area. In theory the thirteen channels could be used over and over again, all around a big country. Nor was there much worry about nearby stations on adjacent channels interfering with each other—though 4.5 million cycles of information would normally generate a "sideband" on each side of the carrier frequency, which would leave the FCC-allocated 6-million-cycle channel insufficient to prevent interference, the AM radio broadcasters had learned to "suppress" one of their sidebands, and the FCC assumed that television broadcasters would be able to do the same.

If any of these technical decisions turned out to be wrong, the FCC thought there would be plenty of time to correct them. Emerging from the war, the nation was still bound by a depression psychology, remembering the sharp economic setback that had followed less than three years after the end of World War I, and the catastrophes of the 1930s. Television sets were expensive, and the medium was seen as a rich man's toy. The FCC announced that any qualified company, including the radio networks, could be awarded up to five local television licenses in the VHF band, but CBS took only one—"an expensive gesture of contempt," as Eugene Lyons wrote later in his biography of David Sarnoff. "In due time CBS would buy these discarded licenses for telecasting stations for sums running into tens of millions." Tom Coffin, now vice president in charge of research at NBC, recalls that in the late 1940s, as a professor of psychology at Hofstra University, he sought the advice of CBS president Frank Stanton, himself a former psychology professor, on whether he should take a research job in broadcasting. Dr. Stanton was enthusiastic—but, he said, Coffin should stick to radio: there was never going to be much money in television.

In mid-1947 only ten stations were on the air, and production of receivers was at a rate of only 160,000 a year. Martin Stone, who then produced *Author Meets the Critics* for NBC, remembers sitting around the two small offices television then commanded, with the boss saying, "We have $150 for Thursday night. What should we do?" Then, very suddenly, television took off. Like the automobile before it, television appealed to a pleasure center deep in almost every man, regardless of age, education, social status or nationality. In 1950 Americans would buy 7,355,000 television sets (in 1971, the record year, they bought 14,860,000, of which 7,250,000 were color sets). By December 1948 there were 127 stations on the air, and it had become clear that the initial FCC technical decisions were wrong.

The nonbroadcast services permitted in the VHF band *did* interfere with television reception; they were barred from Channels 2-13 and given exclusive use of Channel 1. The geographic separations plotted in the original allocations turned out to be insufficient —VHF signals traveled in straight lines, all right, but they could take queer bounces off hills and sometimes off cloud formations and get into other licensees' coverage areas. Worst of all, the sideband suppressors could not be made to work properly. Telecasting on Channel 2 or 10 interfered with the programs on Channel 3 or 11, even some distance away. Through the latter part of 1948, the Commission studied reports from the field as a dog might study a cobra emerging from a basket; and in December the Commission panicked, and "froze" all construction of television transmitters, anywhere in the country.

Ostensibly, the reason for the freeze was the need to consider the possibilities of color television. CBS had a system which it said was ready to go, involving a whirling disc (again) between the cathode ray and the screen; RCA had an idea for a system not yet ready to go, involving fluorescent dots of different colors on the tube, to be separately illuminated by separate cathode rays. The CBS system, unfortunately, was not "compatible" with normal black-and-white television—i.e., a station broadcasting color could not be received at all on a conventional set. And even in the time of the freeze Americans were buying conventional sets at a rate of almost seven million a year. Color telecasts by the RCA system still gave atrocious color (and inferior black-and-white: when RCA actually began color-casting in 1954, its first "color spectaculars" were so deficient in black-and-white contrast that, "in a rare manifestation of the old pioneer spirit," as Gilbert Seldes put it, "millions of Americans walked all the way over to their television sets and tuned out the programs . . . it was the worst case of betrayal of the public interest in the history of broadcasting"). But at least the RCA system was "compatible," and would not deprive customers of the benefits of their investment in monochrome television sets.

No color system should have been authorized in 1950, but under political pressure the FCC approved the CBS system, and authorized broadcasting to begin on November 20 of that year. Fortunately, controls on the civilian use of electronic components during the Korean War gave CBS an excuse not to try to exploit its triumph at the FCC, and late in 1953, without much complaint from CBS, the Commission reversed itself and authorized a slightly altered (but much improved) version of the compatible RCA system. Incidentally, the CBS system, inherently well adapted to computer processing techniques, returned to life in 1970 as the source of color pictures from the moon.

The freeze on new stations was lifted before the end of the color controversy. In April 1952 the FCC announced a full national plan for the allocation of channels to localities—2,000 channels, divided roughly 550 in the VHF band and 1,450 in the UHF band. The first UHF station—KPTV in Portland, Oregon—went on the air in September 1952. But the agency had again miscalculated. Nearly twenty million television sets which could not receive UHF transmissions were already in the hands of the public. "All-channel" receivers cost more than VHF-only receivers; in the absence of outstanding UHF programming the public was not disposed to waste its money; and in the absence of receivers advertisers were not prepared to support any programming (let alone outstanding programming) on UHF channels.

At best, moreover, UHF transmission was marginally inferior to that on VHF, more subject to interference, and not only in cities. "I ran a UHF in Ann Arbor," says David Connell, executive producer for *Sesame Street*. "In the winter, our signal got all the way to Ypsilanti. In summer, with the leaves on the trees, it wouldn't travel six miles." Britain is in process of converting all television to the UHF band to free itself of the constraints a crowded Europe is forced to place on any one nation's use of the VHF range. Forty transmitters covered England and Scotland adequately in VHF broadcast, but for UHF the technicians now estimate a need for at least six hundred and perhaps a thousand "masts." Parliament is forcing BBC to share the towers with its commercial rival ITA for fear that otherwise, as an MP put it, "you'll turn this country into an upended hair-brush."

In 1962 Congress passed a law requiring all television set manufacturers to produce only all-channel receivers after January 1, 1964, but the horse was out the barn door. Even in television's *annus mirabilis* of 1969, nearly two-thirds of commercial UHF stations lost money. The total operating profits of 504 VHF stations in 1969 were just over $500 million; the total operating *losses* of 169 VHF stations were $43 million. It may be worth noting in this age of consumerism that between 1964 and 1971 the American people were forced by Act of Congress (incited by the FCC, the Ford Foundation and assorted critical commentators) to spend about $1.2 billion for UHF tuning capabilities the vast majority of them did not want and have never used.

5

Of all the differences between radio and television so carefully noted by scholars, one is in practice especially important: making television programs is enormously more expensive than making radio programs. Neil Borden noticed this at the very beginning, and pointed out that the *Ford Television Theatre* required "orchestra, producer, director, associate director, talent buyer, editor, writer, two sound men, two engineers, four camera men, two men on audio, one dolly-pusher, five men in control room, thirteen stage hands and helpers." The total weekly cost for the hour show in 1949 was . . . $20,000. Milton Berle's *Texaco Star Theatre*, proportionately the most popular program in history (it had a rating of 75—of the nation's first million sets, 750,000 tuned to Uncle Miltie every Tuesday), cost $15,000 a week to produce. In addition, there was a time charge for a network of twenty stations, covering roughly the quadrant Boston–Chicago–St. Louis–Washington, on Du Mont, then the most extensive network, of $6,750 an hour. For about a million dollars in 1949, an advertiser could have his own one-hour show thirty-nine weeks a year.

Even that was high by radio standards, but at these prices it was possible to retain the radio network procedures by which sponsors bought time and filled it with programs. In addition to the *Ford Theatre* and the *Texaco Theatre*, there were in 1949 a *Kraft Theatre*, a *Philco Television Theatre*, a *Colgate Comedy Hour*, a *General Electric Theatre*, an *Alcoa Playhouse*. Even the news had a proprietor: it was the *Camel News Caravan*. In television as on radio, the *Gillette Cavalcade of Sports* owned exclusive rights to the World Series—and here, in fact, the sponsor hung on until 1965.

But it couldn't last. Between 1949 and 1959 the costs of producing and transmitting nighttime television shows rose by 500 percent (and they doubled again between 1959 and 1971). By 1960 sole sponsorship of television programs was becoming a rarity. "Even if you developed a show that was a hit," says Don Durgin, president of the NBC television network, who used to be and of course still is a salesman, "it was too much money to throw against one operation." Most advertisers preferred to split the risk with someone else. When two advertisers "sponsored" a show, neither could produce or control it.

The motion away from sole sponsorship and control was led by two very different men, very differently situated: Sylvester L. (Pat) Weaver of NBC and Leonard Goldenson of ABC. Weaver, a jug-eared stringbean from California via Dartmouth, had been Young & Rubicam's man on the Fred Allen show, and went to NBC in 1949, first as director of programming and then as president of the network. Weaver saw television as structurally rather like a magazine, with commercial minutes to be sold individually, like magazine pages. ("If you don't sell all the spots," one of his assistants said rather bravely in 1956, shortly before Weaver was ousted, "it's like a thin issue of a magazine.") Goldenson, by contrast, was a custom-tailored lawyer from Pennsylvania via Harvard, who had arrived at Paramount Pictures through work on the company's bankruptcy in the early 1930s, and had been head of the company's theatre division when the Justice Department separated studios from theatres in the late 1940s.

In 1953 the newly independent United Paramount Theatres, with Goldenson as its president, acquired the starving American Broadcasting Company, a television network with a grand total of thirteen affiliates. ABC could not have sold much time to advertisers on the old you-supply-the-programs basis: signals from the ABC affiliates reached only about one-third of the nation's television homes, and advertisers with programs just naturally took them to CBS and NBC, where they could get nationwide coverage. Anyway, Goldenson and his colleagues saw television as structurally rather like a movie house. "The great appeal of the motion picture business," said Simon Siegel, financial vice president of the merged Paramount Theatres and ABC, "was that if you knew your area, knew your theatre, knew your public, knew your product, you would do all right. Well, the same thing is true of television."

Weaver in the early 1950s put together three daily "magazine concept" shows—*Today, Tonight* and *Home* (a midday women's mag that failed to catch audiences)—plus a weekend afternoon catchall called *Wide, Wide World.* Nobody "sponsored" any of these: dozens of advertisers bought individual minutes once, twice, three times a week—at simple per-minute prices which combined time and program charges. Goldenson went to Hollywood, looking for product which ABC could use to get time on stations already affiliated with another network. Both Weaver and Goldenson wound up in the offices of Walt Disney, probably the only really sure thing in show business. Goldenson needed him worse, and got him, in 1954, by putting up some of the investment capital for Disneyland. Disney gave ABC a show called *Disneyland* and then *The Mickey Mouse Club,* which devastated all opposition in after-school television; in the two-station markets which had locked ABC out, the NBC and CBS affiliates scrambled to get a "secondary affiliation" with ABC and access to Disney. Then Goldenson moved across town to Warner Brothers, which agreed to make forty hour-long films based on the characters in three Warner movies (*King's Row, Casablanca* and *Cheyenne*), other shows which "primary affiliates" of the other networks would wish to carry. The price to ABC was $3,450,000 (forty films at $75,000, twelve to be rerun at additional payments of $37,500 each).

These were network shows: ABC could put them on the schedule whatever day and hour met the network's needs. By the late 1950s all three networks were learning the hard way that the placement of programs was something that could not safely be left to advertisers or other amateurs. Some shows were best suited for early evening, when children and teen-agers greatly influenced where the dial was set; others for later, after the children were (mostly) in bed. Some shows were failures in one time period and successes in another (*Bonanza, Hawaii Five-O, Dick van Dyke*); others lost hit status on moving (*I Spy, Ben Casey, Red Skelton*). Moreover, the audience for any show was to a considerable extent a function of the audience for the show that preceded it. "I can't abdicate to *Flip Wilson,*" CBS program chief Fred Silverman said nervously while making out his 1971–72 schedule. "I've got two hours after nine o'clock that could be killed." *The Voice of Firestone* lost its air time, though the tire company was quite prepared to continue sponsoring it, because its presence on the schedule was thought to diminish the audience for other things.

It does not seem to have occurred to anyone at either the networks or the agencies that network program control would be incompatible with the sponsorship model of buying. Once the network controlled the show, whatever a salesman might say, advertisers could no longer buy time on a network: the time was not separable from the program. Then the networks' power to move shows around on their schedules destroyed the "gratitude factor." Advertisers who could not count on getting this show at this time next year could not assign much value to the steady viewer's wish to reward a sponsor, which was clearly a long-term asset.

Meanwhile, across the table, all packaged-goods advertisers had been influenced by Procter & Gamble's decision that on television "reach" (the number of different homes to which the message was delivered) would be more important than "frequency" (the number of times each home saw the message). Television was at its most effective as an advertising medium when introducing new products. Weaver argued in vain that his magazine shows (including night-time programs like *Producer's Showcase*, which mixed comedy, classics and contemporary drama) reached everyone in the country over a period of a few months. Advertisers seriously interested in maximizing reach did not want to be in one time slot week after week; they wanted to be in as many different places as their budget could cover.

With minutes all over the schedule, P&G could guarantee that its new brands were exposed to virtually everybody in America within a few days of their introduction. Moreover, it was safer. "If you're in twenty shows, you can't be hurt," says Paul Huth, who buys time for all P&G brands. "If you spend it all on five shows and two go bad, you're in trouble." P&G, while retaining a few franchise time slots (most notably, in the evenings, 8:30 on NBC on Sunday night, which it has controlled for donkeys' years), shifted most of its budget, the largest in television, to the purchase of individual one-minute commercials on a widely scattered collection of programs.

One by one, with few exceptions, the other advertisers followed.

While the advertiser was producing the show, he ran the risk of audience size. He paid the network the same price for the time he used, whether his show was a hit or a flop. And though most advertisers obviously wanted huge audiences, some were willing to support less popular shows—perhaps because they sought prestige, perhaps from corporate pride, perhaps because (like U.S. Steel and Du Pont and Alcoa) the audience they most wanted to reach was a relatively small community of relatively serious-minded men influential in politics, in corporate purchasing and in the stock market. But if advertisers were not really identified with the prestige program—if they merely bought a minute in it—then the prestige values were clearly much diminished.

And the risk moved. Now the network, not the advertiser, would have to calculate the costs and benefits of smaller audiences. Instead of collecting a flat price per hour from an advertiser who chose his own program, the network would have to negotiate a price per minute—and over the long run the price of that minute would tend to be a function of the size of the audience to the program it interrupted. The network's revenues became a reflection of the popularity of its programs. In spring 1970, CBS had slightly higher average nighttime ratings than NBC and slightly higher receipts from prime-time sales to advertisers; in spring 1971, NBC pulled ahead in ratings and also, according to the continuing study of these matters by Broadcast Advertisers Reports, in prime-time revenues.

Attitudes at the networks suffered an inevitable sea change. In the early days, a network had not been disgraced by a failure to sell some time periods to advertisers. Jack Cowden of CBS, a graciously cynical veteran whose official function is television public relations but who serves everybody unofficially as archivist of the industry, remembers "a feeling that you *shouldn't* be more than 80 percent sold. If you were, your prices were too low, and you should raise prices until sales fell back to 80 percent. Then you had the other 20 percent for your own 'sustaining' shows." This was never popular with comptrollers or with affiliates, who didn't get paid to carry sustaining shows, but it gave the leaders of the networks the certainty, now much eroded, that they were running something more important than a public convenience for industry.

Whatever virtue there might be in an unsold hour that allowed your own people to produce their own programs, there was clearly nothing but vice in an unsold minute, or a minute that could be sold only for disreputably low prices because the audience size did not justify higher prices. Cant at the networks says that ratings are used only as one piece of information among many, and do not dictate program decisions, but the evidence is to the contrary. In the 1969–70 winter season, the average Nielsen rating for prime-time programs was 18.7. Of the 89 prime-time programs regularly telecast by the networks in that season, 53 had average ratings of 18 or higher, and all but 5 of the 53 returned in 1970–71. The other 36 had ratings of 17 or less, and 26 of those did not return. In 1971–72 seven of the ten shows at the bottom of the Nielsen list for the fall were dropped at mid-season.

Once a network's income became a function of the ratings of its shows, the tendency to seek the highest possible audience for each minute became a compulsion, irresistible, ultimately seen as "natural." The "magazine concept" by which Weaver was going to liberate television from advertisers became, instead, the instrument of its enslavement to ratings.

CHAPTER 2

I Call Hello Out There
But Nobody Answer

The real degradation of the BBC started with the invention of that hellish department now called Listener Research.
—LORD REITH

The problem with this business is that there's no coupon for people to clip.
—OLIVER TREYZ when president of ABC-TV,
quoted by MARTIN UMANSKY, general manager,
KAKE-TV, Wichita

Network television lives and dies on Nielsen data. Other sources are interesting, enlightening, nice to see, but you don't live on them.
—EDWARD I. BARZ, vice president
and director of media research,
Foote, Cone & Belding, New York

My radio plays almost all day every day as I like to wake up to it and hear it as I come in the door at night. And my dog enjoys it as much as a dog can.
—Comment by a lady keeping a Nielsen radio diary

No other major advertising medium can match television in the scope, accuracy, detailed nature and usefulness of the information published by Nielsen Television Audience Research Services. The practical results and value derived from the use of this research are suggested rather clearly by the fact that, in the United States, it is used continuously by every national television network and every advertising agency handling network television, and by every network television advertiser—either via direct subscription or through its advertising agency.
—ARTHUR C. NIELSEN, SR.

1

Broadcasting is, in a way, the quintessential modern activity. It is a man-made force which acts invisibly at a distance, and the distance is so great that even the proximate consequences of an action are undiscoverable. The arrow is shot into the air, and God alone knows where it comes down. Ham operators gleefully keep logs of those who respond to their calls (indeed, once they have made contact, they typically have nothing to say to each other save self-identification). But nobody responds as a matter of course to the broadcaster's call. If he is alone in the world, he stands in a dark and soundless room where no amount of groping will bring sensation to his fingertips. Obviously, he needs ancillary services.

"Ratings," which describe the size of the audience for a broadcast, are an almost universal feature of television in the Western world—indeed, the presence of audience research seems a necessary (clearly not sufficient) criterion for deciding whether or not a society is "democratic." Yet a television system in which all program decisions were determined by the vote of the ratings would be the worst kind of democracy, in which minorities always lose. For the ratings are soulless and simple-minded—one man, one vote. A ratings service knows no way to reward intensities of feeling on the other side of the screen. Few common modern comments are so heartfelt as the housewife's lament, "Every time I like something, they take it off the air." The essence of democracy in operation is logrolling, the swap of support by nonconflicting interests to produce transient majorities. But there is no mechanism by which television viewers can logroll.

None of these difficulties is absolute. Even in the complete absence of rating services observers can acquire some notion of how a broadcasting system is working. When Amos 'n' Andy first began in Chicago, in the early 1930s, the telephone company noted that the number of calls in that city dropped by 50 percent between 7 and 7:15 P.M., and even today most American city water departments find that the pressure in their mains drops at station breaks, every hour on the hour. Nor is the measurement of intensities entirely beyond the wit of man. Sometimes gross measurements occur by accident, as when NBC cut short the telecasting of the New York Jets–Oakland Raiders football game to present the opening moments of *Heidi,* and the Circle 7 telephone exchange in New York was put out of business by angry callers. Equally gross is the persistent habit of counting the unsolicited mail (which two decades ago, in the days of *Red Channels* and blacklisting, gave a letter-writing supermarket owner in Syracuse power to destroy the careers of all but the most talented television performers). Somewhat more refined, and for a while quite popular with network researchers, is the Tv Q Service, which sends questionnaires to a panel of about two thousand respondents, asking them to say whether a given program is or is not "one of my favorites." A more sophisticated variant of this approach is the W. R. Simmons study of "attentiveness," in which viewers are asked to say whether during a show they were: (1) in the room paying *full* attention; (2) in the room paying *some* attention; or (3) out of the room most of the time. But advertisers use these measurements very rarely, and since the mid-1960s the network's haven't used them at all.

Each country handles these problems its own way. In Austria the broadcasting authority (ORF) is required by law to gather ratings and to publish the results. (Management, which is deeply conscious of a cultural potential in the medium, is not entirely happy with this provision. "The worse the program," says Intendant Gerd Bacher a little irritably, "the greater the revolution if you try to get rid of it, because people know it is popular.") In four ratings sweeps ranging in length from four weeks (in November and January) to eight days (in June and September), ORF interviews 2,800 people a week to find out how many were watching what, and how much they liked it on a scale from +10 to −10. The second most popular regularly scheduled entertainment program in 1968 was *Solo für O.N.C.E.L.;* the best-liked over the course of the year was *Daktari* (consistently better than +8), closely followed by *Flipper.*

In France, by contrast, there were no ratings at all until l'ORTF began broadcasting commercials in 1964, and though the state-controlled company now does buy a rather elaborate continuing service, it is strictly against the law for anyone to publish the results. The procedure involves three French market-research firms which maintain panels of respondents to answer questions about purchasing habits. Each supplies diaries of television viewing to random samples of its panels, the diaries (about 1,400 a week) to be returned to and processed by l'ORTF. Though a great deal of French television is in fact rather vulgar, theory insists that public opinion shall not have a determining effect on what is broadcast. Speaking of its nightly three-minute cartoon strip *The Shadoks,* a member of the team at l'ORTF's experimental center reports a comment by director Pierre Schaeffer: "If people don't like *Shadoks,* they are stupid anyway."

In Britain there are two competing ratings systems, as there are two competing broadcasting systems. The noncommercial BBC, broadcasting over two channels, employs year-round some 700 part-time interviewers, who interrogate some 2,250 adults and 450 children every day of the year, asking them what they watched yesterday. There is also a panel of 2,000 homes, one-sixth of them new every month, who supply what Brian Emmett, head of BBC audience research, calls "quick and dirty feedback on television programming." The corporation mails them a questionnaire asking them to mark on a five-point scale (A+ to C−) how much or little they liked certain programs they had had a chance to see in the previous week. In the case of international sporting events, liking correlates very well with how Britain did. In the case of variety shows, it tends to parallel ratings. "But in a series like the Wednesday play," Emmett says, "new dramatists, lot of kitchen sink about it, the correlation between liking and rating is effectively zero." Though an early study by William Belson indicated that the arrival of television tended to drive one or more members of large families out onto the streets of an evening, in general British researchers feel every program will be seen by many people who don't like it— if only "because," as a BBC pamphlet puts it, "the 'set' is often in the one warm room in the house."

Commercial television in Britain is supervised by the Independent Television Authority, which actually owns the towers over which the programs are telecast. ITA is required by law to "ascertain the state of public opinion concerning the programs (including advertisements) broadcast by the Authority." The ratings method used involves a panel of homes in which the use of the set is measured by an electrical recording device. The great advantage of the machine, of course, is that it tells the advertiser how many homes (as a percentage of the panel, for sure; as a percentage of the country, by extrapolation) were tuned to the channel carrying his message during its brief mortality. "Otherwise," says Ian Haldane, director of research for ITA, "I must say I prefer the BBC interviews. Machines don't watch television; people watch." The panel housewife also keeps a diary identifying individual viewers of each program, allowing the service to estimate for advertisers the age, sex, income, etc., of the viewers.

In addition, Haldane runs his own, rather sophisticated surveys of affection and disaffection among viewers. "Partly," he says, "it's the law. The act lays down that we shall broadcast nothing that offends good taste or decency or conduces to crime. So we do studies asking people, 'Have you seen anything recently that you thought should be taken back of the woodshed?' Then, we want to help programmakers. The act lays down that on Sunday evening from six to seven-thirty nothing can be broadcast without the approval of an external religious board. We call it the Holy Hour. Our research shows that pure religion gets no audience and no liking, but that religion related to personal and world problems *does* interest. We make suggestions. For a while, drama was full of tramps sitting in dustbins and talking of eternity, and our research showed the drama boys that Mrs. Bloggs couldn't be asked to take an interest in that. The TV audience is not the audience for a West End theatre. They listened, and it improved our audience for drama a good bit."

When ITA adopted a machine system in 1968, BBC considered going along; but the asking price was about a million dollars a year, and there were all those employees at Broadcasting House who would be displaced. Thus the two broadcast services measure different things and the results cannot be compared. The BBC 1970 annual report shows the family comedy series *Not in Front of the Children* as the corporation's most popular program, with 13,780,000 *viewers* (many of them, no doubt, children: the title was pretty irresistible to a child, and the show went on at 7:55 P.M. on Fridays). ITA gives this program an audience of 5,353,000 *homes* and shows it trailing no fewer than thirteen ITA programs (three of them being different nights on *News at Ten,* the ITA half-hour nightly news service).

British advertising men seem to agree that when ITA was new, in the 1950s, it outdrew BBC by nearly two to one. This produced a considerable shift in BBC programming policy, which had been perhaps a little stodgy ("The BBC," Fred Allen once remarked, "begins its program day with a lecture on how to stuff a field mouse, and continues in the same vein for the rest of the day"). Tom Sloan, "Head of Light Entertainment" for BBC, remembered his arrival there in 1954, when the director of the television service felt "frivolity on television was a literal waste of time. . . . We really were regarded as red-nosed clowns and strolling players." Fear of commercial competition for audience drove BBC to import large numbers of American westerns and action programs (now greatly diminished in volume and appeal), and to develop its own collection of domestic situation comedies, two of which have recently been adapted for American production as *All in the Family* and *Sanford and Son.* By the late 1960s the two BBC channels between them were pulling about as much average nighttime audience as the one commercial channel. Whether this condescension to public tastes was a good thing or a bad thing is a matter of continuing debate in Britain; the most articulate speakers in the debate tend to believe it was awful.

"We have an obligation to maintain some share of the audience," says David Attenborough, inventor of *The Forsyte Saga* and *Civilisation,* the BBC's director of programs. "It's no good going out on the blasted heath and speaking in the most marvelous aphorisms if nobody is listening. We want to have between 40 and 60 percent of the audience, and I don't care if it's 60 in our favor or in theirs.

But if week after week we got only 30 percent on our two services combined, I would know we were not the national broadcasting organization. And if we consistently got 70 percent, I would know we weren't braving enough." What will happen to this way of thinking when ITA gets a second channel is still a matter for speculation.

2

Ratings in the United States started in March 1930, when a group of national advertisers met together with opinion researcher Archibald Crossley to form the Cooperative Analysis of Broadcasting. The technique employed was the "telephone recall"—interviewers would call people and ask them to run down a list of what they had heard yesterday.

The first improvement on this system was Claude E. Hooper's Hooperatings. His technique was the "telephone-coincidental," in which an interviewer calls a home and asks the person answering the telephone whether the radio (or television set) is playing, and if so what's on. This method is still used, on special order from a customer, by the American Research Bureau, to provide an overnight answer on how well a network show seems to be doing nationally. Its weaknesses are that it can't provide ratings for the early-morning and late-evening hours; not everybody has a listed telephone; and some people lie—a famous *New Yorker* cartoon in the 1950s showed a man telling a telephone he was watching *Omnibus* (the Ford Foundation's first effort at high-prestige programming) while the screen behind him showed men on horses firing guns at each other.

Next came the method still used by the BBC, the "aided recall" interview, where the interviewer comes in the flesh, armed with a roster of yesterday's programming. The weaknesses here were the possible influence of the interviewer's personality, a tendency for people to say they had watched something yesterday because they "always" watch that show, even if they had in fact missed it the night before, and, again, the little ego-building lie when the roster in the interviewer's hand revealed a show the respondent felt he should have seen. It will be noted that because the average interviewer is socially upscale from the average respondent these biases would tend on balance to boost the ratings for the BBC, but not for any American network.

The self-administered questionnaire, as researchers who ran household "panels" for advertisers were discovering, saves the cost of interviewers at little loss of accuracy, given an appropriate subject matter; and listening to or viewing broadcasts seemed most appropriate. And this is, indeed, the most widespread way of measuring broadcast audiences in America, used for local station ratings everywhere except in New York and Los Angeles (which are big enough to support their own machine installations). Diaries are distributed (at the rate of 85,000 a month by the American Research Bureau in the three rating months—October, February and May), and the recipients are asked to note in appropriate places, every day for a week, what members of the family saw what on the tube.

This technique is subject to a whole slew of possible errors, depending on who makes out the diary and when. ("If you look at those diaries," Peter Langhoff said while president of ARB, "and you see how sloppy some of them are, you can't imagine how we get ratings out of them; but the fact is that we do, and good ones.") One suspicious investigator distributed a hundred diaries, telling recipients he would be back to pick them up in seven days—and then returned in five days instead. He found a number of diaries still blank, the householder having put off till tomorrow what he forgot to do today—and an almost equally large number already complete for all seven days, the helpful respondent having gone through *TV Guide* at once and written in the shows the family "always" sees. (To counter this bias, ARB diaries have a special place in which people can write the names of programs they always see but somehow missed this week; those citations are not tabulated, of course, but the diary filler has no way of knowing that.)

For national network purposes, however, American ratings, like the British ITA ratings, come from machines—specifically, the Audimeters of the A. C. Nielsen Company, which began offering the service, for radio, in 1942. This machine, variously and aggressively patented, consists of a timing device set to local time, a cartridge of photo film, and a lamp that exposes the film in an appropriate place to show the time when a set is turned on and the channel to which it is tuned. Such machines are placed in a relatively small sample of homes—just under twelve hundred for the nation as a whole—and most homes stay on the panel about three years. Participants are paid 50¢ every week in the form of two quarters that fall out of the film cartridge when it is inserted in the Audimeter. Some cartridges don't work right or get lost, but between 950 and 1,000 usable film cartridges return in the mail every week to the Nielsen Company, and from the company there issues every other week the magisterial Nielsen Television Index, not just a rating but a kind of revised and annotated bible. (In some key months the NTI arrives every week, with fresh numbers.) For these national reports and associated local reports, Nielsen seems to be receiving—the figures are confidential, except for a world-wide total including receipts in four other countries—at least $10 million a year, probably $2.5 million from the networks, at least as much again from stations, and most of the rest from the advertising agencies, some of which pay upwards of a quarter of a million dollars a year for Nielsen services.

Nielsen's primacy in national ratings traces back to a committee of the Advertising Research Foundation, formed in 1952 to bail the broadcasting industry out of a situation where every time salesman used figures from a different rating service. With an insistence rare in social research, the committee supported "tuning" rather than "viewing" as its "standard of exposure," and "households" rather than "viewers" as the "unit of measurement." In the words of Norton Garfinkle, a former economics professor whose Brand Rating Index dabbles once a year in the measurement of audiences for television programs, "ARF's report said, as rat psychology does, let's not do what has to be done but what we know how to do, which is an accurate report on whether the set is on." Another, perhaps more immediately compelling reason was the fact that "households" reached by a program could reasonably be compared against magazine circulation numbers by advertisers—each magazine sold presumably goes into one household (doctors' offices and barbershops being roughly offset by television's hotels and bars). By opting for tuning and households, the committee walked into Nielsen's bag; and Art Nielsen, a tall, American Gothic, tennis-playing, fiercely upright gutfighter, drew the string on the bag.

All these rating techniques, of course, rest on the proposition that one can describe the activities of a large "universe" through the analysis of what is done by a small "sample" of that universe. This proposition is, bluntly, true. It is hard to believe that there are literate people who in the Year of Our Lord 1972 still deny belief to information because it is gathered by sampling techniques. (As Kenneth Baker of the Broadcast Ratings Council liked to say, "Next time the doctor wants to make a blood test, don't let him take just that smear—make him take all of it.") The mathematics of sampling dates back to Pascal in the seventeenth century, and television ratings are the least of an immense list of kinds of information we have only from sampling—about employment and bank deposits, air and water quality, glandular secretions, public opinion, etc. And the logic of the situation is such that television ratings should be among the more accurate examples of information gained by sampling. The behavior to be measured is common (about 98 percent of American households have a television set and 94 percent turn it on at some time during the course of a week) and extremely simple (is the set on or off, and to which channel is it tuned?).

For many purposes a research sample doesn't even have to be very carefully chosen: a man trying to find out the incidence of left-handed people in the population can station himself at the 125th Street subway exit in Harlem and get results just as accurate as he would get from a statistically impeccable selection of respondents. But a researcher trying to estimate the proportion of the population with incomes under $5,000 would produce a most inaccurate report from respondents clustered at 125th Street. Television viewing would appear to be more like income than like left-handedness—children, old folks, low-income families and Negroes watch more television than young-adult upper-income whites. To put together a truly "projectable" sample of television viewers is an extremely difficult job. One of the reasons the industry was dissatisfied with telephone-coincidental, interview and diary techniques was a suspicion of the validity of the samples used.

The perfect sample would be one chosen in such a way that every home in the country had an equal chance to be part of it. This can be done, in theory, by giving a number to every residence in every census tract (the maps used by the Census Bureau to make sure the whole country is counted once every ten years), and then applying a computer-generated "random-number table" to the resulting list of homes. In practice, census tracts are "clustered" and the random-number tables are applied first to choose clusters, then to choose tracts within clusters, and then again to choose residences within tracts. This sort of thing takes time and money, and even more time and money are required to secure the cooperation of the families whose homes are ultimately picked by the computer. Because Nielsen would use every home in his national panel for several years, he could, presumably, spend the money to make his "sample design" perfect, and to persuade everybody to come along. Among the persuasions Nielsen offers, incidentally, is payment of part of the repair bill if the set goes out of whack.

Whether in fact Nielsen was following such procedures was until the mid-1960s a conclusion that had to be drawn from premises more than from evidence. Everything about Nielsen breathed science (he had started as a chemical engineer, and his firm had been in engineering prior to the Great Depression). Headquarters was an utterly unswanky four-story building looking much like an old suburban high school, in the wilds of North Chicago. The man himself had a unique gift for telling you what crooks his competitors were without making you wonder about him, being in a business like that. ("Nielsen," says F. Kent Mitchel of General Foods, "has proved that you *can* be honest in this business.") He couldn't, really couldn't, tell you too much about his panel, because each member of it represented fifty thousand homes' worth of viewers. With advertisers spending an average of $50 per household per year to put commercials before the public, control of a single Nielsen meter gives significant influence over the expenditure of $2.5 million a year. It was obviously in everyone's interest to have the identities of Nielsen's Audimeter homes kept a dark secret, and to allow details of key procedures to remain a riddle wrapped in an enigma.

Warren Cordell, Nielsen's chief of statistics, was if anything even more convincing than his boss. A Midwesterner with a round face—one could almost say apple-cheeked—he was obviously a plain, blunt man who gave straight answers to questions. (There was a bit of spice about him, too, because in private life he was a bibliophile of standing among collectors of incunabula.) It does not seem to have occurred to anyone before 1963 that the same qualities that made Cordell so admirably trustworthy an informant might lead him to feel, in a pinch, an overwhelming loyalty to his organization—until finally he would write in an internal memo that "Government investigations were trying because we preferred not to let these people learn and publish some of our vital weaknesses." Then the House Commerce Committee began an investigation of broadcast ratings services. In March and April of 1963, under the hammering of Congressmen and committee staffers, over ten very long days of testimony, Cordell and Nielsen executive vice president Henry Rahmel very reluctantly revealed that the trust Nielsen had demanded from his subscribers had been significantly abused. As Kenneth Baker of the Broadcast Ratings Council said shortly before his death in 1971, "We all found out they weren't doing the things they had said they were doing."

Reading the transcript of these hearings is an odd and unsettling experience. At the beginning the Congressmen seem rather stupid, asking questions that first-year statistics students could answer— and asking them, often, with that special nastiness permitted in our culture only to those whose office boasts a flag in the corner. Then, slowly, the reader begins to realize that the Congressmen, thanks to classy staff work, know something the reader doesn't, that Rahmel and Cordell are being evasive, that Nielsen's claims are not going to stand up under this kind of scrutiny.

The splendid sample on which no expense had been spared turned out to be based still on a design made in 1947, before probability techniques were perfected; replacements had been rather casually introduced, and there had not been nearly enough of them: some homes had been in the sample ten and twelve years. In the one market thoroughly studied by the committee's staff (Louisville), a disturbingly large fraction of the Audimeters were on the fritz— five out of eleven, for example, during July 1962. In apartment houses, a statistically improbable proportion of the machines turned out to be in the super's flat, obviously the easiest one for an inter- viewer to reach.

Perhaps worst of all, the Nielsens themselves disappeared for the period of the hearings—board chairman Arthur, Sr., to Paris, president Arthur, Jr., to Australia. (Junior managed to get back for the last day, and it was advertised that he would make a statement to the investigating House committee; but when he was called, it turned out that some other piece of business had taken him away.) Art, Sr., today waxes wroth over the memory of these hearings, and likes to talk in some detail of how one of the Congressional investi- gators subsequently abused his inside knowledge of Nielsen's opera- tions, but the misbehavior of others scarcely sweetens the taste of his own failure to testify.

"We, of course, received quite a jolt from the ratings hearings," Hugh M. Beville of NBC told the House committee about ten months later. In response to that jolt, the networks formed CONTAM —the Committee on Nationwide Television Audience Measurement —to look at practical questions growing out of the shadow the committee had cast on the ratings. Meanwhile, the Broadcast Ratings Council, in effect an accrediting agency for ratings services, was formed by a joint effort of broadcasting companies, advertisers and advertising agencies. And the end result—net net, as they say on Seventh Avenue—was a bonanza for Nielsen. For it turned out that the deficiencies in the national ratings were, like the placement of the researcher investigating left-handedness, random to the problem being studied.

When CONTAM compared the Nielsen rankings from 1,100-odd Audimeters to national rankings that could be drawn from the 55,000 diaries in an ARB local ratings sweep, the correlation was .99.

("This is extremely high," Julius Barnathan of ABC told the House committee, one hopes a little gratuitously. "In fact, to give you some idea of how high it is, if you measured a person's blood pressure at two points of his body at the same time, and compared the correlation between the two results, you would get a correlation of around .92.") And when Nielsen permitted Ernst & Ernst to audit its Audimeter placements and results (and the Justice Department gave a specific okay to the Broadcast Ratings Council as a *legitimate* conspiracy in restraint of trade)—the auditors found the Nielsen operation less good than Nielsen's sales literature had claimed but not beyond the normal discrepancy between promise and performance in commercial operations.

In 1969 a fourth CONTAM study nailed the surviving doubts about the reliability of the Neilsen system. These doubts derived from the reasonable proposition that people who refuse to have a meter on their sets have viewing habits different from those who go along with a ratings service. A giant telephone-coincidental survey found proportions of Homes Using Television (HUT) virtually identical to the "HUT scores" in the Nielsen Index—provided the telephone girls let the phone ring eight times (and then called back the Don't-Answers for one more set of eight rings) before giving up and marking the family absent and its set off.

During the years since the House committee hearings, Nielsen has of course greatly improved its procedures, updating and replacing its samples, dropping its radio ratings (which were the main source of damaging material in 1963) and spending substantially more on both its national and local television services. His customers believe that for every dollar Nielsen spent, they paid him about a dollar and a half, to the point where broadcast ratings, a loss item for the company in the 1950s, are now a considerable contributor to the profits of the world's largest market-research operation (the basic business continues to be the Food and Drug Index, which audits the inventories of supermarkets and drugstores to tell manufacturers how fast their goods are moving out to the public). Other firms could not afford similar investments, and today Nielsen has no competitors in national television ratings.

Nobody in television really doubts that the numbers are roughly right: "A couple of years ago," Henry Rahmel says a little complacently, "NBC moved *I Spy* to Monday night and it bombed. They called me—'Something's got to be wrong, Henry.' But there wasn't anything wrong. We haven't had any other complaints like that in quite a while."

In fact, somebody who wanted to go finding flies in ointments could still fish out a few (a family in Harlem, for example, that was counted in for months even though its picture tube was not functioning: the audio was enough to trigger Nielsen's meter). New problems of definition quite as difficult as the old war between "tuning" and "viewing" have been created by the spread of multiset households (now more than one-third of the country—and from an advertiser's angle the best third). "Nowadays you can't get the home with a kid show at seven-thirty," says ABC's Barnathan. "The kid's got his own set in his own room." And the sports people are unhappy about Nielsen's failure to measure places like country clubs, where a lot of golfers watch golf matches of a late afternoon. An old-timer observes that the ratings are universally accepted today in part because Harley Staggers—who succeeded Oren Harris as chairman of the House committee in charge—"isn't interested in these things." Whatever the reason, the fact is that Nielsen national television ratings—the results of the popularity contest—are no longer a source of controversy.

3

"I was once introduced at a meeting of advertisers," Norton Garfinkle of the Brand Rating Index said reminiscently, "as the man who made us understand that people don't use products; people who use products are products. But the fact remains that there is a general law here: one-third of the users account for two-thirds of the sales." And, of course, what interests advertisers primarily is not the number that pass through the turnstiles but the market potential they represent.

"In the past," says Jay Eliasberg, research director for the CBS Broadcast Group when he is not out in the swamps and forests making the most implausibly beautiful photographs of birds ever hung on an office wall, "there was a kind of democratic one-man, one-vote thing. You could complain that the form of voting wasn't a proportional-representation thing, that people outvoted in one half-hour were also outvoted in others. But that was taken care of in part by the crazy advertiser, the man who wanted to reach Cadillac owners, and by the demands of the intellectuals, and by the fact that the people who run networks are intellectuals. Now we're saying that some people are more equal than others, on a demographic basis: if they buy more, they vote more."

The cant word here is "demographic." In general, the ideal viewer is a married woman with four or more children living home and a husband with an income over $15,000 a year. She needs and can afford to buy lots of soap and toothpaste and frozen orange juice, and doubtless wants plenty of soft drinks, cosmetics, minor appliances and aspirin. But there aren't many such women: a representative sample of a thousand or even five thousand Americans would not turn up enough of them to give a statistically valid sample of their group.

To get decent ratings for subgroups in the population, the "cell" containing that subgroup must have enough members to permit projection. In 1962 the American Association of Advertising Agencies urged Nielsen to provide a three-way breakdown of the population —18-34, 35-49 and 50+. To supply such figures, Nielsen added to its Audimeter sample an additional revolving nationwide diary panel which runs in about 600 homes a week, 1,200 over the two-week period which is the normal reporting schedule. Exactly how this Nielsen Audience Composition panel is meshed with the Nielsen Television Index group is one of Nielsen's secrets, but it seems to be okay with Ernst & Ernst, so it's got to be okay with us.

Now, the group of women 18-49, which is the breakdown most advertisers use, represents 62 percent of all adult women (and 71 percent of all supermarket expenditures). If the sample is large enough to give accurate ratings for the market as a whole, the 18-49 "cell" of that sample will be large enough to give good ratings for this audience segment. To call success with the 18-49 contingent a "youth orientation," of course, merely flatters those of us who are in the upper part of that range without saying anything real about the program. Moreover, as Eliasberg has pointed out, the 18-24 sector of the "cell" is "not an economically homogeneous group" from an advertiser's point of view—that is, some women in the group are indeed "homemakers" with households to supply and children's noses to wipe, but lots are in colleges or stewardessing in airplanes or living the swinging, fearful life of the single in the big city. Anyway, ABC does much better with the 18-24 group than CBS does. In fall 1970, CBS and NBC persuaded Nielsen to make some new breakouts from his data, and to publish a 25-64 category as well as the old 18-34 and 35-49; and now, in the oldest tradition of the advertising business, time salesmen from different networks can present contradictory arguments, both accurate, because the same information has been broken into different categories.

Age categories are only one of the more obvious breakouts that can be made from audience surveys. Also important to the advertiser may be family size, income level, education, occupation, place of residence, product usage—even, ideally, brand usage. Probably the most important product-usage data are those from W. R. Simmons & Company and the Brand Rating Index, both of which make annual fall sweeps of 15,000-18,000 homes, gathering information (by diary) on product usage, brand preference, magazine and newspaper reading, television viewing, ages, sexes, incomes, occupations. The replies to these questionnaires are massaged in the computers in dozens of different ways, and tens of thousands of pages of correlations are spewed out to advertisers, agencies and media salesmen.

BRI material tends to be used in index form; Simmons' material, relatively similar, tends to be used in big brute numbers. Indeed, with help from a partly owned affiliate called Interactive Marketing Systems, Simmons has put all its stuff into a time-sharing CDC-10 computer, which will play into a terminal available in the subscriber's own offices at a base charge of only $75 a month (*using* it, of course, costs more). Helped by a coolheaded young Chinese-American named Ed Lee, I myself queried this machine on how many Pepsi-Cola drinkers I would find in the audiences for several network shows, and how I could with three or four minutes to be bought on a scattering of programs maximize the number of soft-drink users my words would reach. And the machine paused briefly, then banged out answers in very specific, very authoritative-looking numbers.

4

Much of the data available from television audience studies can be stimulating stuff. Some of the information is just inexplicable. Why, for example, did a television special with Diana Ross and The Supremes draw lots of women over fifty but very few older men? Why do most movies draw heavier audiences in the last half than they do in the first half? When all the movies on television were rereleases after theatrical exhibition, it was believed that the new-comers to the audience for the latter part of the show were people who had already seen the picture at the movie house and were tuning in to catch some remembered high points at the end—but the same phenomenon seems to appear in the made-for-television *World Premieres* and *Movies of the Week*. Perhaps people just enjoy a denouement even if they don't know the situation that sets it up.

Friday night is the weakest night on television—one almost never finds a Friday show among the top ten. The reason, probably, is that Friday night is when the high schools across the country stage their athletic events. It's also the night of the paycheck: people go out more. In any event, the Friday audience even more than others tends to be composed of the young and the old, with fewer than usual representatives from the 25-49 group. Sunday night, on the other hand, is the heaviest viewing night, and gets all ages.

At any hour of any weekday, there are more women watching television than there are men. The age group over fifty watches more television than any other. But viewing by older people is strongly patterned. They constitute more than half the audience for the nightly network news shows, for westerns and for variety programs, but less than a third of the audience for movies. Among other very popular shows with a relatively low component of over-fifty viewers are the doctor shows like *Marcus Welby, M.D.* and *Medical Center*, presumably because the subject matter cuts a little near the the bone. One of the oddities of the fall 1970 burst of "relevant" shows, which were supposed to draw younger audiences, is that they did well with the over-fifty group, but bombed with everyone else.

The best-educated and highest-income audiences in television are those for football and for movies, though many specials and *Laugh-In* also do well at the upper end of the income distribution. It is not true that light viewers look for "better," more serious programs; studies done for Group W, the Westinghouse stations, indicate that light viewers tend to choose much the same shows the heavy viewers choose, and it is the moderate viewer who makes or breaks a more serious effort. Jim Yergin, who runs the Group W research operation, explains these rather tentative findings on the theory that the heavy viewer has a need for entertainment; the light viewer turns the set on only when he has nothing better to do, so he, too, seeks entertainment; but the moderate viewer has to decide whether to look or not and thus can be tempted by something out of the ordinary. Maybe so; maybe not.

The first factor in determining whether a show will do well or badly in the ratings is the popularity of its competition. The foundation-supported *Omnibus*, an effort to bring upper-middlebrow tastes to television, drew a larger audience on Sunday afternoons than it did when moved to prime time, because the competition in prime time was much stronger. (Except perhaps for Sunday morning, there is no "cultural ghetto"—as the pro football games demonstrate, the small audiences drawn by more ambitious programs on Sunday afternoons reflected not an unwillingness of the American people to look at television on Sunday afternoons but the slight attraction of the shows.) When CBS on October 21, 1971, programmed three full-hour documentaries one right after the other in prime time, two new opposing ABC shows that had not previously penetrated the Nielsen top forty received the third- and fifth-highest ratings of all shows telecast that week. ABC, knowing the CBS plan, promoted the evening heavily, and linked one of the shows to a plot line in *Marcus Welby*, to help nature along.

Public-affairs shows in general do not draw much audience; indeed, live news coverage at a time when people expect entertainment is likely not to draw much audience. On the first night of the Six-Day War, all three networks carried the UN Security Council emergency meeting. In New York, where a quarter of the television homes are Jewish and presumably have more than the normal interest in the fate of Israel, a rerun of *Alfred Hitchcock Presents* on an independent station drew more viewers than the three network stations combined. Nightly, habit-forming news shows do better, but they are by no means overwhelming winners. In Washington, which lives on news as Milwaukee lives on beer, the nightly national news shows were staggered in fall 1970—Howard K. Smith at 6, David Brinkley at 6:30, Walter Cronkite at 7. At 6 in the Washington market, according to the November 1970 ratings sweep by the American Research Bureau, the most widely watched program in Washington was reruns of *I Love Lucy*; at 6:30, reruns of *Petticoat Junction;* at 7, reruns of *Dick van Dyke*. In New York in early 1971, reruns of *I Dream of Jeannie* on an independent station drew a bigger audience than any of the three network news shows in the same 7 o'clock time slot. During its last two years as a regular Tuesday night feature, the *CBS News Hour* averaged a rating just under 9 points, a share of roughly 15 percent, an estimated audience of between 5 and 5.5 million households—a lot of homes and a lot of people, but less than the circulation of either *Look* or the *Saturday Evening Post* at the time of their collapse. Moreover, the audiences drawn by both news and documentaries tend to be slightly below average in both education and income, a fact that always shocks people who have not thought much about television.

It is hard to see how matters could be otherwise. Leland Johnson of the RAND Corporation, who did studies for the Ford and Markle Foundations on the prospects for cable television, was apologetic about his failure to watch the medium at all. "My problem is," he said, "that television is a very low-rate data transmission system, and I just don't have time for that." Despite much assertion to the contrary, television for most reasonably well-educated people is an extremely inefficient way to learn about anything. People really do learn at their own rate, and television is the most hopeless of lockstep classrooms, insisting that everyone in the audience work on the same time scale. As Wilbur Schramm and his associates put it in their book *Television in the Lives of Our Children,* "Watching television, the viewer cannot set his own pace. . . . This quality, of course, makes for good storytelling, good fantasy, because in those forms the storyteller *should* be in charge, and the viewer *should* surrender himself. But it makes learning harder. That is why the child, after he learns to read well . . . tends to seek information more often from print. With print he is in greater control."

None of this is to deny that documentaries have been artistically among the most satisfying and socially among the most important contributions of television, or to accept the idea that the poor ratings and minimal audience quality of documentaries give networks an excuse not to make and air them. But it does suggest that among those who insist Middle America is very stupid there are some who may not be so bright themselves.

5

What the advertising agencies and network sales departments break out of the ratings data becomes more sophisticated every year, correlating the geographical location, age, income, family size and education of people thought to buy the product with the geographical location, age, income, family size and education of people thought to watch each television program. NBC especially likes to play with BRI index figures—"AUDIENCES OF 12 RETURNING NBC SHOWS," says the forty-first of the network's "Product Usage Highlights" in 1970, "REVEAL HIGH USAGE OF DRY DOG FOOD." BRI computers calculate the average consumption of each of several hundred product categories, then measure the consumption in that category reported by houses that also reported watching each television show, and put the latter as a percentage of the former— *Adam-12*, in the NBC announcement, showed an "Index of Usage" of 131 for dry dog food, which means that average viewers of *Adam-12* buy 31 percent more dry dog food than the average American family. NBC then multiplies that index figure by the most recent Nielsen ratings figure for the show to get, in this case, a "Dry Dog Food Usage Rating" for each program on the air. "It gives the salesmen ideas for places to look for customers," says Sam Tuchman of NBC research, who puts together this material, "but do the salesmen use it? We don't know."

Advertising has a flourishing trade press, including publications as specialized as *Media Decisions,* and this very detailed "information" makes copy for the publications and sermon texts for speakers at the many, many advertising and broadcasting trade lunches, dinners and conventions. There is no secret about the fact that many of the detailed breakouts from audience survey data are not much good. Queried about his computer's answers to my questions about Pepsi-Cola drinkers, Bill Simmons said, "Well, we sell our clients tapes of our data, and they can go to tabulating firms, and there's the tape on the tabulating company's machine, and what can we do? I thought we'd better have a service ourselves." Henry Rahmel of Nielsen cheerfully admits that many of his customers are chasing a will-o'-the-wisp: "When you refine your demographics, you're putting in more and more statistical dollars to buy thinner and thinner cells and more sampling error." But the figures are there, and in Banks' Law (named for Seymour Banks, research director of the Leo Burnett agency), "Available data drive out necessary data."

When people in television sales talk about their work, however, about what actually happens day by day, they do not in fact stress these arcana. What they talk about is "$4 television," which means an expenditure of $4 to reach a thousand households, as measured by Nielsen ratings. "We can tell you," says Burnett's Banks, "about the viewing habits of beer-drinking housewives whose husbands are professionals. We do cross-tabs all the time on everybody's data. But none of this is as important as ratings, simply because the variations in cost-per-thousand households are greater than the variations in kinds of households and purchasing habits. We try to get the thing that is the best buy, with minor qualifications for kind of audience. The qualifications are necessary, but not sufficient. Cost per thousand is both necessary and sufficient."

What it comes down to is that advertisers are reluctant to pay more because the audience is younger and richer, but will gladly knock down the price if the audience is older and poorer. All any medium has to sell is access to its audience (for some reason, this is regarded as shameful in television, though not in, say, the *New York Times Book Review* or *Variety*), and the nature of the audience must affect the price. The shake-up of the CBS program schedule in fall 1971—the elimination of trusty stalwarts like *Beverly Hillbillies* and *Green Acres* and *Family Affair*—was caused partly by the feeling that the network's salesmen were having a hard time getting the right price for minutes because "the demographics are bad." But ABC salesmen, offering much younger audiences and total ratings only 12 or 15 percent lower than those of CBS, were almost always forced to settle for considerably lower prices per minute than CBS got. In the end, all information about audience, brute popularity numbers and subtle demographic breakouts, gets melted together in Mime's forge to make Nothung for some Siegfried in the sales department.

CHAPTER 3

The Mystical Business
of Selling Time

When I came to NBC, I brought the radio-television heads of nine major agencies with me. Our sales department never sold anything—I sold it, I and my fellows. We were agency men selling to other agency men. By 1952, NBC was sold out from three in the afternoon to midnight.
—SYLVESTER L. (PAT) WEAVER

Network is very competitive, and every day there's a spot left on the shelf you've got a loss. Like a store full of grapes when you're not selling grapes.
—EDWARD R. SCHURICH,
president of H-R Co., station representatives;
formerly vice president, station relations, CBS

The critical decision in TV is, when do you break price, and how far. Unhappy salesmen may make that decision badly, which hurts you.
—EMANUEL GERARD,
Wall Street institutional broker,
specializing in entertainment stock

There are no secrets in this business; if somebody makes a horrible deal, we hear about it in half an hour and roar.
—WARREN BOOROM, vice president
and national sales manager, ABC

1

Among those who sit on the program committees when a network is exploring its next season's schedule is the vice president in charge of sales. His is not the controlling voice by any means, because he has no special expertise in predicting whether or not one show or pattern of shows will draw more audience than another, and sooner or later the network's success in drawing audience will determine his success as a salesman. But there is also a sense in which, to quote Paul Klein shortly after he left NBC's research department, "A hit is what you can sell." *Monday Night Football* was a hit for ABC in the fall of 1970, even though it was often outdrawn by NBC movies and Bob Hope specials and by one of the strongest CBS prime-time schedules, because automobile makers, airlines, oil companies and beers were willing to spend heavily to reach football's special audience of high-income men. The rash of "relevant" programs that broke out on the ABC and CBS schedules in fall 1970—junior lawyers and interns and such, Helping the (clean) Poor—spread from a demand by salesmen for programs that would draw younger audiences, which such shows were supposed to do and advertisers were supposed to want. The increase in movies on all the network schedules in 1971–72 reflected the previous years' experience that movie audiences were weighted toward younger families. No one can begin to understand how programs get to screens unless he gets a reasonably good fix on the nature of the business qua business.

The business today is selling minutes, and it is complicated. First off, there are so many of them: for each network, six an hour in prime time, up to twelve an hour in the middle of the day (both figures by custom: there is no law here); all must be sold. And nobody can write specifications for the merchandise that changes hands when the sale is made, because the goods will not be manufactured until the instant of use. What a minute is "worth" is a question like what a common stock is "worth," answerable only by the market report of what somebody is willing to pay for it today. A ninety-minute movie will have nine network commercial minutes embedded in it, all presumably equal in value, but there may be a 25 percent spread between the highest and lowest prices advertisers actually paid for these identical minutes, depending on when each of them was sold. Add to this the fact that a given audience has different values for different advertisers: the makers of razor blades will not pay much for minutes on *The Partridge Family*, while the makers of depilatories will pay even less for minutes on *Monday Night Football*. Stir in the further fact that usually there is no "price" at which any given minute was sold, because the buyer has paid a lump sum for a "package" of minutes scattered through a number of shows. The result will be less a marketplace than a witch's caldron.

A further source of confusion is the obviousness of the "minute" as the unit of sale.* In fact, the networks at the start of the selling season do lay down a line of suggested list prices for minutes on different shows. But the unit the advertiser is buying is not the air time; it is the audience watching that network at that instant. If *All in the Family* has an audience of 20 million homes and *Dick van Dyke* an audience of 16 million (which were about the numbers given by Nielsen in November 1971), an $80,000 minute on *All in the Family* and a $64,000 minute on *Dick van Dyke* are in reality identically priced—both cost $4 per thousand homes. In 1971 that was what advertisers wanted to spend for the undifferentiated mass television audience—$4 per thousand homes per minute. Movies, with their younger audiences, might command a premium; westerns and variety shows, with older audiences, might go at a discount. The salesman's art is not in the pricing of minutes on individual shows, but in the assembly of a package especially tempting to an individual advertiser, moving a lot of minutes at the highest possible average price. "You can't really say there's *a* price for anything," says CBS sales chief Frank Smith, "because there are never two

---

*The term "minute" is used as a convenience. Since late 1970, when CBS forced the issue, the unit of network sales has been the thirty-second spot. It was not a big change for the viewer, because most larger advertisers had already taken to using a minute commercial as a "piggyback," with two brands sharing the time, usually on a basis of thirty-seconds each. CBS went to the new system mostly, one suspects, because it had the most elaborate computer installation in the business, and felt that in a depressed selling season it could expand its proportion of industry sales through sheer computerized efficiency. It was also true, however, that the Federal Trade Commission was investigating network selling practices, and some FTC staff members had professed themselves disturbed by the idea that a one-brand firm like Wilkinson Blades would have to put the price of a full minute on one item, while a multi-brand firm like Gillette could divide a minute between razor blades and hair spray, paying only half as much per product for each household delivered.

advertisers buying the same thing under the same conditions." Or, as an NBC executive put it, "All our sales are like tailor-made suits, for a humpback."

When selling began for fall 1971, NBC was asking $86,000 a minute for *Flip Wilson*, ABC was asking $84,000 for *Marcus Welby* and $71,000 for *The FBI*, CBS wanted $72,000 for minutes on *Hawaii Five-O* and *Medical Center*. The theory behind Section 5 of the Federal Trade Commission Act would require networks to sell just these minutes to anyone who asks for them, but the world doesn't work that way. "Do you think," asked James Duffy, president of the ABC network, formerly head of its sales division, "that there's any advertiser on *Flip Wilson* who buys only *Flip Wilson* from NBC? Neither do I." The fact is that minutes on the most popular shows serve essentially as "sweeteners" in packages of minutes. Only packaging makes it possible to sell for apparently satisfactory prices the minutes on the new shows (two-thirds of which, by television's saddest iron law, will fail). *Marcus Welby* undoubtedly helped ABC sell *Owen Marshall*; *Flip Wilson* undoubtedly helped NBC sell *The Funny Side*.

"If a man comes in here and says, 'I want to buy movies,'" says a CBS salesman, "I can't tell him to jump in the lake. But I can tell him I'll book it, and he can wait; I need those minutes for packages." A buyer of television time at one of the nation's half-dozen largest advertising agencies says, "When you're talking with a network salesman, the thing you really want to know is what he's got in the bottom drawer"—in other words, how many of the most-prized minutes have not yet been sold. It will never be possible to say whether or not the networks sold these top shows for their advertised top prices (all well above $4 per thousand homes, though the amazing popularity of *Flip Wilson* in its second season brought the $86,000-per-minute price for that show, the highest ever asked for a regularly scheduled program, down to about $4.25). But the packages in which they are buried will be expected by both sides in the transaction to price out to an average not far off the going rate. The fact that nearly all minutes in prime-time network television are sold in packages, by the way, makes it next door to impossible for the network comptrollers to say whether or not any given program is profitable to the network—you never know for sure how large a contribution to the sale of the package was made by any one of the shows in it.

2

Network minutes are sold in two quite separate markets, divided by time—an "up-front" market six months or so before the season begins, and an "opportunity" market a few weeks (or less) before the minute will go on the air. If he is willing to buy time in every program in next season's series—a commitment of $3 million for a minute a week in a single show—an advertiser prepared to come to a decision in March can probably gain a weekly presence on a show of his choice for the succeeding September to June. This is, of course, show business. The advertiser making an up-front investment in a television program is like a businessman deciding to invest in a new Broadway musical. Even decisions to continue advertising on one of last year's top twenty require judgments on the new shows opposite it in the same time slot: the ratings for the current year don't prove much about what is going to happen next season.

When the networks first asserted exclusive control over programs, in the late 1950s, they still tried to sell on something like a sponsorship basis, and show-business rituals grew up around the late-February announcements of what would be on the air next fall. Peter Bardach of the advertising agency Foote, Cone & Belding recently looked back, unwistfully, to "the old days when everybody went to CBS on Washington's Birthday, the days when I would sneak over to NBC to see something which theoretically nobody else had seen, while the guy I was faking out was secretly at ABC." That high-pressure blasting out of advertisers' money has stopped, because advertisers no longer buy shows; but even in the 1970s, with the overwhelming majority of sales made in packages, advertisers must attempt some show-business judgments on how well each program will do.

They don't get much help from people who are supposed to know better. In September every year, the trade press—especially *Variety* and *Broadcasting*—publishes estimates from the agencies about how well they expect the new season's programs to do. A chimpanzee pointing a stick at a board listing the programs would predict about as well as the published estimates, which are often grotesquely wrong. In the fall of 1970, for example, *Variety* quoted a consensus of agencies as believing *Flip Wilson* would run third best in his time slot, that Andy Griffith's *The Headmaster* would "do particularly well . . . one of the new season's heavyweights," that *Storefront Lawyers* would do well for CBS and *Mary Tyler Moore* would do poorly. Everyone was "bullish" on *Bill Cosby*, and there was "fairly wide agreement" that *The Partridge Family* would be the worst of a bad lot for ABC on Friday nights. "*Johnny Cash*," *Variety* reported, "is projected by all at a healthy 32-33 share. . . ." It was thought by the agencies reporting to *Variety* that *Jim Nabors* would hold *Ironside*'s level, that *Red Skelton* would do well on NBC, that either *Don Knotts* or *Beverly Hillbillies* or both would top ABC's *Mod Squad*. . . . All wrong; in sum, about as many wrong as right.

Most of the action in the first selling weeks, in March, comes in the shows scheduled between 8 and 10, in what Hal Tillson of the Leo Burnett agency calls "the gut time periods, the hours when people are watching television rather than doing something else." And it comes from relatively few purchasers. Broadcast Advertisers Reports, which monitors all network commercials, says that half the network revenues come each year from twenty to twenty-five companies; and these companies do much more than half the buying "up front." Television is their prime sales tool, and they cannot risk being without it. They must operate in highly competitive markets, fending off enemies marketing products so similar that in many instances the contents of the packages could be interchanged with nobody ever the wiser. They want their commercial "protected" against commercials for brands of similar products in the same program. And, not quite consciously, the buyers for these companies want the big hits, even at the highest prices: "Nobody ever got criticized," says a Procter & Gamble agency man, "for buying the number-one show." All these goals require long-range, early commitments.

Even up front, however, most buys tends to be a scattering of minutes on a number of different shows. "I've had them come into my office," says a man who was once president of a network but has sailed into calmer waters, "and sit across my desk and treat my program schedule as though it were the menu in a Chinese restaurant—they'd take three from Column A and two from Column B and four from Column C." Over the course of a calendar quarter, most nighttime network programs carry commercials from at least a dozen advertisers. On the other side of the same coin, most heavily advertised brands appear on as many as fifteen to twenty shows during a single quarter. Commercials for Procter & Gamble products have appeared on as many as 140 different network shows (day and night) during a single quarter.

The shift from the sale of hours and half-hours to the sale of minutes has had one salutary effect on programming: it has made much easier the substitution of "specials" for regularly scheduled shows. In the old days an advertiser whose time had been pre-empted had to be reimbursed for the talent payments he was contractually committed to make to the people who normally put on his show. Since the mid-1960s network contracts for minutes have routinely provided for pre-emption on ninety days' notice, with no penalty; and the network contracts with talent have been for specific numbers of shows, always less than the number that would be necessary to fill the time slot all year.

Specials can be planned many months in advance—indeed, at NBC, which does more of them than the other two networks combined, the schedule of specials is known only a few weeks after the schedule of regular weekly programs. Because specials offer the hope of a "gratitude factor," specials are often sponsored (and not infrequently supplied) by a single advertiser.

The other side of this coin is that the program departments of the networks are not able to control their specials schedule anywhere near so completely as they control the series schedules. American nighttime television in 1970 left unremarked Beethoven's two hundredth birthday, although CBS had put a good deal of money into filming Leonard Bernstein following the footsteps of the Meister round and about Vienna. No advertiser was willing to sponsor the Panther-scarred Bernstein, and CBS was not prepared to air the show for free. (It was broadcast on schedule in England, and on Christmas Eve 1971 in America.) Every once in a while an advertiser finances and controls a blockbuster special, guaranteed to pull a giant audience, and can shop the networks for a place, as Budweiser did with its 150 minutes of John Wayne. "We thought," says sales vice president Jack Otter of NBC, which won it, "that they might like to be on Sunday night."

3

Before 1971 the premier customers for up-front minutes were the cigarette brands, which wanted reach and frequency both, had lots of money, and for public-relations reasons felt restricted to time slots after 9 in the evening, when presumably the kiddies were in bed. More than $150 million of network business came from this source (chunks of it, of course, on weekend sports presentations), and when it was banned halfway through the 1970–71 season, it left the networks with only about 60 percent of their minutes sold up front. This reduction in up-front sales, moreover, was merely an acceleration of a trend line that had been running through the second half of the decade. More and more advertisers wanted to buy on a quarterly rather than an annual basis, held back on their commitments, waited for the "opportunity buy."

"The whole thrust of business today," says Bern Kanner of Benton & Bowles, "is to be flexible. And top management in a company is no longer involved in decisions about television buys. The brand manager is measured by his profits and volume of shipments, not by what shows he sponsored. If you ask him to decide in March what shows he wants to be on a year from next month, his first thought is: 'Oh, hell. I'm not going to be here in a year. . . .'"

What had kept the networks' up-front books close to filled before

1970–71 was much the same set of factors that enabled the Metropolitan Opera to sell out, or nearly so, by subscription: prestige, plus the fear that if you wait for just the things you want you won't be able to get them. The removal of cigarette advertising pulled the plug, and the networks came into the first quarter of 1971 with literally thousands of unsold prime-time minutes. The big advertisers waited for their quarry to weaken, then moved in for the kill: in February and early March, all three networks made some sales on successful shows at prices below $3 per thousand homes delivered. In the sellers' market before 1971, the time salesmen had talked rather glibly about the market for minutes becoming an auction market, with prices controlled entirely by the flow of bids and the supply of minutes. In early 1971, like stockholders living through a Wall Street crash, the salesmen—and everybody else who worked for a network—learned how dangerous an auction market can be.

CBS vice president and Eastern sales manager Bud Materne describes the process: "An agency guy comes in with half a million dollars, and says it's for a cereal company that wants to reach eighteen-to-forty-nine-year-old housewives. He's not expecting something quoted from some rate card—he could do that himself. So the salesman goes to Dana Redman in research and says, 'I think NBC will come in at three-seventy-five; what can we do?' And Redman says, 'We can't do better than four dollars—you've got to *sell*.' So the salesman goes and says, 'Here's *Mission: Impossible, Lucy* . . . nothing on Sunday night, when you don't sell crackers, anyway.' And the guy says, 'Nothing doing. ABC comes in at three-sixty, NBC at three-fifty-five. I'm going to recommend we give half of it to each of them. You're wiped out.' *Wiped out?* So the salesman goes back to his boss. . . . Every year they sell less and less of the schedule up front, which leaves more of the pie for me. Aren't I lucky?"

Minutes are offered as long as they are alive. Jerry Jordan remembers cheerfully from his days as advertising boss at American Airlines that "I bought lots of minutes this morning for tonight." Jim Shaw, a tall Irishman with a fringe of red hair who presides over all sales operations for ABC, says, "Sure. The sales proposals department makes the shoe to fit on Cinderella's foot before midnight.

But the law of supply and demand here is exotic. You have to look at what does the early buyer get and what does the late buyer get. You have to look at the longevity of the relationship, and you have to stay out of childish auctions." The great advantage of the late buyer is that he already knows this year's ratings, while the early buyer may be struck by a bomb. Help in these situations is part of the "longevity of the relationship." An agency man says cautiously, "CBS takes care of you. ABC, too, though it may be next year. NBC may say, 'That's show biz.'"

Suddenly, in early spring 1971, the market for minutes turned around, fed by improvements in the economy and a fear that there was going to be a shortage of minutes under a new FCC rule which would eliminate eight network half-hours, seventy-two commercial minutes, every week starting October 1971. Within three or four days, prices rose by almost a dollar a thousand for prime-time network minutes. Encountered in the elevator at the ABC building, national sales manager Warren Boorom confirmed rumors in the trade press. "Absolutely," he said. "All those horse-shit projections I was making—they turned out to be true."

Among the reasons for the improvement in the market was the networks' success in developing new customers to bid for the time necessarily abandoned by the cigarette companies. Some of this demand came from companies that previously advertised on local stations—Avon Products, Continental Baking, Coca-Cola—more than they did on networks; they liked the new low prices. Some of it was hard sell to businesses like motel chains, McDonald's hamburgers, insurance companies that had begun television advertising relatively recently and were ripe for demonstrations that their expenditures had worked for them. But some of it was from companies that had never thought of seeing their logos on a home screen.

"I started in 1969 from scratch," says William Firman, ABC's vice president in charge of marketing, "from ground zero. The idea was to create new dollars for the medium. Television had been delinquent here, because people didn't have the time to devote to it; these were difficult dollars to come by." Firman is a faintly military man with blue eyes, a lined face, long gray hair, wearing a narrow regimental tie; a little older than the rest of the ABC sales staff, occupying a corner office. "When you walk into one of these companies that have never been on television, they'll say, 'I only want to reach one hundred people in the whole country—that's all, just one hundred people.' 'Well,' I'll say, 'are you interested in recruitment? in morale? in Congress? in stockholder relations?'—and by the time I get through it makes a hell of a lot more than one hundred people. But they know the editors of the trade publications, and they don't know us from Adam's off ox."

Firman smokes cigarettes; here he takes a puff. "I made a presentation at lunch in sixteen major advertising cities across the country," he says. "Subject: 'Televised sports as a way to reach the nation's influentials.' That starts the dialogue. Sports is only one thing we do—we have talk shows, news, documentaries. Then I show them a reel we've stolen, of corporate commercials—Clark Equipment, 3M, North American Rockwell, GE for recruitment—and a slide presentation, *Fortune's 500*, with success stories.

"But the whole game with these cats is *extensional thrust*. If you can make it work for him off the air, you keep him stimulated. Say a guy buys a documentary. We can arrange to have key people preview it at nine-thirty in the morning at studios all across the country. Elmer Lower [ABC News president] gets on camera just before it rolls, and says how pleased we are to have this company associated with it. Or he buys *Issues and Answers* and he can come to the studio for the show, have a cocktail or lunch with the guest. Then he gets home the next day and he says, 'Hey! Do you know who I had a drink with in Washington? . . . *Melvin Laird!*'

"Maybe he buys NCAA Football. We can get him tickets: he can take his key people. Ditto National Football League. Golf. We have a hospitality tent at the big tournaments. He can play a round with Byron Nelson, meet Jack Nicklaus. When it's done, he can have special films, news clips, audio tapes, fliers. . . .

"It's been slow. We started at a time when the economy turned sour. The very first call I made worked, but the guy wound up on CBS. Nevertheless, this can be a very big swing in this decade." You bet.

4

Not everything in television gets thrown into the auction pot. Both ABC and CBS have held the line on pro football minutes, which will be given to the Cancer Society rather than cut in price. (NBC, with the rather more questionable attraction of the American Football Conference, teams in only five of the nation's ten largest metropolitan areas, has had to bargain a bit more. Bargaining on the World Series, incidentally, can be cute: advertisers may be offered a minute for free on the seventh game, if it happens, in return for a full-price purchase of a minute on the third game.) Newsshow minutes go on a rate card, based on a charge of about $2.50 per thousand on the average audience anticipated in the next quarter, for each of the five commercial minutes in the half-hour. Daytime, too, is by formula, the price changing every quarter according to historical viewing patterns during the quarter. CBS, with a sold-out roster of soaps, sells daytime minutes only on a fifty-two-week firm commitment, and there are agencies and advertisers lined up outside the door waiting for a minute to open up. Some daytime dramas (including As the World Turns, the doyenne of the medium, sixteen years on a single story line and top of the ratings every year) operate on something like the old radio basis: Procter & Gamble supplies the program. But the deal is more complicated than radio deals used to be, and CBS sells some of the minutes on the P&G shows.

Prices are also fixed, at least to national advertisers, in the sale of time on local television stations. These sales of local minutes (generically, "spot" television, from the days when the networks sold time periods for programs and only stations sold minutes) are the basic source of income for the stations. In 1970, of total revenues of $1.89 billion earned by all the nation's commercial television stations, only $240 million came from payments by networks to affiliates; $1.09 billion came from national and regional advertisers and $506 million from local advertisers. Every once in a while a local station may sell a time period to a sponsor who slots in a show—

everybody's favorite is Billy Graham, who not only buys his time but counts later as religious programming (*very* good) when the station must tell the FCC how its broadcast year was divided up among various categories. But the normal procedure is for a national or local advertiser to buy a minute or a smaller time period (20 seconds is a common length on local television) for the insertion of a commercial, supplied to the station in the case of the national advertiser, often made by the station itself (at a fee) for the local advertiser.

The highest-priced minutes, on most stations, are the "adjacencies" between prime-time network shows. Almost as expensive on all stations, and even more costly on some (especially in the Midwest, where the local channel's own 10 o'clock news may be the highest-rated show in town), are the "fringe-time" minutes around prime time. A station like WCCO-TV in Minneapolis, with a crack news staff and a dominant market position, may take in nearly $2 million a year from the sale of minutes in its 10 o'clock news. Then there are minutes to be sold in shows presenting the station's own library of movies, in filmed series or reruns of network shows no longer on the network, in "syndicated" shows rented by the local station from their producers (including *David Frost* and *Mike Douglas* and some original entertainments like *Story Theatre, The Mouse Factory, Wild Kingdom, Monty Nash*), in slots left black for local sale within the late-night talk shows, the daytime soaps, the football games.

For each of these time periods, in each season of the year, there is a station rate, published in big paperback books by Standard Rate & Data. Sales of local time to national advertisers are made through one of several dozen firms of "station reps," which handle scores of different stations scattered all over the country, employing large numbers of salesmen (Blair once built an 11-story office building north of the Loop in Chicago) and processing huge quantities of paper. There is no standard charge for reps; their commissions (applied to the net receipts of the sale, after deducting the separate advertising agency's commission) seem to run on a sliding scale from 8 to 15 percent, depending mostly on volume.

Basically, the rep's job is to keep in touch with the agencies (most do not, in fact, maintain contact with advertisers); his worst headache is keeping track of the "avails," the minutes available on the station's schedule at each time period in each day. With few exceptions, spot sales are short-term, often for a month or even two weeks, rarely for more than a quarter.

Like the network salesman, the rep relies heavily on ratings, which are done market by market in periodic "sweeps" (at least three a year in each market) by both Nielsen and the American Research Bureau. Here the dominant factor has not been Nielsen but ARB, partly because everybody (maybe even including Nielsen) wants to avoid a situation where one company has a monopoly of ratings. ARB covers 220 markets, distributing from 250 to 1,500 diaries in each, according to the population in the market. The composition of the population must be considered as well as the size. In the Negro slums, rewards as high as ten dollars a week have failed to produce returns from much more than 10 percent of the diaries mailed out. In addition to sending diaries to the homes in the sample in a dozen such areas, ARB girls (some Spanish-speaking) call the home every day to get the information on an immediate-recall basis; and the telephone interviewer makes out the diary for the family.

ARB cannot begin processing diaries until eight days after the close of the sweep period, but it gets the computer-prepared books for some larger markets (every market has its own book) into the mail only a week after processing begins. These books are the bible for the station rep, as the Nielsen "pocket pieces" are for the network salesman. For national advertisers there is also a fascinating summary volume, an *ARB Network Television Program Analysis*, which gives for each market an index figure expressing the relationship between this market's share of the total television population and its share of the audience for each program. Thus, a rural-oriented show like CBS's late *Hee Haw* might have a New York index of 35 and a Nashville index of 145, telling an advertiser that in New York he was getting only 35 percent of the audience he might expect from a nationwide rating of 20, while in Nashville he had a 45 percent bonus over what the nationwide rating indicated he received. By correlating these geographic indices with the distribution of his brand and its sales around the country, the advertiser can decide where he needs more "pressure" to fulfill his "marketing plan" for the year. Network shares always tend to be a little lower in the largest cities, which have more and stronger independent stations, so the advertiser who wants to achieve as much pressure in New York and Los Angeles as he gets in Des Moines will have to buy extra spots on New York and Los Angeles stations. This is not the least of the reasons why the big-city stations are so much more profitable than any others.

Local television gives advertisers headaches, because from the point of view of a New York advertising agency local broadcasting is something done in private. F. Kent Mitchel, marketing vice president of General Foods, puts it mildly: "One of the weak links in spot is that there's no sure way to know it gets run." Ultimately, this problem may be eliminated by a computerized service which imprints a code on each commercial, automatically picks the code off the broadcast air in each city, and sends the advertiser huge books of data on what ran when where. Such a service has already begun operations for a few large advertisers in some dozens of cities, under the name Digisonics; at the end of 1971 it was still plagued by a few technical and many conceptual bugs. For the time being, most information on this painful subject comes to advertising agencies from Broadcast Advertisers Reports, which covers the fifty largest markets.

BAR's basic tool is the ordinary sound-only tape recorder. For one week each month, television set tuners feed the sound from the broadcasts of 265 stations to tape recorders BAR has placed in private homes around the country. The tapes are then mailed to a one-story brick building in a slummy neighborhood of suburban Philadelphia, where dozens of women (seventy all told, on two shifts) sit at little tape decks, wearing earplugs, listening, noting on a printed data sheet every commercial broadcast on this channel. With practice, the women learn to run the tape fast through the program material and to note only the spot commercials; an experienced girl will note all the commercials from a station's eighteen-hour broadcast day in a single eight-hour working day. A separate book is published for each of the fifty markets (and a confidential separate report on a direct monitoring of the networks' own feed is made for exclusive use of the networks themselves). The value for the advertiser is partly a check on whether his own commercials are being aired, partly a look at what his competitors are doing in the market. The value for the station rep is the data on what the rivals of *his* station in this market are selling, which gives him a guide to where to go to steal business.

A round-faced, efficient, not humorless man named Phil Edwards left a job as a station time salesman in Chicago to start BAR back in the 1940s. "I came to the office one day," he recalls, "and found all the chairs had been taken out of the sales department. The manager told us the salesman's place is on the street; he didn't want guys sitting around in the office. I decided I wanted to get into some other aspect of this business." He reports that stations pay up to $4,000 a year for their monthly pamphlets, while advertising agencies pay on a scale that tops out at $52,000 a year. BAR also reports its own estimates of what advertisers have paid stations and networks for their minutes, information gathered through a network of informers in both television and advertising. Until 1971 BAR summaries of the business at all three networks were printed every week in *Broadcasting* magazine; early in 1971 the separate figures for each network were eliminated from the breakdown by "dayparts."

There is, of course, a limit to how neat local television advertising can be. Many local spots are sold "run-of-station" (like the newspaper's "run-of-paper"), which means they can appear in any time slot in a "daypart" (i.e., prime time, early fringe time, daytime, etc.) and still meet the conditions of sale. Others, in the evening, are sold for reduced prices on a pre-emptible basis—that is, if anybody comes along willing to pay list price, the earlier sale is canceled. Still others, though one doesn't want to talk about it, are bartered: the garage that supplies tires to the station's trucks has been paid off in minutes its owners can peddle to any likely buyer. "TV advertisers who understand the medium," Edwards says, "have learned to live with a little confusion." Stations in the fifty largest markets broadcast upwards of a hundred commericals a day, apart from the network feed, and even if the agency supplies the right piece of film or cartridge of tape for every thirty seconds involved (which is by no means inevitable), a busy back-room crew must do a monumental job of organizing every day if the messages are to pop onto the screen as ordered. And, of course, strange things can always happen when one must deal with people. "Let's say there's a time buyer for some agency," says Hal Tillson of Burnett, "and somebody at a station in Minneapolis can't stand the son of a bitch—sells him fifty-two weeks, pre-empts fifty of them and never tells him . . ."

There are problems about when bills get sent and when they get paid, and problems also, though nobody wants to talk about it, in how some advertising agencies handle the money. Because so many stations are late billers, and advertising agencies live on the commissions they receive when their clients pay bills, advertisers in the 1960s got into the habit of prepaying their agencies for the local spot schedules the agencies had bought on their behalf. Sometimes not all the spots on those schedules were aired, and sometimes agencies forgot to refund the advertiser's money. Some agencies began to use these advertiser prepayments as their own working capital rather than segregating out into escrow accounts the 85 percent they would eventually be asked to pay the stations. If the agency got into financial trouble, it meant others would be in trouble, too. But even these clouds have silver linings. Interpublic, the second largest advertising agency group in the country (built around McCann-Erickson) was saved from bankruptcy only after its clients learned that if the agency *did* go under they would have to pay a second time for local television spots: Interpublic had already used for its own purposes the money its clients had handed over to pay bills from stations, and the clients were still liable to the stations. The clients decided it would be cheaper to keep Interpublic alive, as indeed it was. In 1971, however, advertisers let the well-established Lennen & Newell agency go under, and invited stations and networks to sue for their money.

5

The most likely source of really major change in the American system of broadcasting is the greatly improved organization of local television advertising which can be foreseen for the latter part of the 1970s, when the bugs are removed from the computers and their programmers. A sizable proportion of network business now comes to the network mostly because it's so much more convenient and sure to buy network minutes than to buy local minutes. A change in this situation would move the locus of authority in television.

In real life, such apparently trivial commercial factors usually are the force behind what later look like major societal shifts. Television killed off the mass magazines not because it is necessarily a better advertising medium (the Magazine Advertising Bureau can show all sorts of studies about how ads in magazines outpull commercials on television, just as the Television Bureau of Advertising can show studies about how television works better than print), and not because the numbers are so much bigger (the basic damage was done while the circulations of *Life, Look* and the *Saturday Evening Post* were not far short of household audiences for all but the most popular television shows), but because the traditions of broadcasting gave the time salesman a major competitive advantage over the space salesman.

Magazines historically have had to sell their pages at the same price to every advertiser. If a magazine cuts the price of a page to one advertiser, it is contractually obligated to give all other advertisers in the issue the same reduction. Magazines can be thinner or fatter, depending on ads. But the broadcasting day is invariable; unsold minutes are gone forever and the loss cannot be made up by adding minutes in better times. As the fatal minute nears, then, the time salesman has authority to cut his prices, before the merchandise spoils, while the space salesman can only repeat yesterday's arguments.

People tend to think of economic competition as something that goes on everywhere and all the time. In fact, most business is committed to one supplier or another from early in the game, and meaningful competition occurs only at the margin, among the relatively few who are undecided. Television inevitably won all marginal competition against magazines because it could offer cut prices, and as time passed, the margin receded into the territory once controlled by print.

Newspapers and magazines have been known to cut rates. (Carey Street, home of the bankruptcy court, sits right behind Fleet Street, home of the press, in London; and the cant line of publishing says that the path from Fleet Street to Carey Street is paved with broken rate cards.) But it's an illegal activity, and it can't occur too often. Magazines were unable to compete effectively until the later 1960s, when they began to offer advertisers a chance to buy in a variety of regional editions. Now they had perishable merchandise, too: a page that had to be printed anyway, because it had been sold in one regional edition. If a page was sold for, say, the State of Utah, that same page could be sold at any price in the other forty-nine states without any concern about prior guarantees to the purchasers of other pages. By then it was painfully late in the game for the magazines: the packaged-goods companies had grown a generation of brand managers who had never used print at all. Anyone who cares about magazines must be deeply grateful to a Congress that banned cigarette advertising from the airwaves, moving it to print.

Commentators have been tolling funeral bells for the networks for some years, foreseeing a future when people will be so bored with commercial pap that they'll watch public television, or the multiplicity of channels on cable television will fragment the audience, or everybody will be watching sex manuals distributed via video cassettes. But a much more likely menace is the improvement of efficiency at the local stations. Markets are local; advertisers given a real chance to tailor their broadcasting expenditures to their geographical marketing patterns may grow skeptical about the quantities of waste circulation inevitable in a network buy. Networks can and do sell regional advertising time (viewers in the South would be amazed to learn how much antifreeze advertising appears during football games as broadcast in Northern cities), but they can't cut it very fine. And people who have computers just love to cut things fine.

If the preponderance of the revenues from broadcasting began to go to the stations rather than to the networks, everyone would have to think again about the question of how television programs should be supplied to the American public. Until then, commercial network television very much as we know it now is what the country is going to get, for better or worse, in sickness and in health.

CHAPTER 4

Flip Wilson and Other Prime-Time Phenomena

You know about the African tribe who saw their first movie? It was *King Kong*, and after it was over there were big cheers. The next week they were shown a movie again, and they tore down the tent and the screen and trashed the projector—because it wasn't *King Kong*. I think people want to see the same thing week in and week out.
— ROBERT GOLDFARB,
director of program development, CBS Television

I firmly believe that people do respond to different things. I just wish I knew what they were.
— FRED SILVERMAN, vice president
in charge of programs, CBS Television

The great secret is not to give things pretentious labels. "Satire," a gloomy theatre manager in the provinces was once heard to remark, "satire is what closes Tuesdays." As I see it, entertainment runs forever.
— TOM SLOAN, Head of Light Entertainment Group,
Television, BBC

Of course you have no right to ram things down people's throats or patronize them. On the other hand, you have every right to give them what you think is worthwhile. It is monstrous to give them what you know to be crap on the grounds that you believe that is all they are worth. . . . It is all a question of the respect the professionals have for their subject and their audience. And it is important to understand this distinction because it cuts right across the old divisions of highbrow and lowbrow or middle class and working class.
— RICHARD HOGGART,
assistant director-general, UNESCO

73

1

Down in the basement of the immense shed of the NBC Studios in "beautiful downtown Burbank" a crowd of twenty-five people was gathered in a narrow hall outside a long, narrow windowless room with a mirror along one long wall and acoustical treatment on all other surfaces. Inside the room, as could be seen through the panes of glass in the top half of the doors, Flip Wilson was earnestly talking to and gesturing at a group of half a dozen men standing around a hexagonal table, some in sports jackets, some in shirtsleeves. They were his producer and director and writers, plus his "guest star," Tim Conway. While at lunch that day Wilson had thought of a routine with Conway, involving how to cheat at the poker table, which he wanted to add to that week's show. As the group listened, a man in the back lettered key lines of Wilson's discourse and his colleagues' suggestions onto small placards, which would be displayed to Wilson and Conway during the rehearsal: this skit would go from idea to prompt cues without ever becoming a script.

The twenty or so people out in the hall all had some connection with NBC or the show; they were to be clumped in rows at one end of the room to make an audience for what producer Bob Henry called "the eternal run-through," the last it's-still-taking-shape rehearsal before moving upstairs to the studio. (NBC rents a studio to a show at $450 an hour; the basement room is logged at $10 an hour.) That rehearsal was now being delayed by the serious business of creating a new skit. Henry likes to have an audience around to help sharpen people's timing, but he doesn't want their reactions distorted by in-group fooling around, and a Wilson run-through has a matter-of-fact quality very rare in theatrical rehearsals. ("Sometimes," Henry says, "the cameramen tell me that what the show needs is a little humor; I tell them to wait till we tape.")

During breaks, the other cast members cluster and tell stories, but Wilson sits by himself on a folding chair in a corner, smoking a cigarette, his eyes in the middle distance, the big smile familiar on the screen flashed to anyone who does distract him, but no encouragement whatever to distraction. "He's working," Henry said. "Perry Como once told me, 'I know everything that's going on at all times,' and Flip is a lot like that, too. He never had any formal education, never went beyond eighth grade, but he's a human computer, he absorbs, he remembers, he feeds it back. He puts everything he has into this show, and he doesn't *do* anything else—except the Rolls. The one thing he's done is, he's bought a Rolls. He loves to drive. Sometimes when we get two days off, he'll just drive the Rolls up to San Francisco, and he'll fly back to work while his man drives it home. Or he'll drive it over to the ocean and just sit in it and look at the water. In his own world he's a very experienced guy, and so am I—I've been in television since 1950—and when we started, I told the staff, the best thing we can do is get out of his way.

"I didn't need this show," Henry continued. "I'd done four years of the *Andy Williams Show* and forty-one weeks of Nat King Cole, and for two years I'd been making specials, having a ball and doing very well financially. New York had tried to do something with Flip before, and it hadn't worked. Then I caught Flip at Melodyland in Anaheim, a tent kind of thing, and I'd never seen him perform so well. He'd hear laughs behind him, and he would react with his body—that body is *gold*, it's the best since Chaplin. So I did that set with the stage in the middle and the runways, give him an audience all around him and put him above them, give him the benefit of a low camera angle to make him look taller. We keep a camera slaved on Flip, so even when we plan to take other shots we can always cut him in later, and I've told the cameraman never to cut him off at the waist.

"Back home I've got all but two of the Chaplin films in my own collection. Chaplin is a genius. Carol Burnett is a genius. Flip may be a genius, too. We did a take-off on silent film, in which he's an artist, and he hasn't eaten for six years, he's been kept playing the piano. Don Rickles offers him a cookie and says, 'Play! If you play well, you can have another cookie.' The way he looked at that cookie . . . I think, maybe he is a genius." This was toward the end of the 1970–71 season, and Henry, who is about as short as Wilson but stouter and far from a performer's physical condition, remembered that he was tired. "But it's every week for an hour," he said, rising to go to work. "It grinds you down. Thank God for commercials—that means it's only fifty-two minutes, not an hour."

The producing function is the key to the variety show, and probably to most other television programs. (In the 1950s, when dramatic presentation on television was live from New York, the producers were almost all men who had been or wanted to be theatrical directors, and could handle the problems of putting actors on stage; by the end of the 1960s, when dramatic presentation was on film from Hollywood, the producers were almost all men who had been writers, and could handle the problems of getting a believable shooting script before cameras.) If the system is working properly, the producer guards the concept of the show, keeps the writers from giving its star the wrong lines, keeps the star from self-expression that might alter the image of self projected on the screen, keeps the director from visual effects that change the shape of the package.

On the positive side, the producer, maintaining with the star a relationship both friendly and distant, will make the important choices of surrounding personnel, writers, guests, directors, and will establish the *pace* of the program. He must be able to see what the show will look like to people not engaged in its production; he is the responsible man. Under the most usual commercial organization, each show is separately incorporated (even in the movie studios, ventures are kept separate, to maximize flexibility in bookkeeping, a phrase that can be interpreted in as many ways as the reader has the wit to interpret it). The producer always owns a piece of the show and participates in its profits. The network participates in the profits by its sale of minutes to advertisers at total prices higher than the price paid for the show, and by its rental of facilities at outrageous prices—$450 an hour for time in the studio, $350 an hour for editing time on the videotape machines, 75¢ per tile for the oversized vinyl tiles on the studio floor (which must be rented separately for each use), and much else. Network books are kept in such a way that the program department always loses money (and the news department loses *lots* of money), while the facilities always show a handsome profit. In the old days networks used to own pieces of shows produced by outside companies, but the FCC has now prohibited such investments.

A great deal of the atmosphere of a show depends, of course, on the personality of its star. Back in the days when shows were produced by advertising agencies, Mike Kirk of the Kudner Agency put Milton Berle on television for Texaco. "We decided," he said in the late 1950s, very businesslike, a businessman whose business was show business, "that the day of charades was over, and that the first advertiser who got a real TV show would clean up. I thought, 'vaudeville.'" After that it was Berle's doing: the wink, the huge grin, the leer, the horrid tragicomic grimace all projected onto what was still a very small screen. Berle's theory was that there were no new jokes, and he was deeply resentful in 1952 when the show began to falter and Kirk brought in Goodman Ace to write lines for a "new Berle"—even though the changes Ace wrought in the show (giving Berle a pair of sidekicks, removing him a little further from reality by casting him as an actor playing Milton Berle) put the show back to the top of the ratings for two years. ("He missed the essential Berle," said Berle. "I would have liked to," said Ace.) But for all the frenzy, Kirk looking back on those days had fond recollections: "When there was work to be done, Milton was always there, doing it. You didn't have to worry that he was out boozing or in bed with some babe. I'd rather work eight more years with Berle than two more weeks with Gleason."

Triumphs like *The Flip Wilson Show*, which led the ratings through most of the fall of 1971, are not accidental at all. Good stand-up entertainers are the lifeblood of broadcasting in all countries (except perhaps in Eastern Europe, where the state cultivates drabness with loving care), and even when advertisers controlled most programs, the networks tried to guarantee themselves a supply of comics. In 1948 CBS established itself as a serious ratings rival to NBC by buying up the services of Jack Benny, Edgar Bergen, Red Skelton and Burns and Allen (all of whom continued, however, to be "sponsored" in the usual manner). And ever since the Pat Weaver days of the early 1950s, NBC has tried to systematize the arrival of new comedians on the air. Weaver had a vice president in charge of development, Leonard Hole, whose first responsibility was a stable of six to eight young comics and the recruitment of writers for them. "Kaye Ballard," Hole said in 1956. "I kicked out her entourage, said, 'You're a long way from being a star—bad physique, never studied pantomime because you move badly, your material is lousy. You sing a great song.' We took her over, worked all summer and spring on her. The night club Bon Soir took her three weeks with our material, and she's still there, there's a line outside the club. The next step will be exposure to camera—on the Steve Allen show [the first edition of *Tonight*]. She wants Perry Como. I'll beat her over the head. Then if she goes on Allen, there will be a series of exposures, and next year, her own show. We did it with Jonathan Winters, too; we put George Gobel together with Hal Kanter. And we'll do it with others."

To get writers for these comics, Hole wrote to college papers and drama clubs, and to four hundred radio station managers, soliciting applications for a training program. NBC got almost fourteen thousand entries to its contest, and, Hole said, "We read every one. The big bulk were carbons, stuff submitted before and rejected, and in no time we were down to nine hundred. We asked these nine hundred for more material, and asked them for a comic they felt comfortable with, and then reduced the list to thirty. We took their stuff to Goodman Ace, Sid Caesar, Bob Hope, Weaver, Sarnoff, etcetera. Everyone agreed that nine of them showed great promise, most of them in New York, one in Boston, one in Baltimore. We gave them seven-year contracts. They work for the development department, but they also sit in on sessions with the great comics, and if the stars want 'em, they buy 'em."

Nobody at NBC remembers what happened to these projects, exactly. Mort Werner, a rather lumpy little man in a wrinkled gray suit who was Weaver's assistant in the 1950s and has been NBC vice president in charge of programming since 1960 ("I have worked for this company since I was a child"), now believes that "the idea of the comedians in the Catskills was a big mistake." The fallacy lay in the confusion of ambiance—the night-club and resort-hotel comic must come out and knock 'em in the aisles fast, and most of his audience is not going to see him again for a long time, if ever. A television comic comes back to people's living rooms (if they tune him in) every week. It should be said in fairness to Hole that he contemplated monthly rather than weekly appearances for the comics he was nursing (and worried that "even a show a month is an enormous burden to a staff").

Werner shifted the development effort away from the night clubs to broadcasting, to guest spots on variety shows, summer replacement shows, and especially *Tonight.* "When we decided Flip Wilson was the best young comic on the horizon, a couple of years ago," he said in early 1971 as the new show gathered power, "we put him all over, in the series, on the *Tonight* show, and then we made a special. It was a disaster. I've never let anybody see it. I have the tape here in my office, locked up behind doors, and when I feel bad, I show it to myself; then I feel better. A year after we made that tape, we got Bob Henry to try, and we had our show."

The Flip Wilson Show is a careful blending of established radio and television conventions, tailored to the strengths of its hero, who can dance a little and sing a little (one of the running gags is the despair of the show's musical director, George Wyle, on the question of Wilson's singing), act a little—and time everything to the nanosecond. Guests do their own stuff, join Wilson in a musical number, and are inserted into an episode in the continuing history of a number of characters Wilson has created—Geraldine, the Negro lady of independent temperament ("Don't you touch me! Don't you *ever* touch me!"); the Reverend Leroy of the Church of What's Happening Now; Sonny the cleaning man, etc. What is most fascinating is that all these performances are in an older tradition one would not automatically have believed viable in the 1970s—the tradition of Bert Williams and Bill Robinson and Bubbles, the great Negro entertainers whose stock-in-trade was a lightheartedness that verged on the feckless. It is comedy as a weapon of self-defense, expressing the invulnerability of a folk that survived conditions of pitiless hostility.

"We were worried about Geraldine and the Reverend Leroy," says Herb Schlosser, who for years sat in on conferences as the lawyer attached to Mort Werner's programming office in New York and then was transferred out to run the West Coast end of NBC programs. "But there's been no protest at all about Geraldine, and only a handful on Reverend Leroy." Henry says, "When Milton Berle puts on a woman's dress, he's working in drag—but not Flip. He doesn't play for easy laughs. It's an artistic creation; he does it so well, and so straight. The strength here is that we're all of us concerned all the time about a good show: there's no substitute for that."

The Flip Wilson Show is done before a real audience in one of the four enormous "major studios," each about the size of a football field, at the NBC Burbank building. It is not, however, broadcast "live," as all the big radio shows were and as the early television comedies were. Each episode in every show is separately taped, with breaks between to allow for costume and scene changes, and for moving the cameras into just the right positions for what comes next. In all, the taping requires two to two and a half hours for the fifty-two minutes of program. Even when the show is completed, Henry retains options—tapes made at the dress rehearsal, and tapes from the cameras not chosen by the director at any given instant, can be cut in as desired. *Flip Wilson* takes much less editing than *Laugh-In*, which runs from 200 to 250 cuts an hour. But there are still two days' work of cutting and splicing to be done down in the great dungeon of the Burbank tape room, where twenty-four huge Ampexes whir in the dimly lit, slightly dank air of a controlled-humidity environment.

And if worse comes to worst, the experts in the Post-Production Audio Room in the basement can be called into service to "sweeten" the show, adding laughter and applause as necessary.

2

As of fall 1971, nothing on network prime-time television, with the single exception of ABC's *Monday Night NFL Football*, was being broadcast live; only the variety shows, *All in the Family*, and (in late fall) *The Odd Couple*, which were taped before real audiences and thus produced in a relatively continuous manner, preserved any part of what was once considered the essence of broadcasting. And with the demise of the *CBS News Hour* and *The Ed Sullivan Show,* nothing scheduled every week on a network in prime time was being originated in New York. In effect, nightly network television, except for news, has become a distribution system for Hollywood-made film.

It is a strange and sad story. When television was new, it seemed clear that the picture tube in the box was best suited to what past generations of critics had called "closet drama"—Ibsen and Chekhov, the domestic play in which interest is maintained by the gradual unfolding of the characters under stress. The movie, as John Ford had pointed out, was the domain of the running horse; television would be the domain of people at home, of the kind of play that implies shifting perspectives on character. (Many plays, as some television programmers have never learned, are written for examination through the single perspective of an audience looking at a proscenium stage, and are excruciatingly difficult to revise for film or television, with their demands for changes in camera angle.) One of the more successful examples of television-in-the-box can still be seen every once in a while in early editions of *The Honeymooners,* episode after episode staged in a single room of a grim New York tenement apartment, the window on a court, the front door opening directly into the middle of the living space, the writing often third-rate or worse, but the stage illuminated by the genius of Jackie Gleason as a bus driver and Art Carney as a sewer worker and Audrey Meadows as a wife, the ensemble viable after twenty years.

But very little of what was telecast before the later 1950s survives at all; indeed, the key element in television in its early days was the evanescence of the programs. What was being prepared was a play. Usually, it originated in a theatre (few radio studios were equipped to handle the technical requirements of television). The cast was given scripts to study, roles to memorize, as though the show were to be staged on Broadway. Rehearsals were held in bare rooms, directors maneuvering actors and actresses around the chairs and tables that served as surrogates for the sets and props still being built, and then the company moved on stage for rehearsals of increasing security. Though the stage was in effect considerably smaller (because all the sets used in the play had to be crowded together, emplaced for the entire production) and though somewhat special talents were needed by the director (who would look at scenes through variously placed viewfinders rather than from a single standpoint in Row E center), the process was familiar to anyone who had ever worked in the theatre. When the night came, the company performed before the cameras exactly as it would before an audience, while the director in the booth, ordering cameramen to dolly here and there, cuing in this camera and then the other, determined the perspective from which the real audience—out in millions of homes—would see the performance.

This was nighttime television in "the Golden Age." Perry Lafferty, who now runs West Coast programming for CBS, recalls that when he was directing television plays written for the *U.S. Steel Hour*, "it was one of eleven anthologies running out of New York every week." The challenges were technical and psychological. To some extent, the format inclined writers and producers to a rather lugubrious product, because in the absence of an audience to laugh a comic play would lose timing, and even a synthetic laugh track played onto the stage would not help much. (Anyway, the networks felt they were taking care of comedy in the variety shows.) Though actors were used to having each performance good only for the time it was on the boards, writers were often upset, consciously or subconsciously, by the idea that some months of work would disappear in ninety minutes, never to be heard from again. The intensity of collaboration required created tensions that could—depending on the talents of the actors and the director—heighten the sense of excitement carried through the air to the picture tube in the home.

This sort of television is not dead in the world: British television, both BBC and ITA, does nearly all its dramatic productions as they were done in America in the 1950s, except that the facilities employed are superbly equipped studios rather than decayed theatres. Even when convenience in studio use or individual schedule leads the BBC or ITA producer to tape his show rather than broadcast it live, the tape is not edited and the performance is regarded by all concerned as a once-for-all live presentation. Coupled with the fairly busy London theatre season, this stage orientation in television has produced an extraordinary cadre of quick-study character actors who are happy to take a week's time for a television play. The quality of performance makes the stuff seem much better than it is: neither in seriousness of purpose nor in quality of dramaturgy are most British televised plays obviously superior to most of the made-for-television dramatic films in America, but for those of critical temperament the British product tends to be a much more compelling experience than the American one.

What drove American television production to film was, simply, the possibility of reuse. "Kinescopes"—films taken from television screens, in a process more or less the reverse of the telecine chain that transfers film to television—could be made from broadcast television plays, but the film tended to be grainy and lacking in contrast, and there were further losses when it had to be translated back for rebroadcast. Videotape was not available as a professional production technique until well into the 1960s. In the 1950s some big hits were repeated—*Peter Pan*, for example, and Reginald Rose's *Patterns*—but they had to be completely restaged; even the sets had to be built again.

Television shows on film could be sold abroad, where broadcasters could use them regardless of the television system they had adopted. (Incidentally, the immense popularity of American series abroad is not entirely a function of their quality: they could be and were sold for very little to foreign broadcasters because the domestic market had paid all the costs of making them. The rules all countries have now adopted to limit the use of American programs on their television screens do not necessarily reveal anti-Americanism; they simply protect local products from egregious dumping.) More important, the shift of program control from advertisers to broadcasters had opened up the prospect of much more frequent use of any program in America itself. Advertisers had never been nuts about the idea of sponsoring the same program twice, and they were entirely unwilling to let anyone else sponsor "their" show. But a network could rerun a show at night and then maybe fill time with it during the day, and then sell it to independent stations, one use at a time. Each time, the commercial minutes embedded in the show could be sold again, to new advertisers, at whatever prices the market would pay.

Film was something one made in Hollywood, especially in the late 1950s, when the movie box office had been smothered by the spread of television and both talent and studios were unemployed. Moreover, alas! there were advantages in the move for the average viewer. "The play is a limited, a static and frequently more difficult form than a motion picture," Pat Weaver has written in a still unpublished memoir. ". . . One must work essentially with the characterization of the actors and the story itself becomes much less important than it is in motion pictures." Fairy tales and myths have no characters: they are stories for everyone, of everyone. They exert a deeper and more widespread pull than the more realistic theatre. "These beloved merchants of dreams," Weaver wrote in his discussion of the movie bosses, "have a story fixation that is hard to believe." But the fixation derives from experience in dealing with box offices; film is a more popular art form than theatre.

Technically, film required very different talents. It is, we are told, a director's medium; the director is the *auteur*. Actors memorize nothing, and need not create their own characters. A shooting schedule puts together collections of brief scenes in an order that bears no necessary relation to the order of the script. Scenes are done over and over again until everybody gets everything right: four, seven, even ten feet of film may be shot for every foot used in the program. Decisions on which perspective to take on a scene can be changed in the cutting room. In television production as in conventional movie-making, the film is developed after the day's work, and the producer and director look at "rushes" late that night to see what they want to change tomorrow. After a television program is completely shot, at least six days and maybe months of work remain to get the film edited and assembled and in the can. There are no mistakes: the potter's hand smooths the clay before it goes into the kiln. "You gain gloss," says Howard Thomas, managing director of London's Thames Television, quite disapprovingly, "but you lose urgency."

Television went to Hollywood in awe and fear. ("All those great names of my childhood," says Carl Lindemann, who was in NBC programming before he became head of NBC sports, "why would they even *see* us? But we had the money.") The upshot, surprisingly and very soon, was to establish the networks as not just the dominant but almost the *only* source of new television programs. While the advertisers and the networks had been producing live in New York, low-budget film producers in Hollywood had been making television series for sale to the non-network stations in the large markets and to network affiliates for periods when the networks were not feeding. Once the networks moved to film, their old shows—stale perhaps but produced to a much more exacting standard than the independent or non-prime-time markets could support—began muscling in on the independent American film-makers much as they undercut the native producers in foreign countries.

In the mid-1950s, seeing the handwriting on the wall, Richard A. Moore of the unaffiliated KTTV in Los Angeles petitioned the FCC and Congress to do something that would restrict network ownership of filmed programs and protect the independent film packager. The association of station representatives, who sell the spots on local stations, joined Moore's appeal, suggesting that the FCC might forbid the networks to feed their affiliates from 9 to 9:30, reserving for local stations the choicest time on the schedule and thus the greatest possible revenue potential for the purchase of non-network programming. But in 1956, when the Congressional hearings on this subject were held, the networks could demonstrate that the independent packager still flourished—the shows the networks would later "syndicate" for second and third runs on the independent stations were still enjoying their first runs on the networks, and not even the farsighted Moore could *prove* that the availability of used network product would someday make it nearly impossible for anybody to produce television dramas or comedies without a guaranteed network market.

That is, however, what happened. In the late 1960s, when Westinghouse Broadcasting (Group W) petitioned the FCC to restrict the amount of prime time that stations could devote to network programs, CBS and NBC in reply lamented like the walrus and the carpenter that there were no more independent producers making new material for off-network syndication, and thus the "Westinghouse rule" (as it came to be called) would not do anyone any good. This time the station representatives backed the networks. The FCC adopted the rule nevertheless, forbidding the networks to feed to their affiliates in the top fifty markets more than three hours a night between 7 and 11 P.M., effective October 1971 (with an exception for a half-hour network news show at 7 where that time was customary). The Commission permitted stations to rerun *old* network shows in these newly liberated time slots for this first season (after 1971–72 everything that had ever been on a network or in this market before was to be banned from the air occupied by network affiliates during the forbidden time period), but the stations owned and operated by the networks in the biggest cities felt themselves obliged to follow the spirit of the rule from the beginning and to take new shows from new sources for the four hours a week from which network feed had been eliminated. The independent stations in these markets, unaffected by Westinghouse rules, thereupon took over the top ratings in those four hours by stripping across the schedule (i.e., showing the same series five or six nights a week at the same time) reruns of recently canceled network shows like *Hogan's Heroes*, the old series of *Dick van Dyke*, *I Dream of Jeannie*, *Get Smart* and the like.

The plain fact is that no new programming of the slightest importance emerged in 1971–72 as the result of the FCC's action in clearing time for non-network shows. The shows that had seemed potentially the most interesting going in—*Story Theatre, Norman Corwin Presents, Circus!*—were the most disappointing. "Sleazy product," said Murray Chercover, head of CTV, the Canadian network of privately owned stations, "much of it made here because of cost factors." Hollywood, deprived of nearly six hundred half-hours of program production for the networks, sank into the worst depression in its history as a film center.

As 1972 began, it was painfully clear that the next season would see no improvement in the quality of independent programs available to the stations, and no major effort by the stations to do their own programming. The only part of the "reform" movement that had worked was a prohibition of network ownership of syndication rights—instead of the networks syndicating their old shows, the movie studios now retained (some had always done so) the revenues from resale. But this was a matter of intramural interest; from a public point of view, the fact remained that in the producing economics of the 1970s all significant popular programming still had to be tailored to the asserted needs of the networks.

3

Watching television is a habit. The super-top ratings go every year to a handful of extraspecial programs (three of the top ten ratings in the 1960s went to Miss America pageants, two to Bob Hope Christmas Specials, two to the Academy Awards), but on the average, week in, week out, the most successful shows are the series that present the same people, same setting, same time, same station.

Situation comedy began, the BBC's Tom Sloan has written,

on American radio in the thirties. It consisted of a resident cast, headed always by a well-known entertainer, who played farcical comedy in which recognisable characters—like the next-door neighbour, the country philosopher, the newly married couple, the screwball comedienne and, in less complicated times, the white folks' idea of the comedy coloured servant— all played their part. They were Hollywood films for the blind. . . . The quality of reassurance . . . is an essential ingredient in such programmes, but, above all, they had to be funny. . . . People say there are no rules. Maybe, maybe not. I think myself that you can put unbelievable people in believable situations, or believable people in unbelievable situations, but you are asking for trouble if you try and put unbelievable people in unbelievable situations.

The British regard situation comedy as essentially a writer's form; at BBC, writers get top billing, above directors or stars, in the titles of television shows. Comedy writers are cultivated via a half-hour series called *Comedy Playhouse*, which solicits scripts from the general public. This started as a vehicle for Ray Galton and Alan Simpson, who had been the writers of the earliest BBC triumph, a show called *Hancock's Half-Hour* starring a comedian named Tony Hancock who decided to abandon broadcasting for movies (where he failed). The BBC program people suggested to the disgruntled writers a series that would be billed as "written by Galton and Simpson and this week starring ———." The fourth of these scripts was about a junk dealer of the old school and his modern-minded son, and BBC's Sloan persuaded the two writers to make a series with these characters, freeing the *Playhouse* for tryouts. (*Steptoe and Son*, the widely admired Galton and Simpson series, produced only twenty-six episodes in three years—BBC runs a series only six or seven weeks straight in a situation-comedy slot, and then gives the people creating it a rest until the next quarter, or even longer. NBC's version, *Sanford and Son*, if successful, will have to turn out that many shows in its first season.)

Another series started from *Comedy Playhouse* was *Till Death Us Do Part*, modified and softened in the United States as *All in the Family*. After *Till Death Us Do Part*, writer John Speight signed up with London Weekend, on commercial television, for a series called *Curry and Chips*, featuring Peter Sellers and Spike Milligan as a Pakistani and a Cockney engaged in verbal race nastiness in London. Speight got top billing on the show and in the advertising for it, despite the presence of the stars, and it was Speight's talent for rubbing on the raw spots of British race relations that got the series banned by the Independent Television Authority, which controls the programs on the commercial channel, after its first seven episodes. By no means all British situation comedy lives on such interesting levels, of course; much of it is based on slapstick *Lucy* fans—even *Beverly Hillbillies* fans—might find distressingly vulgar. There is, for example, a saga of two incompetent custom tailors, one Irish and one Jewish, with pictures of the Pope and Golda Meir hanging on the walls of the shop, that runs on and on under the title *Never Mind the Quality, Feel the Width*; and kindly recollection blocks the names of its writers.

In America, situation comedy usually starts with the name of an actor or actress available for a role; writers no less than producers are likely to come into a network executive's office with an idea "perfect for" so-and-so. Many cooks contribute to the broth: "Herb Schlosser and I are on the phone all day long," says Mort Werner at NBC. "We see people. They come in with an idea, and when they leave my office it's a different idea. It isn't 'Mr. Werner, a Mr. Jones, a writer, wants to see you.' I've been seeing Mr. Jones for twenty-five years, we work together." Everything is slotted into place in a "step deal"—so much paid to a writer for a "treatment" of the idea, so much more paid (if the treatment is approved) for a script, and then if the script is approved (as amended), an appropriation for a "pilot" episode that can be tested before a commitment is made to a series. Because the actors involved must give an option for considerable time next year, gambling a piece of their future to make a pilot for a series the networks may not buy, big-name actors are hard to sign for the last step of a step deal. Some, like Jimmy Stewart and Shirley MacLaine, can get commitments from networks for a series of shows without passing a final exam.

The writer's job is regarded as the initial creation of characters for the actors and actresses a producer has signed. Once these characters have been established, other writers are brought in to do other episodes, and yet others may be retained to spice up the dialogue with gags. In the end, nobody outside the business knows or cares who wrote this evening's installment of a series (*TV Guide* rarely mentions writers in its program listings). To the extent that the producer was in at the creation—and these days he probably was, because increasingly the producers are ex-writers—continuity of character will be maintained in the script. ("Today," says Aaron Spelling, who produces *Mod Squad* and a third of all the ABC *Movies of the Week*, "a producer is a rewrite man.") But often enough the job of keeping the series in its own track must be done by the director or even the star, or by whatever network chief assistant to the assistant chief has been asked to keep a distant eye on this particular show.

The same rules apply to the "action-adventure" or "medical" or "western" series: the writer who thought up the series and wrote the first "pilot" episode often continues to get both money and credit though the work is being done by other hands. Merle Miller in *Only You, Dick Daring!* describes the bait used to lure him into trying for such status: "You could turn out this pilot in a very few weeks, and since this is going to be one of the biggest projects ever, every time the series goes on the air you as the writer of *one* script, the pilot, collect a royalty of at least $1,000. Every week . . ." In early 1972, in a gesture revealing astonishing self-deception, a spokesman for the writers who prepare the week-by-week scripts—stories and dialogue tailored to characters, relationships, work habits and life styles already established by other writers—complained at a hearing of a Senate committee that producers, networks and advertisers had too much control over what they wrote. It beats writing an occasional play for a weekly ninety-minute drama anthology series, that's for sure.

4

Probably the most powerful reason for taking television production to Hollywood was that the move to film gave network executives much more security in making decisions about programs. A film series would be bought not as an idea nor as a collection of talent; it would be contracted for on the basis of a pilot which could be audience-tested and improved to suit. Moreover, the fact that shows are made some weeks prior to air time would give the network valued breathing space. David Attenborough, Director of Programs for the BBC, said recently that he knew a show was "losing its zip" some time before the ratings began to drop, and that everybody who runs the programming end of a network must have similar instincts. Mort Werner of NBC denied any need for such instincts: "I've already seen what the audience will be seeing in six to eight weeks. . . ."

Audience reaction is still judged mostly through some variant of the Stanton-Lazarsfeld Program Analyzer, developed in the late 1930s by Frank Stanton, then a psychologist, and Paul Lazarsfeld, later director of the Bureau of Applied Social Research at Columbia

University. "Little Annie," as the tool is known at CBS, consists of a small screening room with perhaps a dozen seats at a conference table, each seat equipped with an odd pair of black doorknob-shaped objects from which electrical wiring runs under the table. One of the black knobs has a red button; this one is to be held in the left hand, and the button is to be pushed whenever the viewer sees something he or she dislikes. The other black knob has a green button; this one is held in the right hand, and the button is pushed whenever the viewer sees something he or she particularly likes.

The button-pushing is graphed in the control room, where a trained observer notes what seems to be getting what reactions from which viewers. Then the viewers, who are simply tourists to New York or Los Angeles caught in the tourist places of those cities and invited to help CBS pick programs, will fill out a questionnaire asking them whether they liked or disliked the show and each of its major episodes and major characters, and why. Finally, the observer will hold a little seminar session at the table, asking people questions about the show and their reactions, calling alternately on people who are known to have liked or disliked different parts of the program, to avoid bandwagon effects (though a really formidable man, especially a doctor, can bollix the panel by imposing his opinion). For each pilot being considered by CBS, enough sessions of this sort will be held to give about eighty respondents altogether; and then, until recently, Robert Goldfarb, a young man with long black hair and a corncob pipe, who has now moved on to be director of program development, would write out a report of a dozen pages or so.

"My boss, Jay Eliasberg," Goldfarb said, "has likened this job to selecting an animal for the zoo, an animal you know people will want to see. The zoo keepers tell me that they've already got an animal with legs like tree trunks and a skinny tail and a nose that picks up fruit, and people come to see that animal—they want more animals like that. I can't see the new animals; I must rely on other people's vision. So I ask the other people about some new animal, and they tell me it's got rough skin and it's skinny like a snake and it's gray and it's like a wall—and I say, 'Hey, that's an

elephant!' and I go to the program department and I say, 'You've
got a winner. . . .' "

By 1970s standards, the Stanton-Lazarsfeld Program Analyzer is
a rather simple device. "It defies a hundred marketing rules," Goldfarb said. "It's lousy research. The sample stinks, it's not representative of any group. It's not large enough. People view in an unnatural
situation. The way they register their opinions is unnatural. You
can go on and on and show why it shouldn't work. But it works—
our batting average is 85 percent." What makes it work, Goldfarb
and Eliasberg believe, is less the tool itself than the experience of
the man running it: Goldfarb was only the second man to hold the
job, and his predecessor Tore Hallonquist had worked Annie for
more than twenty years, starting with radio shows, before Goldfarb
took over in 1963. In 1971 the network program division began
asking Goldfarb to read scripts for projected pilots, to get his personal opinion of their prospects. It was Goldfarb's entirely negative
opinion of the script that vetoed a projected pilot for a series that
would have returned Jackie Gleason to television in 1972.

At the other networks, perhaps because no researcher other than
Frank Stanton ever became president of a network, audience testing
is done outside the organization, in 1971 mostly by Audience Studies,
Inc., which operates a luxurious four-hundred-seat theatre on the
Sunset Strip just outside Hollywood. This is a much more sophisticated operation than the one at CBS. Each of the four hundred
respondents at the theatre is tagged by age, sex, education and income. Instead of pushing a button for like or dislike, each viewer
turns a dial to express gradations of liking or disliking. Each night's
audience is "normed" by the study of its reactions to a Mr. Magoo
cartoon shown before the program to be tested, and if the reaction
to the Magoo material is too far out of line, the sample is discarded.
This observer attended on a wasted evening. The 1971 earthquake
had knocked everybody out of bed that morning. The Magoo story
was about skiing, and its high point was a moment when the peak
above the hero's head split in two with a thunderous roar, and
voices in the distance called "Avalanche!" and Magoo chuckled,
"Avalanche is better than no lanche at all." This always got a wild

rise in the "like" graph from a normal audience, but tonight the line on the graph dropped steeply into the area of "dislike." An ASI man in the control booth said glumly, "Well, I guess we'll have to throw everything out tonight." The floor rumbled and bounced in reply.

The big control booth upstairs has a bank of twenty-odd theatre seats affording a view both of the show on the screen and of a collection of oscilloscopes that display a running record of the "like/dislike" graphs resulting from dial-turning by the different groups in the audience. Analysts from producer and network sit and form opinions while a girl at a mike describes key moments in the show into a tape recorder to help future analysts identify what incidents on the screen triggered what turnings of the dial. All of the four hundred viewers fill out a questionnaire, and a dozen of them are kept after school to discuss the show in a somewhat less structured way than CBS favors. This discussion group is itself telecast on closed circuit for the benefit of the analysts in the building (who have a drink while watching it), and a tape is kept for the use of other analysts later. The quantity of information provided is enormous, and its interpretation is exceedingly complicated.

The central problem with program testing, both CBS and ASI, is its strong tendency toward conservatism: most novelties will produce negative audience reaction on first exposure. "I Spy," says NBC's Mort Werner, "was the worst-testing pilot I ever saw." ASI scores normally show 62 to 72 percent of the audience "liking" what is, after all, a free show; Batman came in with only 39 percent liking it. "Our people almost had a heart attack," says Bill Brademan, ABC vice president in charge of program development. "If they hadn't been committed to it already, it wouldn't have run." CBS couldn't test Hee Haw at all: both in New York and in Los Angeles, the viewers walked out. Perry Lafferty, West Coast vice president in charge of programming for CBS, is not in the least impressed by Goldfarb's .850 batting average. "You'll be right 75 percent of the time," Lafferty says, "if you just say any new show won't work." William Rubens, who runs audience research for NBC, remembers "Reuven Frank [of the news division] saying that if we researched a couple of guys named Huntley and Brinkley they couldn't have got on the air. I don't know whether he's right or not. He may be."

Outside the program divisions, network management tends to talk about ASI and Annie as guides to the improvement of programs more than as determinants of whether or not a show makes the schedule. "Every creative guy looks at these reports for the least-common-denominator ASI score," says Fred Pierce, a lean man with a prize-fighter's look and perhaps a prize-fighter's temperament, who rose from ABC research to a job as second-in-command of the network, largely through a superb instinct for pricing minutes. "But the value is not only in the reaction to the program but in the reaction to characters and story lines." And, indeed, it sometimes happens just that way. ABC's Henry Miller remembers that "Jason, the sullen Negro boy, was supposed to be a one-shot in *Room 222*. But they got such a strong reaction to him on ASI that they made him a continuing character." *All in the Family* was made as a pilot for ABC, and rejected by that network. CBS research looked at the pilot and didn't think much of its chances, but when network president Robert Wood decided he *had* to have it (it was one of relatively few times in recent history at CBS that an individual put his neck on the line for a show), Eliasberg and Goldfarb told him to clean up a scene in the pilot, and to recast the roles of the daughter and son-in-law. "If the audience rejects an actor," says Perry Lafferty, "if they don't like him—that's visceral and you'd better pay attention." Similar analysis of ASI results led ABC to add girls to *The Odd Couple*. Research said the show was hurting in the ratings because women wouldn't watch misogynists.

"The real weakness of this sort of thing," says Ian Haldane, speaking of the program testing he does after broadcast to help ITA contractors improve their performance in Britain, "is in the communication of the results to the people who might use them. They say, 'We don't believe it' or 'We already knew this' or 'What on earth do you expect us to do with such stuff?' or 'The program has gone out and what can you do about it anyway?' or 'I make programs from the guts and I know what people want.'"

Jay Eliasberg at CBS thinks the frequent failure of producers to use his results is systemic. "Even when they really want to do what you tell them to do," he says, speaking of program producers, "when there's the best of good will on both sides, nothing happens. The producer agrees. He says, 'That's right—you've put your finger on something that's bothered me but I didn't know how to say it.' But then he doesn't *do* it. I suppose when he makes the pilot he has an unconscious set, and he can't change it."

Bill Brademan, on the other hand—an earnest, forthright man with short sandy hair and a manner *echt* Middle American, who worked his way up from stock-room boy at Universal Pictures to vice president of program development at ABC—feels his people may take the views of the researchers too seriously. "The producers ASI rough cuts of their pilots," he says. "It's like giving students an exam after they've had time to study the questions. One of the reasons you see dogs in so many shows is that it makes the ASI needle jump. Good storytelling almost always starts quietly, but if you start a television show quietly, the ASI report says it starts slow. What you'll get in the end will be standardized entertainment."

Yet the researchers and the program people agree at bottom about the demands of the medium on dramatic entertainment. "A dramatic show," says Jay Eliasberg, "has to have a real threat and a hero who overcomes it. *Perry Mason* is successful because the guy is going to die in the electric chair if Perry Mason doesn't get him off—and he's not guilty. In these *Young Lawyers* shows, the guy did it—or something like it—and the viewer doesn't know whether he wants the guy to get off or not. He didn't turn on the set for that." On the other side of the continent, Perry Lafferty says, "People ask me why the shows are so much the same, and some of us were talking about it the other day. We couldn't think of a continuing hour show in which the hero didn't have the power of life and death—the only possible exception was *Route 66*, and that was marginal. You have to give him a gun or a scalpel or a lawbook, and a jeopardy situation."

Television drama in countries where the system is responsive to the audience tends to stress primitive conflicts because it is this primary, Aristotelian level of reaction that all members of the potential audience can be assumed to share. Those who complain that television drama is superficial have not thought the question through. Television drama is often stupid; but it is profound.

5

All the production feeds finally into the steaming caldron of the annual scheduling crisis. The time is February. All over New York, in what are truly executive suites, stand "scheduling boards," display easels with a chart showing days on the vertical axis and time periods on the horizontal one. For each day, the easel shows each network separately as a colored strip running across the time lines; and on each colored strip the art department has lettered, in sans-serif capitals very easy to read, the shows that company is expected to offer in prime time next season.

This is something more than guesswork and something less than science. "We've already set up our charts," NBC's Mort Werner said in the last week of January 1971. "We know what will go off the air at the other nets—we know what's wearing out. But we don't know what established shows may move. You get surprised. We had no idea *Beverly Hillbillies* was going to come opposite *Don Knotts*—it hit us like a ton of bricks. But some things you do know. CBS can count on *Bonanza* being on at Sunday at nine o'clock."

Audiences change during the course of the evening: many more children are watching at 8 o'clock than at 10 o'clock (though some children are watching at almost any hour), and to the extent that television viewing is a family activity the 8 o'clock show must be geared to have appeal up and down the age range. "Something comes in that's a great show for ten," says Fred Silverman, the CBS program chief, "but the hole in your schedule's at eight, and you can't use it." Bill Brademan of ABC comments gloomily that "You can have a great idea and you can execute it just right—and then what it comes down to is, did you put it in the right time slot? If you didn't, everything goes down the drain."

There is no doubt at all that the ratings a show receives are a function of its competition, and in all likelihood this is a zero-sum game—that is, the total number of people watching in prime time of a winter's evening is not going to be greatly influenced by the available programs, and any audience added to one network's show means a subtraction from another network's show. Zero-sum games are easily susceptible to analysis on a computer, but Silverman scoffs at such mechanical aids. "How can a computer tell you how popular the ABC show is going to be at eight-thirty?" Still, the computer can quantify predictions of what different levels of popularity of rival programs may mean later in the evening. Anyway, computer games are more fun than committee meetings, and their time will come; at NBC, apparently, it already has come, though no details are available.

To the extent that these matters can be analyzed with existing tools, the decisions made by the network program directors and their bosses are rarely surprising. During the brief period in the 1950s when there were, in effect, only two national networks, the programming choices were just what theory would have predicted— both networks presented very similar programs in each time slot. This was clearly contrary to the public interest, but not avoidable in true competition under the conditions the economists call duopoly. The matter is most easily understood in Harold Hotelling's example of "spatial competition"—the analysis of what would happen in a town strung out evenly along a single avenue twenty blocks long, where there were only two grocery stores.

Public convenience in such a situation would place the two stores at Fifth Street and Fifteenth Street, guaranteeing that no customer would have to walk more than five blocks to buy eggs. But a smart grocer on Fifth Street would move to Fourteenth Street, which would give him an edge with everybody who lived below Fifteenth Street—three-fourths of the market. The equilibrium position in this competition finds the two groceries side by side on Tenth Street, though it means some customers have to walk ten blocks.

Hotelling's law of "excessive sameness" in conditions of limited competition is an interesting lens through which to focus on many situations—a two-party political system, for example—and it explains without need for any further analysis the scheduling philosophy of the networks of the 1950s. But the introduction of a third competitor in this simple model creates conditions of great instability, because there is no equilibrium point (any set of locations for the three groceries will give at least one of them what looks like a good reason to move).

In any event, the rise of ABC in the late 1950s produced different competition: a long list of made-in-Hollywood action-adventure, western, cops-and-robbers and medical series which were, by Lafferty's power-of-life-and-death law, the most likely to succeed. ABC needed them: it had many fewer "primary affiliates" than the other networks, and had to blast its way onto the schedules of stations affiliated with CBS and NBC. It should be noted, however, that so eminent an authority as FCC Commissioner Nicholas Johnson has praised ABC's almost exclusively popular programming of the early 1960s. "To the extent [ABC] is not [substantially competitive with the other two major networks]," Johnson wrote in his 1966 opinion dissenting from the Commission's abortive effort to approve a merger between ABC and ITT, "the evidence supports the view that the public is benefited by ABC's more innovative programming, not harmed."

With three networks competing, theory does argue the victory of diversity: "You may confidently bet the family jewels," writes Paul Klein, William Rubens' predecessor as director of audience research at NBC, "that, regardless of quality, the winner in a given time period will be the network that is *counter-programming* that slot." This is a corollary to Klein's basic theory of the "LOP"—the Least Objectionable Program—which holds that since people are going to watch television anyway, the show which the fewest people find unpleasant will get the biggest rating: "The payoff is really determined by 75 million different thresholds of pain plus the law of inertia."

Klein's argument is not quite right: most of the audience to any really successful show is very loyal and will complain bitterly if their favorite is dropped (the big-mail time at all the networks is in the weeks right after the schedule for the next fall is announced: the dismissal of Lawrence Welk brought 62,000 letters to ABC—"well-considered, impressive letters"—in the months after the press release.) Counterprogramming in itself guarantees nothing: *Glen Campbell* did poorly against *Mod Squad* and *Ironside*. And the possibilities for counterprogramming are limited by the general belief that different shows are best suited to different hours. But in television as in politics (another area where majorities are wanted) it is the floating viewer or voter who will determine the result—and the floater is almost certainly more strongly influenced by avoidances than by tropes.

The shows that come before and after a given program will also be extremely important, perhaps crucial, in its audience pull— Klein's "law of inertia." It takes effort to change a channel: not much, but some. Thus CBS put *Cannon* after *Hawaii Five-0* and *Doris Day* after *Lucy*; NBC led into *Sarge* with *Ironside;* ABC ran *The Brady Bunch, The Partridge Family* and *Room 222* one after the other. A network that takes a bad beating in an early-time period has a hard time catching up later on, though as always there are exceptions—in 1970–71, for example, *The FBI* did fine for ABC despite a virtually nonexistent lead-in, first from *The Young Rebels* and then from local programming when the network gave up in midseason on 7:30 Sunday night.

The sequence of decision is: (1) What do we keep and what do we drop from this year's schedule? (2) What do we expect the other networks to do at each time slot? (3) What kinds of shows do we need to maintain our balance and compete effectively? (4) What's in the can in terms of new pilots? The vice president in charge of programming and the vice president in charge of sales and the vice president in charge of research meet before the scheduling board with their assistants. "The word 'committee' has the wrong sound," says Fred Silverman. "You have intelligent people with different points of view; something should come out of it. Anybody who acts as a dictator is crazy." One thing is invariable, year after year: "No matter how many pilots you make," says Jay Eliasberg, "the moment comes at the scheduling meeting when somebody says,

'God, how I wish we had another hour or two of decent programming.'"

Senior executives may intervene. Chairman Paley at CBS is
notoriously his network's final judge of program. Chairman Goldenson at ABC has been known to recall at program meetings some of
his experiences in the late 1930s, when he was head of Paramount
Theatres and Frank Freeman was head of Paramount Pictures, and
"twice a year I would sit down with Freeman to analyze all the
studio's policies—where they were going wrong—as a retailer tries
to guide a dress manufacturer, and tell him what's selling."

During the course of scheduling a show, and touting it to his network president and higher corporate officers, a program director may
recast it, change the story outline, shift the focus. Producers need the
sale, and very rarely complain (sometimes they should; their writers
will). The agony comes if the program director decides that he
wants what was projected as an hour show to fill what is for a
producer a much less profitable half-hour slot. Thus one end of
three telephone conversations during an hour, all of them about the
prospects for *Bearcats,* as overheard in scheduling week in the
office of CBS's Fred Silverman:

"I've got to know by noon tomorrow . . . by three tomorrow. As
long as you have the kid, I think that part is castable. But I want to
have some alternatives in my back pocket. . . . I can almost promise
you a half-hour, we have a half-hour slot. . . . I want to put it back-
to-back with *The Teddy Bears,* which is my best pilot. . . .

"Look, if it's going to be Chekhov for a half-hour, then we don't
want the show. It's *Have Gun, Will Travel,* it's a show with movement, it's not great drama, it's an action-adventure show. You set
up the problem in three minutes and then you solve it. . . . Look,
Morty, life is a compromise. . . . Call me back. . . .

"No, look, you gotta remember this is *Mission: Impossible,* only
it's two guys in a Stutz Bearcat in the West in 1914. . . . I'm sure
you can get the half-hour. I told it to Paley, he *loved* it. . . . I'm
not saying you can't get a full hour, I just don't know. . . . Morty,
I'm doing this not only for our sake but for yours. I want to see you
get on the air, that's all I want to do. . . ."

The show went on the CBS network in fall 1971, as a full hour, in the high-risk slot opposite Flip Wilson. By January 1972 it was dead as Queen Anne, and so was the "best pilot," *Teddy Bears*. But the CBS prime-time schedule as a whole was easily on top of the ratings.

6

Over the years, most television critics have been most suspicious of the advertiser's role in this process. Erik Barnouw, in *The Image Empire*, explained the decline of the New York dramatic anthology in the 1950s as the result of advertiser pressure: "Most advertisers were selling magic. Their commercials posed the same problems that Chayefsky drama dealt with: people who feared failure in love and in business. But in the commercials there was always a solution as clear-cut as the snap of a finger: the problem could be solved by a new pill, deodorant, toothpaste, shampoo, shaving lotion, hair tonic, car, girdle, coffee, muffin recipe, or floor wax. . . . Chayefsky and other anthology writers took these same problems and made them complicated. . . . It made the commercial seem fraudulent." Barnouw cites a letter sent by an advertising agency to Elmer Rice after Rice had suggested to the agency television shows based on his play *Street Scene*: "We know of no advertiser or advertising agency in this country who would knowingly allow the products which he is trying to advertise to the public to become associated with the squalor . . . and general 'down' character . . . of *Street Scene*."

But control of programming was passing from advertisers to networks through the 1950s, and there is no evidence that the networks ever had trouble selling participations in closet dramas. Chayefsky was being ardently solicited by both NBC and CBS for years after he stopped writing television plays. "I've spent a lot of my career and a lot of my blood trying to build a climate for commercials," Robert Foreman said as he left the television vice-presidency of BBDO in 1965. "It doesn't seem to make any difference. You'd think a period piece would be a bad setting for a modern convenience appliance, but then you look at the way Chevrolet sells on shows full of stagecoaches . . ."

The "relevant" programs of fall 1970 sold out faster than new shows usually do, though they promised to deal with hard social problems; what killed them was their failure to deliver audience. Of course, they weren't *really* relevant—they didn't say the country is shit, which is the only *really* relevant thing a television program can say, I mean, y'know, man? Yeah. But that kind of relevance, while it might do quite well once or twice, would be unlikely to draw an audience week after week, and advertisers would have perfectly sound business reasons for avoiding it, quite apart from image or magic. In the 1970s anything short of pornography that seems likely to draw a big young audience will sell easily: if *Easy Rider* were cleaned up enough to make it acceptable for broadcast, the minutes would go at premium prices. The advertiser's bad influence on entertainment for at least ten years has been not his censorship of program material but his demand for numbers; network censorship, which happens occasionally, reflects not personal or political prejudice but the fear that something distasteful or highly unpopular will keep some of the audience from coming back next week.

The implications of a policy to maximize audience were stated as a global matter by David Attenborough in 1966, when he was program controller for the then-infant BBC-2:

The way to gather an immense audience very quickly is quite obviously to put on the most popular programmes. . . . They are: domestic serials like *Coronation Street, Emergency Ward 10* and *United;* pop shows; quiz shows; spy fantasies, like *The Avengers* and *The Man from U.N.C.L.E.;* serial dramas, like *Dixon of Dock Green* and *Perry Mason.* These are formats that have been evolved, refined, and perfected by networks all over the world. . . . It follows, therefore, that if BBC-2 wants to attract large audiences fast, these are the sort of programmes, among others, that it must schedule. What is more, it must schedule them early in the evening to take advantage of the principle that is well understood by all programme planners, that if you do not grab a large audience early on, you will never get one at all. In short, to implement such a policy means to produce programmes that are largely carbon copies of existing programmes and to schedule them in such a way that they clash head-on with similar mass-appeal programmes being shown on other networks.

Attenborough rejected this counsel, put on *The Forsyte Saga* and *Civilisation*, and became Director of Programmes for all BBC television; American program directors accepted it, put on what they put on, and became rich. By 1971 the only regularly scheduled prime-time exception to the rule of maximum audience was the once-a-month two-hour public-affairs show on NBC and CBS, a minimal gesture toward the obligations of the franchise. What is painful about this is less the loss of programs valuable to the viewer (most of the ambitious stuff on television as on Broadway is dreadful) than the loss of spirit and vocation among broadcasters. At a scheduling meeting in 1961, William S. Paley, near-founder, proprietor and chairman of CBS, was reported to have said, "I thought we were going to talk about programs, and all we've been talking about is deals." No such comment would be made in the 1970s—everybody is used to it now.

Nowhere in prime time in 1971 did any of the networks offer its own people a chance to do something into which they might wish to pour their heart's blood (as Thames Television, the commercial contractor in London, financed a year's work by a big crew on a long, filmed love song to the river after which the company is named); nowhere was there any tribute paid to the higher orders of artistic talent (as BBC presented Schubert's *Die Winterreise* sung by Peter Pears before a background of slide projections of the winter scenery Schubert's tragic hero would have known). At six commercials an hour and a CPM of more than $4, each rating point lost by a show costs the network about $15,000. To put on, say, a Balanchine ballet instead of *Mannix* would cost CBS something not much short of $200,000 in advertising revenues in the one night. It is one thing to say that of course they should do it (and *of course they should*); something else to explain exactly why.

A dozen years ago, both Hollywood and New York felt much greater obligations to material of possible permanent value. The television critic for *Harper's Magazine* in 1959–60 noted that during that season "A New Yorker could see *Medea* as filtered through Robinson Jeffers and *Volpone* as bourgeoisified by Stefan Zweig, Shakespeare's *Tempest*, Ibsen's *Doll's House* and *Master Builder*, Strindberg's *Miss Julie*, Chekhov's *Cherry Orchard*, Shaw's *Misal*-

liance and *Captain Brassbound's Conversion,* Wilder's *Our Town,* original plays by Archibald MacLeish and Reginald Rose, adaptations of stories or novels by Cervantes, Dickens, Turgenev, Conrad, James, Faulkner, Maugham, Hemingway, Sinclair Lewis and many, many others including the author[s] of the Bible." Then he added, and his judgment is one I have always trusted: *"Sub specie aeternitatis,* everything was lousy." But the difference between trying and not trying is more than a difference in degree, and sometimes the efforts bore fruit. Gian-Carlo Menotti's *Amahl and the Night Visitors,* for example, was commissioned by NBC and telecast —year after year, to good audiences—in prime time.

The experience of England's commercial ITA ("an extraordinary combination of nineteenth-century liberalism here," says a man who works at ITA headquarters, "and medieval robber barons out in the contracting companies") does indicate that profit-seeking television networks can be made to try harder and keep trying harder. Whether the FCC has such powers in America is at best an open question and at worst an easy answer, because the Communications Act does in so many words deny the FCC authority over programming.

Over the short term, the best hope for the return of somewhat less gassy entertainment in American television would seem to lie in the growth of what Attenborough has called "loose boxes"— program slots where a single name like *Movie of the Week* in fact represents a variety of programs. This is very tricky, because wide departures from what an audience expects from this title will diminish the habitual viewing which is the basis of any large audience—yet the fact is that all three networks in their made-to-order movies have offered a variety of moods under a single title. At ABC, which has been the most successful, movie supervisor Barry Diller, just over thirty and confident, attributes the loyalty of the *Movie of the Week* audience to "merchandising. This isn't a series, and it isn't movies; it's a merchandising campaign."

Though the comparisons are not in Diller's head, the same sort of statement could be made about an opera or repertory theatre subscription or a Community Concerts membership or a Sol Hurok Presents series: the arts are sold through merchandising campaigns.

Diller's campaigns rest on what he calls "concept testing," six annual research studies ABC is genuinely reluctant to describe, which seek to measure public reaction to program ideas. Other networks scoff at this sort of thing—"We have enough trouble deciding whether people will like a show after it's in existence," says Jay Eliasberg at CBS—but it's hard to quarrel with a man whose ninety-minute made-for-television movies command a pretty steady 38 percent share of the viewing audience.

At NBC, the unprepossessing, almost mousy Mort Werner ("most underestimated man in television," says ABC's Bill Brademan) has been trying to steer the network toward a more flexible set of program categories—*The Bold Ones, Mystery Movie, Four-in-One*, all titles for groups of programs, sharing some elements of audience appeal but not cradling the audience in the cushions of eternal sameness. "You know what I want?" Werner said rather dreamily not long ago, reflecting on the longest career as a program director anyone has ever had. "I want the American householder to get up and say, 'I wonder what's on NBC tonight.' That's what I want." He looked out his window at the winter scene of Rockefeller Center, and he snorted. "Twenty-six originals and twenty-four repeats," he said. "Is that really the destiny of television?"

CHAPTER 5

Other Dayparts, Other Customs

I vas in America für zexteen moants. Iss much besser dere. In Austria iss no goot country—iss not eefen any television here before fife o'clock.
—Young male hairdresser to American lady customer,
Hotel Sacher, Vienna

1

"I heard one day from the wife of somebody I know in Maine," said Stuart Schulberg, producer of *Today.* "She told me that some mornings at seven-fifteen she'd see the lobster fishermen coming down the street to the docks, and other mornings there was nobody. One day she stopped one of the lobstermen and asked him why this was so—how did they all decide that one day was going to be good for lobstering and the other wasn't? He said, 'We all watch *Today,* the weather show. Nobody in Maine can predict weather like the *Today* show.'

"Funny part of it is," Schulberg added, "that's true. They won't get their own late weather forecast till twenty minutes after we have it. The Weather Bureau is in this building [Rockefeller Center], on the mezzanine, and they love Mark Davison, our meteorologist. It's like a three-million-dollar weather central we have for nothing."

Today is the most important of Pat Weaver's legacies to NBC, and maybe the most important regularly scheduled television show in the country. Since 1951 it has been on the air every weekday morning from 7 to 9 (in the Central Time Zone, too, for a wonder—it is the only network show broadcast at the same clock hour in both time zones), mixing news and weather and many commercials with serious interviews and funny interviews, self-help features (for a long time there were setting-up exercises) and real live entertainment. The average audience of about 3.5 million homes must reflect viewing of some of the show by at least 7 million homes—and they are the right homes, people of rank and substance. NBC did a survey of cabinet members, Senators and Congressmen, and found that 75 percent of them looked at *Today* before leaving the house in the morning. Once when Congressman James Scheuer of New York was interviewed on the show, he stayed to look at a playback of the tape. "For a Congressman," he told his host, "happiness is watching yourself on the *Today* show."

For an author, too. "Look at the nonfiction best-seller list," Schulberg said in spring 1971. "Cut out the sex books, which I won't put on the air, and every one of those best-sellers was on our program. Over two years, 98 percent of the books on the list will be books that were on our program. Authors are ideal guests—they come in and discuss in nine minutes something that would take our staff a year to develop. And we're washing each other's hands; we're selling their book while they bring us problems worth trying to understand. I could do the whole ten hours on books. But I hold it to one a day; if I do eight in a week, I get a note about it from Julian Goodman."

And of course Goodman is right, because the special flavor of *Today* is its mix of topics and guests and kinds of entertainment. Every three weeks or so, Schulberg has an artist or a group from the world of serious music for a ten- or fifteen-minute interlude that will be, say, a Mozart violin sonata or a long Verdi aria or a movement from a Schumann quartet—not culture-is-good-for-you with long lectures, but the real thing. No other network program presents as much serious music over the course of a year. Without *Today*, which devoted its entire two hours to the occasion, the bicentenary of Beethoven's birth would have gone unremarked on American television.

Schulberg invites pop groups, too, and dancers, and actors. "For entertainment," he says, "we try to do what the nets are not doing at night. I won't put on Steve Lawrence and Eydie Gormé—they're available every night. And we don't do much from Hollywood because they don't come to us—they don't watch us. Hollywood sleeps until nine, and we're off the air at nine." Then there are doctors and lawyers and Indian chiefs, foreign diplomats, college students—the whole mixed bag of active Americans. During the wage-price freeze of 1971, the only organized guide offered to the perplexed of the country was a weekly appearance of the chief administrator, answering questions on *Today*.

Every guest on the show is chosen by Schulberg himself: "There is no way to produce a program like this on a committee basis. Day to day it has to be based on my own curiosity. If you listen to the agents or read the mail too earnestly, you'll wind up with pap. But as you read that mail and listen to those telephone calls, about once in twenty times you'll say, 'I'm going to respond as an individual and put it on the air.' My predecessor had a meat market here—once a month thirty or forty singers and dancers would come down, and he'd sit smoking his cigar and put thumbs up or thumbs down—but I can't do that. I'll listen, and we'll scout around, send people to Boston and Philadelphia. And a marvelous idea may come from anybody—from a small university in Idaho, or BBDO, or the White House. One day I came back from lunch and my secretary said, 'The Rumanian Embassy just called, to see if you'd like to do some shows from Rumania. . . .' "

There is some reason to believe that Weaver lucked into this one —that he was thinking of something much bigger and brasher than he got. He loved the gimmicks—the clocks that told you what time it was in Glocca Morra, the schedules of events at Acapulco, the predictions of today's news that made the first few shows back in 1951 a kind of three-ring circus on a muddy footing. Among the hard-to-handle charms of the early *Today* show was its location, in the RCA Exhibition Hall in Rockefeller Center, where the people on the street could look in at the creatures in the show, and the cameras could look out at them. Dave Garroway, whose easy manner and authentic gentleness gave the show its most endearing (and enduring) quality, was by no means Weaver's first choice for the job. When the first shows got in trouble, Garroway was given a chimpanzee as a sidekick on camera, which was charming for the audience and added a visual dimension Weaver had felt was lacking, but was considerably less charming for host and guests. Eventually the chimp was allowed to depart, and the NBC top echelon accepted the fact that *Today* would often be mostly a radio show on camera. For one brief and nearly disastrous period some interview and entertainment sections of the show were taped the afternoon before, but that wasn't the spirit of *Today,* and since 1961 the show has again been live every morning.

Producing *Today* takes a staff of sixty-six people in New York plus ten in Washington. Four cameras on wheels are used in the New York studio, one of them doing double duty for a newscaster sitting in a corner of the studio and for any titles that have to be superimposed on the screen during an interview, another rolling to another corner for the commercials, which the regulars in the show are still expected to handle themselves, showing the people the bottles of glycerine suppositories or whatever else it is the viewer is presumed to need this morning. This, too, is a Weaver legacy, and if members of the cast sometimes regret their extra role as pitchmen, the extra payments for the work soothe the pain.

The first members of the company report for work at midnight, when a five-man news staff clocks in to handle what the newspapers call the lobster trick, keeping up with the world while all around them are asleep. The production staff starts work at 1 in the morning, and the director arrives at 1:30 to get the lighting right for the events of the morning, rehearse the technical end of the commercials, establish cues. The talent comes in at 6, more than a little bleary-eyed; guests are picked up by limousine and delivered to the studio between 6:30 and 6:45. Interviews are not rehearsed— one of the writers on the show has read an author's book or a politician's speeches or an actor's reviews, and has prepared sample questions for whoever is handling the basic interview (in 1971–72, Frank McGee or Joe Garagiola or Barbara Walters)—and there will be a few minutes during a commercial or a weathercast for the two sides of the interview to make some minimal human contact (and for the audio man to adjust his levels in the booth) before the talk rolls out, live, in front of the cameras. An assistant director crouched beside a camera gives finger signals to alert everyone to the approaching end of this section of the show, and life goes on, not unexpectedly. The news writers go home at ten, and the talent is free as of one in the afternoon, after planning their parts for tomorrow's show, but Schulberg has no such luck: "The problem with producing this show is that you have to be there at six in the morning, and you don't get off until six at night."

Schulberg himself is a man watching his weight, with a large head of gray hair descending to fluffy gray sideburns and pointed beard. The son of one celebrity (the Hollywood producer B. P. Schulberg) and brother of another (the novelist Budd Schulberg), he is not to be awed by the ordinary notorious character. Like almost any successful producer, he feels his first responsibility is to his show and its audience, not to the great issues of the age. In summer 1971 the *New York Times* acquired a new television critic, who gave *Today* the back of his hand because it failed to ask hard, penetrating questions of the guests on the show; but it seems clear enough to Schulberg that people waking up in the morning don't need any more fights at the breakfast table than they already have.

In Rumania, Schulberg worked with a local lady television producer who was terribly upset. "She said," he recalls, " 'How can you do Rumanian history in eight minutes?' I told her, 'Of course it's superficial; we've got one week in Rumania. And we're not the intelligentsia talking to the intelligentsia, we're just talking to people.' " Barbara Walters remembered the lady well, and laughed at the recollection of how simple she had thought the *Today* people were. Schulberg nodded. "But at the end," he said, "she said to me, 'I'm beginning to learn that television can be more casual.' "

The secret of the talk show—and the luck of NBC in Garroway and in Steve Allen (who started *Tonight* at about the same time)— is of course the appearance of ease, the interviewer as surrogate for everyman though inevitably a celebrity himself. (A celebrity, Daniel Boorstin once pointed out, is a person well known for his well-knownness.) Barbara Walters is probably not very much at ease as an ordinary matter, and Dick Cavett runs on immense expenditures of nervous energy—but both of them *seem* casual in the middle distance of television. Except for Jack Paar, whose reign over *Tonight* was considered at the time an almost Houdini-like escape from the defects of his personality, abrasive characters have not worked out on these shows. Mike Wallace cost Westinghouse Broadcasting almost a million dollars on the short-lived *PM East,* and Joey Bishop put ABC's late-night talk show into a ratings nosedive from which Dick Cavett was a long time emerging.

Such attitudes are culturally determined, and while talk shows not very different from those in America do well in England, the hosts are wildly different in temperament. David Frost functions in both countries, flying to London every weekend (perhaps because he is the largest single stockholder in the station for which he does his English show), and the English have been complaining that he seems infected by American blandness. Certainly he was a much less amiable fellow on the air before coming to New York—on one of his English shows, he had as his guest a London doctor accused of abusing his prescription privileges to supply illegal drugs to patients, and he salted his audience with other doctors and ex-patients who rose to denounce the guest of the evening and accuse him of ruining people's lives for profit. It must be said, though, that to those who meet him for the first time in New York the American on-air personality is entirely consonant with that of the man across the table at the Algonquin (antique home of wits: the British look for tradition wherever they can). There is a real desire to please: a favorite Frost story concerns the time he went to the Montreux Festival to receive the Golden Rose of an international first prize, and said "Thank you" in all the fifteen languages represented at the festival.

Frost is the premier example of a type pioneered by the British shortly after World War I—the university-educated club comic. (This is not to say that comics elsewhere are necessarily uneducated, but the fact that Jack Lemmon is a Harvard graduate has so little to do with Lemmon's career that most people are surprised to hear it, while Oxbridge is central to the public reputation of Britons like Frost and Jonathan Miller and Peter Cook and Flanders-and-

Swann.) He is also among the first major television figures to have got his start on television—in 1959, while still a student at Cambridge, where he was head of Footlights, the theatre club, and editor of *Granta,* the literary publication. "Anglia," he recalls, referring to one of the commercial television stations, "had a program they called *Town & Gown,* and they wanted to do a jolly Christmas edition, and they asked Peter Cook and me to do a program." From that program came a summer interview program on Anglia in 1960, and in 1961 a job with Associated Rediffusion in London, where Frost devoted his days to an apprenticeship in television ("a ten-week training tour, learning about VTRs and advertising") and his nights to night-club skits.

It was through the club work rather than through television that Frost made contact with the man who had the idea for the topical revue *That Was the Week That Was.* "July 1962," Frost recalls, "just a year after I'd left Cambridge, I made the pilot for that show. It can honestly be said that that pilot kept its audience on the edge of their seats; it ran two and a half hours and we'd neglected to give an intermission. . . . BBC offered us thirteen weeks' salary for 'preparing' to do the show, and then another thirteen weeks if they decided to air. Rediffusion offered us four-year contracts, guaranteed. But there was no doubt which was the best to take."

The American version of *TW3* was Frost's debut vehicle in the United States in 1963–64, and he considers himself much more an entertainer than an interviewer ("I do two-liners, not one-liners"), but he is proud of certain pioneering efforts on the talk show. "When I started," he says, "everybody told me that you can't interview any one person for more than eight or nine minutes, or you lose your audience. Now we do ninety minutes. The average viewer is much more intelligent than people want to admit. When we first did light conversation on serious subjects in England, the Hampstead coterie *resented* it—resented the idea that the Archbishop of Canterbury might have a bigger audience than a wrestling match at the same time on the other channel."

The Frost show is taped, the tapes are duplicated in Pittsburgh on a bank of thirteen machines that run twenty-four hours a day and seven days a week, and the result is distributed through the mails and the express companies: Westinghouse does not operate a network, because AT&T line charges would cost the company $70,000 a week for three hours of interconnection. Different stations carry the show at different times, and it seems to work well in any "daypart": viewers who see it in the morning feel it is a morning show, those who see it in the evening feel it is an evening show. (But anyone who is used to it at one time will suffer a kind of seasickness if he experiences it at the other: Dick Cavett, who lives in New York and used to look at Frost in the evening, also has a summer home in Long Island, where television arrives from Hartford, Connecticut, and Frost was on in the morning. Finding Frost on his morning screen subtly wrecked Cavett's day.) Most stations carry Frost in lieu of an early movie or kiddie stuff in the 4:30-to-6 slot. It is the simplest sort of show, Frost and guest sitting with a mike between them on stuffed chairs that can be wheeled away, a carpet beneath their feet, musical entertainment in discrete pieces. Interviews are not rehearsed, but they are prepared, a staff of auditioners finding out for Frost what subjects people speak on well or ill. The show runs ninety minutes before an audience at the Little Theatre in New York, and, as Frost says, it occasionally gives all ninety minutes to a single personality; but the more usual show mixes four or five, of different kinds.

Stations pay Westinghouse for the Frost show, and slot commercials they have sold themselves into blank spots left for the purpose. Frost shares in the proceeds, Westinghouse being a firm believer in low overhead and good participations; Donald McGannon, president of Group W, as Westinghouse calls its broadcasting operation, says that in 1971 Frost took out of this program alone about $800,000; and he also had a weekly half-hour revue in the second half of the year, and his program in England, and personal appearances at summer tent shows and the like.

At that, Frost is not Group W's biggest winner: Mike Douglas in 1971 took home, by McGannon's estimate, a full million. Douglas started in Cleveland in 1961, as a sort of late-afternoon Soupy Sales slapstick for grownups, with songs and cooking recipes, a "co-host" to share the burden of being on camera for ninety minutes, and guests from the entertainment business. This show is rehearsed, to smooth the routines between Douglas and his co-host and to polish the musical numbers; rehearsal call is 11 A.M. for a 1:15 taping. Among the early co-hosts, Douglas remembers, was Barbra Streisand: "We paid her a thousand dollars, for five ninety-minute shows, and we had the kind of contract where we could have reused the tapes forever; but somebody in Cleveland erased them to make editorials." In 1963 the Douglas show was offered nationally, and as of 1971 it was being produced in Philadelphia, with guests mostly from New York. There have been more than two thousand Mike Douglas syndicated shows; in 1971 more than 140 stations were buying the program at prices ranging up to $260,000 a year paid by WCBS, Channel 2 in New York.

The rationale for the Group W talk shows was that the five Westinghouse-owned stations were programming movies at 1:30, 4:30 and 11:30. "Simple arithmetic showed you were going to run out of movies," says Chet Collier, now director of operations for all Group W stations (but manager of the Cleveland affiliate in 1961, and Douglas' discoverer). After the first failure with *PM East* and *PM West*, Group W grabbed Steve Allen, who had left NBC's *Tonight* to go on prime time on ABC, and after Allen and some false starts the show went to Merv Griffin, who had been a game-show host, among other things, on daytime television. Only *Tonight* was in late-fringe (i.e., after-prime-time) network distribution at that time, and a number of stations bought Griffin for the 11:30 slot (where most Group W stations ran him). Then ABC went into competition against *Tonight* (the first ABC host, which nobody now alive remembers, was Les Crain), and Griffin's market disappeared—most of the possible purchasers were either NBC or ABC affiliates, and if they weren't, the sensible counterprogramming to the talk shows was a movie. The Griffin show moved to the 4:30 slot, and Douglas retreated to 1:30 (in some places, including Cleveland, where the station is now owned by NBC, rather than by Group W, to 9 or 9:30 in the morning). *Then* CBS decided (mistakenly) that there was money to be made in a third network talk show at 11:30 P.M. and hired Griffin away from Westinghouse in spring 1969. "I told him," Group W's Collier says mournfully, " 'Don't get into the bind between Carson and Bishop—you're too smart for that.' But he did."

Group W had brought David Frost to America to interview the Presidential candidates in 1968, and now rushed him back to take over from Griffin. The show lost markets (Frost by fall 1971 was on 87 stations, Griffin had been on 136), and its ratings dipped where it continued to be carried—"This is a high-risk business," Collier says. "You're hung up on one man. The only reason somebody turns on a talk show is that he likes the host." But then the Frost show got talked about a good deal and built an audience both loyal and of highly salable composition, either in its usual afternoon slot or in its large-market (New York, Los Angeles, Washington) evening slot. Douglas, whose show continues to be as much pickup entertainment as it is conversation, continued mostly in the early-afternoon slot, with ratings higher than any of the late-night shows. In spring 1972, CBS dropped the late-night Griffin show, and Metromedia, an entertainment conglomerate that owns television stations in New York, Washington and Los Angeles, began to package Griffin à la Group W. As the three big Metromedia stations had been the ones that carried David Frost in the evening, Group W was left looking for outlets in the big markets. At press time, it was unclear how badly a revived syndicated Griffin show would hurt Frost, but nobody thought it could help him.

One of those considered by Group W for the Griffin job was Dick Cavett, a small, tense, bright Yale man who had been a writer for Jack Paar on *Tonight* and had turned stand-up comic (doing his act on, among others, the Westinghouse Griffin show). Despite some protective Broadway coloration, Cavett is a good deal less theatrical than Frost in his basic temperament: Frost enjoys doing his show, Cavett on the whole does not. "I remember when I was a writer," he says, "I used to picture myself as a guest on the Jack Paar show. I was sure I could get laughs, and I wanted to do something that would involve *appearing* on a panel show. But not as a host.

"These things look very different," Cavett continued, munching on a banana, "from the stage itself, on the monitor screen, and again on the screen at home. I can be exhausted and drained after a show, and I'll look at it later and I'll wonder why. Only rarely can I be moved sitting there as I might be if I were a member of the audience. I have to keep thinking—'Must break after the next two sentences, for the commercial,' or, 'There are other guests backstage.' It's stupid to believe that four people of different weights and values are worth the same—everybody to get two segments of eight minutes each—but the show sets up that way. Once when I did a show with the Green Beret assassin I just left the other guests backstage, because there was no way I could break that off. And you've got to think, there are some guests going through hell just to come on the show, so you can't depart from the plan worked out by the people who interviewed him before the show. Or there are other cases, where people will be more interesting if they're jogged. Or I'll be more interesting if they're jogged."

Cavett employs four people to do "pre-interviews" with guests and prepare outlines for him on the sort of thing they say well. On the whole, he prefers not to get too involved with the choice of people—"I've been entirely into it, and entirely out of it, and I don't like either." Shows are usually taped that day for air that night, and if there's something big in the news, Cavett may intervene to see whether he can interview somebody about it. In fall 1971 when the New York State Prison blew up in Attica, he rescheduled everything to get a prison guard on his show that night. (Frost also rescheduled, to establish a panel discussion on prisons and prison revolts.) Relations between host and guest on a talk show are an unexplored subject for psychology. "Sometimes you sit there during the commercials," Cavett says, "and you don't say a word to each other. But the ordinary man on a talk show continues to talk in an ingenuous way through every break. And I remember how pleased I was once when Robert Mitchum was on and he kept going because he thought I was interested in what he was saying. And always after the show the world backstage, the world off camera, seems very unreal. . . ."

2

Prime time runs three or four hours a night, depending on who calculates what. Some stations will carry as much as five and a half hours of talk shows a day—Douglas, Frost, a network late-night show and a local hour. Nearly every commercial station runs three to five hours of movies a day (apart from the movies fed by the networks), most of them from a stock bought from the studios in the 1950s and early 1960s, which can be used over and over and over again. (There were no "residual" payments for reuse in movie contracts made prior to the 1960s.) And there are syndicated reruns of old network shows, of which the supply increases annually. But "daytime television" for most people means game shows and, especially, soap operas.

Network daytime television began when Chicago first came on line in the late 1940s. Lou Cowan, who would later operate *The $64,000 Question* in prime time and would be president of the CBS-TV network until the quiz scandals made him too hot to keep (though he was never personally implicated), put together a game-show package called *What's My Name?* with Bert Parks as the host three times a week and Bill Goodwin the other two, and sold the lot to General Foods. At one point Cowan and his partner Al Hollender (who also produced the one-minute Eisenhower spots in the 1952 election, when Cowan was working for Stevenson) had thirteen shows going on the three networks, most of them in daytime slots, nearly all of them games. In 1958, thinking back on a period that already seemed prehistory, the Shakespeare scholar and television personality Bergen Evans, who had worked on many of the Cowan shows, remembered *Down You Go, Superghost, Of Many Things, It's About Time, Top Dollar, RFD America, Pet Shop*—"He's done at least twenty," Evans said.

Game shows come in two varieties—the panel, which presents problems of varying degrees of difficulty and cuteness to a small committee of celebrities, and the quiz, which asks ordinary people to answer questions or do tricks. In their book *How Sweet It Was,*

Arthur Shulman and Roger Youman credit Mark Goodson and Bill Todman, "the maharajahs of paneldom," with the discovery that "the audience was probably more interested in the players than in the game. They assembled their panels with loving care, and their shrewdness is apparent in the fact that few loyal viewers of *What's My Line?* or *I've Got a Secret* can recall last week's lines or secrets, but most can remember Bennett Cerf's pun or the gown worn by Bess Myerson."

A few of these shows had a genuinely educational nexus—Cowan's *The Last Word* (for Bergen Evans) dealt with etymologies, and *What in the World?* showed off the expertise of archaeologists. But mostly the shows used parlor games as a frame within which celebrities from the world of light entertainment could have limited conversations with each other and a "host." As such, the panel show was a direct ancestor of the talk show, and there is probably no criterion by which it could be called better or worse. Television was able to find employment for some of the nation's best comic talent—notably Fred Allen and Groucho Marx—only in the context of game shows. In New York City in fall 1971 some fourteen game shows were still on the air, all before 3 P.M. or after 11 P.M., and my younger son tells me that five of them could be called panel shows with celebrities.

The game shows that involve "ordinary people" have been the blackest marks against television programming, daytime or prime time, from early in the history of the medium. On shows like *Queen for a Day* and *Strike It Rich* women competed to see who had experienced the worst tragedy; Shulman and Youman print a picture of a horribly grinning man and woman, *Strike It Rich* master of ceremonies Warren Hull and contestant Mrs. Eleanor Kane, who "struck it rich for her five-year-old daughter, who was born deaf, dumb and blind." Shows like *Truth or Consequences* and *Dollar a Second* and *Beat the Clock* encouraged people to make jackasses of themselves for the viewing audience.

But the worst of all were those that were supposed to make people look good, by answering questions on subjects remote from their daily concerns about which they had acquired a hobbyist's expertise—a cobbler who knew about opera, a policeman who knew about art, a grandmother who knew about baseball. In 1958, when Lou Cowan was president of the CBS Television network and his *$64,000 Question* was still riding high, Richard Salant, then general counsel to CBS, said that Cowan in producing the show had been motivated "by a desire to show that ordinary people were great." Those who knew Charles Van Doren when he was in apotheosis as the regular guy and universal genius of *Twenty-One* remember his strong feeling that he was striking a blow for the intellectual in an America that had been prejudiced against eggheads.

Ordinary people, unfortunately, are never quite good enough for the purposes of mass entertainment; reality has to be improved. "It does no demonstrable harm," Gilbert Seldes wrote in defense of *The $64,000 Question* in 1956; but he was wrong. Nothing television has done in America has had so great an impact—or so evil a consequence—as the revelation that the ordinary people millions had come to admire were in fact common cheats and liars.

Seldes' tolerance of the game shows (even the agony show: "In its dreadful way it reaffirms one of the greatnesses of television, its capacity to transmit the truth") was not matched by any sympathy for the soap opera—or "daytime dramatic serial," to use the approved description. "During one six-month period when I watched it steadily," he wrote, "I found the incidence of crime very high and the suffusion of hatred almost unbearable; perhaps because the writers and actors lacked skill, the presentation of love and affection, rarely tried, was never convincing, and I felt something sinister as well as hateful coming off the screen." That this savage attack was the result of actual viewing seems clear enough, because six years earlier Seldes had been willing to cite (if not quite accept) the conclusion of W. Lloyd Warner and William E. Henry of the University of Chicago, that the radio soap they had studied "functioned very much like a folk tale, expressing the hopes and fears of its female audience, and on the whole contributed to the integration of their lives into the world in which they lived."

There is in fact a great deal of malice (as well as a great deal of bad luck) displayed on the daytime soap. This is sometimes difficult for the actresses who play the female hostiles, because they may be insulted on the street by viewers who recognize them, and who have accepted the characters of the soaps as part of their own real lives. Letters arrive at the networks, asking where some newly married couple on a show bought the furniture the viewer sees on the screen. "When I tell a woman I'm in daytime television," says Michael Eisner, who heads this division at ABC, "she asks me why so-and-so died." The world of the daytime serial is a separate place, with its own fan magazines, economics and technical procedures. It is a big world. Though most soaps have ratings below 8, and it is only from 3 to 4 P.M. that the networks consider the soap-opera audience large enough to sustain three such programs at once, a Lou Harris poll commissioned by *Life* found 26 percent of the country willing to say that they considered "soap operas" (so labeled) as shows "meant for me." Though *Life* put an "only" before the 26 percent, what the poll said was that the soap-opera audience was two and a half times the circulation of *Life*.

CBS is the home of the soaps, which begin on the network at 11:30 in the morning with *Love of Life* and continue (broken in New York only for a half-hour Australian cooking lesson) until 4 in the afternoon: *Where the Heart Is, Search for Tomorrow, As the World Turns* (the perennial No. 1 in daytime ratings), *The Guiding Light, The Secret Storm, Edge of Night*. Not all of these are CBS properties: four (including *As the World Turns*) are owned by Procter & Gamble and produced on its behalf by advertising agencies. Taken as a group, they are undoubtedly the largest "profit center" on the network. Fred Silverman told *Life* that they cost only $75,000 each to produce each week, and brought in $330,000 a week in revenue. CBS Television was the nation's largest advertising medium in 1970 solely because of its daytime sales: according to BAR figures, NBC outsold CBS in prime time by $349 million to $339 million, but CBS clobbered its rival from Monday to Friday, 10 A.M. to 6 P.M., by $160 million to $100 million. And in fall 1971, CBS was airing more soap operas (seven) than NBC (four) and ABC (two) put together. Interestingly, though, NBC for the first time pulled even with CBS in daytime in that fall's ratings books, and for one or two rating periods *General Hospital* nosed out *As the World Turns*, putting ABC in the race, too.

"The first thing to remember about the soaps," says Irwin Segelstein, vice president in charge of program administration for CBS, "is that their sociological function is to employ actors in New York City." They also keep in being the collection of skills necessary to put on live dramatic television: two CBS soaps actually are broadcast live, and the other five are distinctly live-on-tape—an actor could knock down a wall of a set on camera, and provided he did not accompany this action with an inadmissible expletive the show would air as taped: there is no budget whatever for editing.

Home is the CBS studio building on Fifty-seventh Street on the far West Side of Manhattan, very ambitiously built in the late 1950s when management expected live drama to continue to be the staple of prime-time television. Each soap has the use of half a studio for five hours a day. Solidly built sets are semipermanently emplaced for each show (which is why only half the studio is available for each), and behind the sets are huge metal cabinets full of the props of everyday life—tea services and crystal china, flower vases, watering cans, equipment for all the gracious things American women do every day. Sitting about in the unused portions of the studio are members of the show's permanent crew of stage hands, carpenters, electricians, cameramen, etc., about thirty men in all.

Actors and actresses are also sitting around on available furniture, going over their lines: CBS soap operas do not use idiot boards or teleprompters, because the characters are supposed to be looking at each other rather than at the camera. Parts must be memorized, though only in a rather informal way. The cast is not held to the lines as written (partly because the senior writers of these shows don't write the lines: they do plots, others do dialogue), and there is a certain Stanislavsky feeling that if the actor really gets into the character and knows roughly what's going on he will say the right thing. The actors and actresses live with their characters for a long, long time—some contracts for soaps run three *years*, though the producers have unilateral options to cancel every thirteen weeks; and there are actresses who have played the same role on a soap opera for fifteen years.

The key operatives in planning the show are the director, the associate director and the technical director. The actors have had the script for several days, and they run through it in the morning (for a show to be shot in the afternoon) in an upstairs rehearsal room. The director works with them on intonation and expression and gesture, and plans the camera angles he will want. When the show arrives in the studio, the three men on the directing staff position themselves in the control room, communicating with the cast mostly through a loudspeaker on the set, and with the camera and mike personnel through headsets. In general the director studies the picture that would be going out to the public if this were the broadcast; the associate director, who has made a notebook full of notes on what the director said he wanted during the morning rehearsal, studies the picture on the camera that will be called next, to make sure everything is ready when needed; and the technical director flips the switches that move the various cameras on line. The audio man sits in a separate booth across a glass partition, controls the mikes and blends in sound effects as indicated.

On this set, the afternoon show is *The Secret Storm*, and the director is David Roth, an energetic man with a mane of gray hair, wearing a dark jacket and belligerently checked gray pants. The basic outline of what he has ahead of him can be read on a sheet pinned to the operator's panel of a mike boom:

> Secret Storm   VTR 5/18 Air Wed. 5/26
> Opening Film
> Act I: Val's Living Room
>   Commercial TOFFER/NOODLES WITH CHICKEN 16° COLOR FILM
> Act II: Clayborn Patio
>   Commercial LIQUID PLUMBER/CLOROX LIQUID 16° COLOR FILM
> Act III: Clayborn Patio
>   Commercial STARCH FAB FIN/HOME STYLE BEANS 16° COLOR FILM
> Act IV: Frank's Office
>   Commercial PEANUT BUTTER/NIAG. FAN FINISH 16° COLOR TAPE
> Act V: Frank's Office
>   Commercial WIZARD/EASY-OFF WINDOW CO. 16° COLOR FILM
>   SHOW FILM AND CREDITS

Buried in here is a forty-second stretch when the network is dark and the local station inserts its own commercials. At CBS the bridge music between the acts and the commercials is played by an organist (the other networks have secured union permission to use orchestral recordings), and the organist may be the most highly paid member of the company. He has published his bridge tunes, and every time he plays one of them the cash register at ASCAP clicks off a supplement to his salary. . . .

At the beginning of the rehearsal for each act, the members of the cast set themselves in the bright lights in the places they will occupy, permitting the cameramen and studio engineers to get focus and color just right; and Roth works mostly with the cameramen through their headsets, composing the picture that will appear on the screen. The huge fishing pole of the mike boom hangs over the actors' heads. There are three cameras on dollies (one mounted high and operated by a man perched on a seat, relying on a helper to push him around); all move from one set to another during the commercials. If the picture does not set right, Roth will come bustling down himself to move people or props ("I'm sorry, kids: I go to pieces easily"). The cast is addressed by the name of the character played rather than by the name of the actor or actress: "Deirdre, get around the coffee table as soon as you can, and sit; Marta, watch it, I'm coming around; then I'm going to want three across to Deirdre and Kevin." Deirdre is a pretty girl with freckles, wearing a gingham ankle-length hostess gown for her part. The mind attuned to nighttime television recognizes her as a heroine, but in fact she is the villainess, a selfish, grasping, willful *younger* woman. "I'm really a singer," she says, chewing gum. "I do this on the side." Kevin, the young lawyer for the other side in the cockamamie dispute that is being taken through the labyrinth of *The Secret Storm*, replies to a question with "I really don't know; I'm a new character. But I'll be here a long time." A passing stage hand, XL tee shirt stained with honest toil, comments, "You *hope*."

But the fact is that the theatrical process is not spoofed, and in the end people do put together a play in which the troubles of the characters are far from unreal to the actors who play them and to the characters who direct them. Bob Myhrum, who shares *The Secret Storm* directing duties with Roth, remembers a day when he was directing *The Doctors*, and at the end of a morning scene, a wife bidding farewell to a husband from a pose in bed, the actor playing the husband felt a charge of sexual electricity in the situation and unexpectedly turned around and jumped into the bed as the scene faded. "It would have been impossible at night," Myhrum says, "but in the day it's no problem, because people know the character." Starting a series once for ABC, Myhrum told the newly assembled cast, "You have to remember that this all goes back to the wine-dark sea. The one thing we all have in common is Dickens; we're storytellers. There are very few storytellers in the world; anybody can write dialogue, but storytellers are rare."

This afternoon's *Secret Storm* proceeded, neither efficiently nor inefficiently but on pattern, from run-through to dress rehearsal, Roth incessantly on the line to his cameramen: "Go right in the pocket where you were, Frank—fantastic!—that's beautiful—stay there—now, step to the left. . . ." The cameraman's voice comes into control booth: "I have a chair there. . ." Roth says, "I'll be go-to-hell. . ." Then: "You should bring him to a two-shot with Kevin baby. Keep coming down to the desk. You got it, coach." And the cameraman says, "Okay, but that's not the cue I got last night." At CBS daytime, the cameramen do not have written cue sheets, and depend entirely on instructions from the director plus their memory of what he's been saying. "That way," Roth says, "they can concentrate completely on what's in the viewfinder."

At the dress rehearsal, in Roth's words, "there was a certain amount of chaos left over." He goes down and talks over some sticky moments with individuals in the cast, then returns to the booth. "A bad dress gets the adrenalin flowing," he says cheerfully. "Keeps everybody on his toes." And the tapes roll: "Ready two. . . . Take two. . . . Easy on the birds in the garden, too loud, my producer's getting nervous. . . . Beautiful! Just the way we planned it. . ."

3

Some observers (including Paul Klein) find in daytime the inevitable future for prime-time television, because daytime controls its costs. Game shows cost from $5,000 to $10,000 an hour to produce.

The Frost and Douglas shows work out to about $8,000 an hour; the Cavett show, because the unions hit networks harder, probably costs about $13,000 an hour. Soaps come in at about $30,000 an hour. The average prime-time show, repeated once (daytime shows are rarely repeated), costs in effect about $115,000 an hour, and there have been worries that it may go higher.

For the short term, prime-time costs can probably be controlled. When commercial minute prices dropped in early 1971, New York turned the screws in Hollywood, to some effect. "Price cuts on minutes," says Phil King of the CBS West Coast program department, "filter down on us, like a rain of bricks." But it is also true, in the words of Alan Wagner, whose office is next to King's, that "Nobody will give me kudos for bringing in a cheap disaster."

The pyramid of producers and executive producers and associate producers that loads the costs of Hollywood production will be eroded during the 1970s. "There have been a lot of payoffs to somebody's brother-in-law," says Murray Chercover of Canada's CTV, "and that's going to stop." Stars (who seem increasingly unable to guarantee audience) will doubtless have to make do with a quarter of a million rather than half a million dollars a year. An equal economy will come in the move from 35-mm. film ("which involves," says a Hollywood executive producer, "a cameraman who is a crew foreman and the operator and the guy who turns the lens and the people who push the camera") to 16-mm. film ("which gives you hand-held cameras, small crews, lower film costs"). Direct taping of Hollywood-style productions, which would at the least eliminate the time delays of processing "rushes," has been vetoed in America till now "because there is no three-headed Moviola for tape"—i.e., no machine by which tape can be edited easily and inexpensively, blending separate sound tracks with different picture tracks at will. CBS in early 1971 announced the invention of such a machine (called RAVE, for Random Access Video Editing), but the price tag was $300,000 a unit, and as of the end of the year the studios were still very wary of it. The Japanese were using it, though.

Much of the doubling of costs for prime-time shows in the 1960s was the result of yielding to union requirements for crew sizes and special pay for odd hours worked. "Everybody conspired to drive prices up, because then the same margin gave you a higher profit," says John H. Mitchell of Screen Gems, who has put a hundred series on the air. "The producers had their hands in the till, the unions had their hands in the till—and, of course, the medium was performing." The burden of overhead became insupportable. "Every time I go to a Hollywood studio," Murray Chercover says, "I could throw up."

Nobody knows whether any of the featherbedding in picture production can be eliminated. "The unions here," says Willis Grant of Screen Gems, "are like the coal miners. They see the business eroding, but they want to protect those still working, they're afraid to give anything." (This analogy will not do, because what killed the mine workers was their *acceptance* of mechanization; it will work, though, with union printers at the newspapers.) But Hollywood had a terrible scare in 1971, and demands from both the Writers Guild and the Screen Actors Guild in summer contract negotiations were much less than producers had feared they might be. "Things that couldn't be done a year ago," said CBS cost controller Don Sipes in spring 1971, "can be done and have been done now." It is not impossible that the technical unions—Jack Cowden of CBS says that on one production his company had to deal with seventy-eight separate unions—will also lend a helping hand to the cost controllers.

But after all the fat is cut away, television production, too, is likely to become a victim of what William Baumol and William Bowen of Princeton have called *The Economic Dilemma* of the *Performing Arts*, which is also the economic dilemma of education and medicine and the other service trades. In an economy where all wages are pegged to the productivity of industrial workers, the costs of services rise rapidly, because service workers increase their productivity only slowly, if at all. At some point, then, there will be heavy pressures on the costs per hour of television production and only three possible ways out of the squeeze—to produce fewer hours, each of which would be more important in itself and would run more often; to reduce the production quality of prime-time product; or to increase the amount of money spent for television programming, either through a form of pay-TV or through an increase in industry's expenditures for advertising. There are, after all, questions of national priorities here. The nation in 1971 spent more than $73 billion for education, but only about $3.5 billion all told for the operation of the television broadcasting system, which for most people is clearly a more important activi——— . . . but now we have entered into realms of philosophy, and must retreat.

In any event, it seems unlikely that producers will be able to escape the cost burden by fleeing the country. *The Men from Shiloh* went down to Mexico to make several episodes, but each one of them took so much longer to make that the extra time ate up the savings from exploiting peons. Canadian production does save money, but—so far—only at a cost in quality. As befits a frontier country, Canadian television production is a little rough-and-ready, which does not harm the product locally (and may even increase the charm of a show like *Pig 'n' Whistle,* a recreation of an English working-class pub as a setting for old music-hall numbers), but causes obscure irritations in America, where people expect a glossy finish.

And over the long run the British unions are going to be at least as tough as American unions. "We were in the middle of rehearsal," said a veteran actress on *Crossroads,* an English soap opera which runs four times a week on the commercial stations (at 6:30) and is in some regions the most popular show on television, "and this workman walks by and gives me a civil hello, and he says, ' 'Ere now! Wot's that?'

"And I say, 'It's an electric fire.' [She means an electric heater; one can't help what the English say.]

"He says, ' 'Ow'd it get there?'

"I say, 'I don't know. I suppose a prop man put it there.'

"He says, 'That won't do. That won't do at all. That there's helectrical hequipment.'

"Well, there was the most terrible row, all the time while we were trying to learn our lines, and the only way it could be resolved was that two men carried the electric fire onto the set—the prop man carried the fire itself, and the electrician carried the wire. . . ."

CHAPTER 6

Sesame Street and
Saturday Morning

Television is peculiarly the child's companion, from an earlier age and to a degree not matched by any other mass medium.
—J. D. HALLORAN

Children who have been taught, or conditioned, to listen passively most of the day to the warm verbal communication coming from the TV screen, to the deep emotional appeal of the so-called TV personality, are often unable to respond to real persons because they arouse so much less feeling than the skilled actor. Worse, they lose the ability to learn from reality because life experiences are more complicated than the ones they see on the screen, and there is no one who comes in at the end to explain it all. The "TV child" . . . gets discouraged when he cannot grasp the meaning of what happens to him. . . . If, later in life, this block of solid inertia is not removed, the emotional isolation from others that starts in front of TV may continue. . . . This being seduced into passivity and discouraged about facing life actively on one's own is the real danger of TV.
—BRUNO BETTELHEIM

When educational programs like "Science Review," "From Tropical Forests," or "Have You a Camera?" appeared on BBC, children who had access to both BBC and the commercial channel would almost invariably turn to the commercial one, where they would find usually a choice of cartoons, Westerns and the like, much like American commercial television. But if the child had access only to BBC—that is, if the commercial service had not yet reached his community—then there was a choice only of turning off the set or of watching the educational program. Under these circumstances, the English investigators report, "quite a number of children chose to see such programs and *in fact enjoyed them.*"
—WILBUR SCHRAMM (paraphrasing
and quoting HILDE HIMMELWEIT)

Sunday morning in this market was notable by its lack of kids' programs. When the nets a couple of years ago began carrying Sunday cartoons, the stations here wouldn't clear. So we offered the time, and came out with, oh, a fifty percent share. We were UHF, unaffiliated, and new, and we were the No. 1 rated in this market in these time periods. Then, one by one, the network programs were no longer available to us; the local affiliates picked them up.

—LOREN W. MATHRE, program director,
WTOG-TV, Channel 44, Tampa–St. Petersburg

1

Who should write the criticism of children's television shows? The children themselves cannot be trusted with the job, because every piece of research ever performed in this field comes up with the conclusion that children will express a favorable attitude toward any show seen in isolation, without direct or immediate comparison. The specialists in Early Childhood Education are ineligible, because a critic must keep up with the literature of his field, and there are very few specialists in Early Childhood Education who can both read and write. And the conventional television critic, always happy to see the well-intentioned, is an absolute sucker for seriousness of purpose when the anticipated viewer is juvenile.

A few years ago, an international symposium of social scientists came up with the pleasant idea of testing in various countries children's reactions to the programs that had won the UNESCO-sponsored *Prix Jeunesse* as each year's very best children's television show. The winners were voiceless puppet, cartoon and mime features, presumably cross-cultural in their appeal; one was Swedish and two were Czech. Despite the planning, the social psychologists in the different countries did not follow the same testing procedures, and results were varied (as indeed they might have been under identical procedures). But few of the research teams disagreed with a central finding of the British group, that in each case "the children themselves would not have chosen the film as a prize-winner." Several teams tried to enlarge the do-you-like-it question to provide a basis for comparison (i.e., by presenting a picture of a chest of drawers, each drawer labeled "my favorite" or "one of the programs I like best," etc., in a descending scale, and asking the child to pick the drawer into which he would slot the *Prix Jeunesse* show). Such tests produced a median result somewhere below the second-best drawer.

Developmental psychologists tell us strange things about children's relations to television. "Except for the more intelligent child," says one of the English reports on the *Prix Jeunesse* winners, "comprehension of the story line, when using photograph sequencing tasks, is not possible until seven years"—that is, children below the age of seven (unless very bright) see the frames of a film as discrete units and cannot understand that they fit together to tell a story. Moreover, children below six or seven years are said to regard everything they see in a film as "real," because they lack the ability to distinguish fantasy from fact.

To the extent that these assessments are correct, the task of making a program for children is entirely different from that of making one for adults. But only a social scientist could begin to believe that four-year-old children do not sense the telling of a story in a film that does in fact tell a story—and while children are clearly different from adults in their sense of "reality," the disparate definitions are not necessarily *contradictory*. Children's preferences among *adult* programs, which most of them begin to watch before their first attendance at school, do not seem wildly different from those of their elders; and where there are differences, subject matter (i.e., science fiction v. domestic comedy, adventure v. variety) seems more important than differences in technique.

For a long, long time, it has been known that children love ghost stories and slapstick comedy and tales of heroes and heroines (preferably heroes and heroines a few years older than themselves) beating off challenging, mysterious and dangerous adversaries. The building up and releasing of tension is for them as for grownups an essential aspect of "entertainment." On re-examination, the great "children's " writers of the English language—Defoe, Swift, Dickens, Twain—turn out to be, more often than not, the great writers of their time. There are, clearly, *fewer* major creators of material for children than of material for adults. The notion that the government can order "better" children's television programs and the world will go off and make them is a remarkably simplistic proposition even for the early 1970s.

Yet the question of who shall define "better" and by what standards—the question of who shall write the criticism of children's television—is an extraordinarily interesting one. Adults are entitled to choices, even if their choices are "wrong," and it is arguable that the sum total of human happiness is increased by the adman who makes products look more attractive than they might seem in the harsh light of a consumer organization's laboratory analysis. But pushing goods to children is an abomination, and allowing the most popular shows to float alone on the surface, buoyed by ratings, clearly abdicates adult obligations to children.

And there is, of course, a possibility—much publicized as more than a possibility—that individual programs may harm children. In the language of the National Commission on the Causes and Prevention of Violence, "Violence in television programs can and does have adverse effects upon audiences—particularly child audiences." Harvard's James Q. Wilson, reprinting that sentence, adds the comment, "The blunt truth is that there is almost no scientific evidence whatever to support the conclusions of . . . the Commission." J. D. Halloran, R. L. Brown and D. C. Chaney of Leicester University in England have noted that while writers on television always talk about the way it promotes crime, writers on crime almost never mention television. In early 1972, very surprisingly (because all the good guys were on the other side and the politics of the situation argued for viewing-with-alarm), the Scientific Committee on Television and Social Behavior, appointed by the Surgeon General at the request of a Senate subcommittee in 1969, reported that television seemed to be a relatively minor element in the causation of juvenile misbehavior.

No doubt art can be disturbing—Hilde T. Himmelweit found in the 1950s that a televised dramatization of *Jane Eyre* gave nightmares to many children whose tough hides were totally impervious to the gunfights of western and crime shows. And Dr. Leon Eisenberg, chief of psychiatry at Massachusetts General, has found some evidence that viewing the news is upsetting to children, partly because some of the scenes show terrible things happening to children (i.e., juvenile victims of the Vietnam war, starving Biafrans or oppressed Bengalis), partly because children will tend to watch news shows, if at all, with their parents, and parents while watching the news are incomprehensibly different from what they are at other times. Many of the most successful and apparently acceptable children's shows are sources of trouble: Lassie in danger has disturbed the recent generation of children much as Snow White's encounters with the witch disturbed their parents.

Portrayals of sadism, knives, lingering torment make for bad dreams; so do screaming fights between adults, especially husband and wife; and Dr. Fritz Redl has argued that for "disturbed and delinquent children" the sweet situation comedy with its portrayal of the happy family they have never known can be "traumatic." The one thing that seems safe is what the critics worry about most—the ritual, patterned violence of the cartoons, of Batman and Robin, of western gunfighters. Just as the ladies who watch daytime television will accept without offense an explicitness about sex that would anger them in a nighttime show, because they know to the dotting of the last "i" the rules of the soap-opera game, children can accept without concern the boulder Road Runner drops on the pursuing Wily Coyote, or the mowing down of the bad men in the Bad Lands.

Worries about what entertainment will do to the young are among the atavisms of the tribe of Western man. Victorian parents deeply feared novel-reading by their children (especially their daughters), and Fredric Wertham's hysteria about the comic books made him a famous psychiatrist. Halloran and his colleagues quote a fifty-year-old essay by G. K. Chesterton on *The Fear of the Film:*

> Long lists are being given of particular cases in which children have suffered in spirits or health from alleged horrors of the Kinema. One child is said to have . . . killed his father with a carving knife, through having seen a knife in a film. This may possibly have occurred, though if it did, anyone of common sense would prefer to have details about that particular child rather than about that particular picture. . . . Is it that the

young should never see a story with a knife in it? It would be more practical that a child should never see a real carving knife, and still more practical that he would never see a real father.

None of this defends what children are actually given: junk is junk, whether it is produced for an adult market or a childish market. Nor does the scientific disreputability of most of the studies indicting children's television detract in the least from the perfectly reasonable, common-sense distress mothers feel when they look in at what the kids are watching. (And this distress is well documented: when the FCC requested comments on a possible rule-making in this area, the Commission got eighty thousand letters.) Unfortunately, the most vocal and influential critics are humorless, doctrinaire educators and busybodies for whom no amount of priggishness on the tube could ever be too much. Children's rights do not include the right to be pandered to, but they do include, please God, some access to entertainment, even if their parents find the entertainment vulgar.

James Duffy, president of ABC-TV, told a meeting in June 1971 that it might be a good idea to eliminate Saturday morning ratings; and perhaps it might be. The point about the importance of *not* giving people a choice of things to view if you wish to influence their behavior applies with greatest strength to children's television —and it is of course perfectly legitimate for adults to wish to influence children's behavior. But the American commitment to competition and freedom of choice is deeply held and, really, fairly easy to defend. There is, after all, one triumph on the record of American children's television: *Sesame Street*. And it is quite impossible to believe that this show would have been anywhere near as good as it is if its creators had been given a monopoly of all channels —a captive audience—for their program.

2

Perhaps the most remarkable thing about *Sesame Street* is the fact that few schoolmen take it very seriously. It is by any standard the largest educational experiment ever. Children's Television Workshop, which runs it, estimates on the basis of some breakouts from Nielsen surveys that eight million children see the show every week; and there may be six million who see three or more shows every week. Though the cost works out to only about one cent per exposure per child, the total budget for the show runs nearly $6 million a year, of which $2 million comes from the Corporation for Public Broadcasting. (CTW as a whole has a budget of $13 million a year, which also pays for *The Electric Company,* a new daily half-hour reading show aimed at second-graders; a separate unit makes profits to help support the show by developing printed material, puppets and games built around the *Sesame Street* substance.)

The sources of support are big guns—the Office of Education, the Ford Foundation, the Carnegie Corporation—none of which has historically been shy about claiming significance for the end product of its grants. But among the many things thought on ahead of time by those who planned *Sesame Street* was the publicity the project should receive, and the Carl Byoir agency was involved from the start. The appropriate drum rolls were struck, of course, when the Educational Testing Service of Princeton reported that *Sesame Street* watchers learned much more of this and of that than tots who didn't watch. On the whole, however, Byoir has followed the line laid down for the psychologists by Dave Connell, executive producer of *Sesame Street,* at the earliest meetings of the committee that planned the show: "Our purpose is both to instruct and to entertain, but always to entertain."

The show entertains spectacularly—lots of people, not only children, are devoted to some of the cloth "Muppet" characters created for and growing with the show: to Ernie, the cheerfully persistent little fellow who drives big brother Bert right up the wall; or Oscar, the nice grouch who brings negativism to the block and lives in the garbage can and yells at people for daring to forget that an "R" has two legs and a "P" only one; or Kermit, the pitchman-frog who sells numbers and the letters of the alphabet. Some of the songs from the show, bouncy, brightly orchestrated, funky yet flowing, have appeared on the best-seller charts—Frank Sinatra himself recorded Kermit's "It Isn't Easy Being Green." Running gags and black-outs, in the *Laugh-In* tradition, enchant adolescents and adults, which helps keep the set tuned to the channel on which *Sesame Street* appears. The youngest in the family, after all—this is another thing thought on ahead of time—doesn't control the set.

But almost every moment in almost every show starts from a severe, strictly non-show-biz "theme sheet," a prescription from the team of six (it has been as many as eight) educational psychologists who work full time for *Sesame Street* and are responsible for planning and testing everything that goes out on the air. A *Sesame Street* hour is likely to be divided into thirty to fifty separate bits, very rarely longer than three minutes, occasionally as short as twelve seconds. Every bit must fit into one of four categories:

- Symbolic Representation (the letters and numbers, sometimes words and geometric shapes—15-20 minutes scattered through the show)
- Cognitive Organization (matching and discovering shapes, discriminating between subjective reactions to and objective properties of a thing, sorting and classifying objects by size, position, function, etc.—about 15 minutes)
- Reasoning and Problem Solving (inferring antecedent events and predicting subsequent developments, generating and evaluating explanations—about 10 minutes)
- The Child and His World (recognizing body parts and mental functions, understanding social groups and social interactions—15-20 minutes)

Within the four big categories, the psychologists will make up a list of perhaps twenty specific targets, specifying the letter and number for this day, the relational concepts to be taught, with a period of time for each (i.e., "Classification. Time 3:00. Sorting According to Quantity").

Sesame Street repeats a great deal of material within shows and from show to show—in the 1970–71 season, only about 25 minutes of every hour were created fresh for that day's program—and the psychologists specify what is in the can that they would like to see used on this installment. Thus the letter for show No. 249 was "P," and the psychologists' instructions began with "P-Pin :12 [seconds]; P-Painting 1:43; P-Puppy :28; Skywriter [during the planning phase,

Sesame Street had all the letters written in the sky and photographed] :38 (V. O. [meaning "voice-over," no sound on the film as supplied]). Do not repeat P-Painting." The 25 new minutes get tested on live children in day-care centers by the house researchers, and are occasionally vetted by an advisory council of about twenty headed by Gerald Lesser of the Harvard School of Education. "Every once in a while," says Jeff Moss, who heads the program's team of six writers, "they may say, 'This stuff which is supposed to teach such-and-such really doesn't,' and we'll say, 'We know—but it's funny. If we teach 95 percent of the time, let us be funny 5 percent.' But if they say it's teaching the *wrong* thing, then, of course, it goes."

Even the Muppets are not just beasts or people thought up by their very talented creator Jim Henson (who since the late 1950s had been making puppets for commercials and Ed Sullivan, and special items like Rolf the Dog for Jimmy Dean). "I sat in on the early educational meetings," Henson recalls, looking like a slim Pirate King with pointed beard, heavily brocaded black jacket, silk scarf, wide belt with ornamented heavy silver buckle. "We were thinking about what new characters might be nice for the show. We wanted to get out all the things kids usually hold in. The concept of a large goofy character, an outlet for making dumb mistakes and being silly, grew into Big Bird [a seven-foot-tall canary]. The idea of a character that was wholly nice but completely negative became Oscar. There were solid learning reasons, you know, for all these things to be built."

3

To some extent, *Sesame Street* is a tribute to the patterns of work that John Gardner built into the Carnegie Corporation while he was its president. In general, foundations worry about the relation between the social contacts of their officers and the grants those officers propose; Gardner encouraged his people to consider themselves on duty all the time. The University of Illinois Arithmetic

Project, for example, grew out of the chance that Fred Jackson, then a Carnegie executive assistant (later president of Clark University), saw David Page inventing math games for his son while the dinner company had a drink. *Sesame Street* grew out of a conversation at the home of Timothy and Joan Cooney.

Mrs. Joan Ganz Cooney had come east from Arizona with a degree in education, but her working experience had been in the newspaper business and in public relations, and in New York she moved to noncommercial television, rising through the ranks at Channel 13. She was producer of *Court of Reason*, which was—largely because it commanded the services of Columbia sociologist Robert K. Merton, on camera—the best of the long string of essentially empty-headed efforts by educational television to gain audience for serious discussion by giving it the fake drama of staged adversary proceedings. Among her guests at dinner was Lloyd Morrisett from Carnegie, one of very few men in the foundation game whose appearance leaves a striking memory—tall, lean, fanatically austere, long narrow head, black suit. Morrisett has since moved on from Carnegie to become president of the Markle Foundation; his Rockefeller Center office has three dark wood-paneled walls and a white wall behind his desk, and not an item of ornamentation, not even a diploma, on any of them.

Childless herself (she and her husband now have a black foster son), Mrs. Cooney was involved in various antipoverty efforts to do something about the condition of black and Puerto Rican children in New York. The discussion at dinner touched on the amount of time children spend watching television, and on the wasted educational possibilities of the medium, especially in the preschool years. The more Morrisett thought about it, the more interested he became. He called Mrs. Cooney and invited her and another of the dinner guests, Lewis Freedman, then director of programming for Channel 13, to come to Carnegie and talk about the question seriously. From this meeting came a grant to Channel 13 to free Mrs. Cooney for three months to make a "feasibility study."

The result was a workmanlike, not very imaginative document that suggests *Sesame Street* only in its recognition that what most children like most on television is the commercials, and that the commercials seem to teach: "Parents report that their children learn to recite all sorts of advertising slogans, read product names on the screen (and, more remarkably, elsewhere), and to sing commercial jingles. . . . If we accept the premise that commercials are effective teachers, it is important to be aware of their characteristics, the most obvious being frequent repetition, clever visual presentation, brevity and clarity. Probably, then, their success is not due to any magic formula." But neither in the feasibility study nor in the subsequent proposal to Carnegie (dated February 1968, two years after the dinner party) did Mrs. Cooney propose the straightforward use of advertising techniques in the show.

She planned to start, for example, "by devoting ten to fifteen minutes . . . to story and conversation. The discussion would take place between three 'regulars'—a woman who would do the reading, an intelligent child of twelve or so, and a little puppet who would provide humor in the form of wrong answers, simplemindedness and general clowning." There were to be other puppets, but their function would be to deal "with the problems of everyday living encountered by young children." These episodes would be little morality plays, "situations involving feelings of possessiveness, rivalry, aggression and fear which could be dramatized effectively in this manner." Science would be taught "by performing little experiments on camera in the studio." Though there is no real doubt that Mrs. Cooney planned animated-cartoon sections of the show from the beginning, she did not mention them in her report or proposal to Carnegie: the do-good community from which support for the show would have to be solicited considers "cartoons" the root of all evil in children's entertainment.

Though in fact all her specific ideas for the show could have been done for little money, Mrs. Cooney insisted that a show good enough to draw children away from professional productions on other channels would have to be budgeted at professional costs. The networks buy Saturday morning cartoons at a price of about $50,000 for 22 minutes (for that price, they get the right to air the half-hour six times; and the price, incidentally, does not cover the animation studio's costs: the studio relies on subsequent reruns for its profits).

Mrs. Cooney was planning 130 one-hour shows—26 weeks of 5 programs each—and there clearly wouldn't be enough money to match the networks dollar for dollar. But a collection of experienced television producers, summoned to conferences, said that much probably wouldn't be needed, especially for a program with as little animation as Mrs. Cooney said she wanted. Morrisett finally budgeted Children's Television Workshop at $8 million, to cover a year's preparatory work, initial publicity and a year of program on the air.

From Carnegie and Ford, Morrisett raised $4 million, to be matched by gifts elsewhere. The broadcasting companies looked through Mrs. Cooney's documents and were not impressed, and no rich individual backers appeared. Both Ford and Carnegie had worked closely with Harold Howe, Lyndon Johnson's Commissioner of Education, while he was Superintendent of Schools in Scarsdale —and, indeed, earlier, when he was principal of Newton High School in the Boston suburb. Howe committed the Office of Education to Morrisett's project, then wandered around the government scrounging pieces of change from the Office of Economic Opportunity, the children's division of the National Institutes of Health and the National Endowment for the Humanities. All grants were made, formally, to NET (National Educational Television), to establish the Workshop. In March 1968 the proposal was unveiled to the press.

Among those reading the story in the papers was Mike Dann, who had just finished the annual agony of establishing the next year's CBS prime-time schedule, and wanted to turn his mind to higher things. He wrote a letter to Mrs. Cooney (whom he had never met) congratulating her on the idea and the grant, and urging her not to get trapped by her own noncommercial television background and the noncommercial sponsorship of her program: the sort of thing she had announced would need a commercial producer experienced in handling the crises attendant on daily production of television shows. Mrs. Cooney called Dann to thank him for the letter, and learned that Dann had a man in mind: David Connell, who at the age of thirty-six was a veteran of eleven years as producer of *Captain Kangaroo,* the daily CBS morning kiddie show.

Connell, however, was not interested. "I'd left children's television," he said recently, "given it up, never wanted to see children's television again. I'd gone into the film business, and I was very content in it. I'd read the announcement in the *New York Times*, with the big list of educational advisers and the sponsors, and I thought that like other projects this one was going to be advised to death—I thought they were going to blow eight million dollars. But I met with Joan—hours and hours and hours of conversation— and I was totally charmed by her."

Connell activated what the English would call an old-boy net, based on *Captain Kangaroo*. Sam Gibbon had been his assistant in planning and producing *Kangaroo*, and had also left the show; he was hired to be one of three day-to-day producers. Jon Stone, a product of Williams and the Yale Drama School, had written and helped to produce *Kangaroo* (and had shared an apartment with Gibbon when both men were bachelors). Stone had given up on television a little earlier than Gibbon, and had gone on to Vermont, built a house for his family with his own hands, and "was hoping to find some way to make a living there, to do some writing." Stone was persuaded to return to New York, to become the chief writer for the new program, with the simulated rank of producer. The third producer hired was Matt Robinson, who had been writing and producing black-oriented public-affairs shows for a Philadelphia local station.

Mrs. Cooney's initial proposal to the production staff was, in her words, "We're going to do something like a *Laugh-In*, for children, with commercials to sell them things they ought to know." Consultants were brought in by the dozen to help flesh this out— among them, the children's book writer-cum-illustrator Maurice Sendak, Young & Rubicam's president Stephen Frankfurt, cartoon producer Chuck Jones from Warner Brothers (later to be head of children's television for the ABC network); George Heinemann, who had produced *Ding-Dong School* in television's early days (a show Mrs. Cooney had specifically identified in her Carnegie Report as one that would no longer go down with the more sophisticated children of twenty years later; Heinemann later became head of children's programming at NBC).

The consultants produced what consultants usually do produce, and the staff undertook to map their own show. By far the most important of the staff ideas was a permanent set, an old neighborhood in a city, the residents to provide "hosts" for the *Laugh-In* format and day-to-day continuity. The name found for the street and for the program continues a grand tradition of contributions to American culture from *The Thousand and One Nights*: it comes from Ali Baba's "Open Sesame." Another example of the sophistication of the planning is that, although the symbolism of the name is charming to all concerned, neither the show nor its publicity has ever linked the word "open" to the street.

In addition to the life on the street, the producers planned a spoof of the television action-adventure shows, which have always been among children's favorites (often to their parents' horror). This juvenile *Get Smart* would be entitled *The Man from Alphabet;* its bumbling superspy hero (who entered by falling through plate-glass doors) would solve his cases with the help of a smart little boy who ran a newsstand for its absent proprietor, and could fill in from his reading all the gaps in the hero's knowledge. Connell, Gibbon, Stone and Mrs. Cooney were all utterly sold on it, and the educational advisers liked the idea, because it gave opportunities to teach something about ratiocination as well as something about reading. (A Muppet named Sherlock Hemlock was later created by Henson to serve these functions.)

From the beginning, Mrs. Cooney had planned on puppets, specifically the puppets of Marshall Izan, a friend who had had a much-admired show on WCAU-TV in Philadelphia; in the formal proposal, she had also suggested the possibility of employing Burr Tillstrom, inventor of *Kukla, Fran and Ollie,* television's first big children's hit. Stone had a better idea: Henson, with whom he had worked two years before on an ABC production of *Hey, Cinderella.* Henson's "monsters" and other unrealistic puppets would give writers a wider field for imagination than they would have with the rather realistic creations of other puppeteers. Henson, moreover, was a man of many talents, who had written and filmed prize-winning experimental shows for NBC using live actors as well as his Muppets, and was the secret weapon of many "creative" advertising men. It was understood that the street episodes and the puppet episodes would be kept wholly separate—educational psychologists would never approve an indiscriminate mix of reality and fantasy.

4

   With the decision to go to Henson came a parallel, equally important decision to place the musical side of the show in the hands of Joe Raposo, another veteran of *Hey, Cinderella*. Trained by his band-leader father, by Walter Piston at Harvard and by Nadia Boulanger in Paris—and still under thirty-five in 1971—Raposo is a butterball of a man with a great aureole of curly black hair, crazy about little children (of whom he has two), exuding the sweetness of the talented man whose talents are being used to their utmost. If any one person should be credited with the near-universal appeal of *Sesame Street*, it must be Raposo. "He's the only one in this whole outfit," says Jon Stone, including himself in the comment and not running anybody down, "in whose taste I have absolute confidence all the time."

   Raposo wishes this affection and respect could be translated into money for his people (for himself, there's almost too much money: under ASCAP regulations, he could not sign away performance fee rights even if he wished to do so, and every time anybody other than *Sesame Street* plays one of his songs the cash register at ASCAP clicks off more royalties for him). "When people are drawing up budgets," Raposo says, "music is the last thing they budget. Then they think, 'Gee, we've got this forty-thousand-dollar set—how about five hundred dollars for music?' With the budget problems, we thought we would use popular children's records, and the Muppets would mimic; and there are many fine children's records, but by the time you got them on the show they'd be dying. We could have done the show with a standard music man, piano and celesta, but that's not what kids like; they like stone-cold hard rock.

   "So I decided I would write the music—eight pounds of it a week, the equivalent of a month's output from a jingle house, or a whole musical comedy, every week. And I turned out this little sports car of a band, seven men. We're up to our ass in Emmys and Grammys and Gold Records, and I keep pointing out to people, who won't listen, that the famous things these men do are playing on *The David Frost Show* and *Tonight*. They work for us at educational-television scale rates—forty-four dollars for the first hour, thirteen dollars for each subsequent hour; and we get the show done for a week in two three-hour sessions." For the benefit of those not arithmetically minded, *Sesame Street*'s cost for executant musicians works out to $980 a broadcast week, or just under $200 a show, which is less than one percent of the production budget. The total music cost for the 130 hours of program in a year, including studio fees, tape, Raposo, office space, etc., runs less than the cost of forty minutes of animated cartoon.

Raposo grew up writing for available instruments: "My father had orchestras of all descriptions. He would tell me to write for four violins, two mandolins, trumpet and flute, and I'd do it. Then I started conducting star packages of Broadway musicals in a summer tent theatre in Wallingford, Connecticut, when I was nineteen. The show would arrive with music for a standard twenty-six-piece Broadway orchestra, and there you were facing twelve men, because a tent budget could never afford more than that. When I finished the last performance Saturday night, I'd have to go through next week's score in about three hours. The experience has served me in tremendous stead."

When he finished Harvard in 1958, Raposo took a job at WCBH, the Boston educational television station, while he taught at the Boston Conservatory, wrote musical background for plays at the Loeb Student Center, collaborated with Eric Bentley on productions of Bertolt Brecht, acted as music director for both of the city's resident experimental theatre companies. "Good actors," Raposo says, looking back. "I thought I would polish my craft in Boston and be issued an invitation to practice it in the Pantheon of New York. Nevva happen. So I just came to New York, with wife and one baby, in November 1965. A week after I came—I was never worried; I'm such a whiz-bang pianist I can always work—an off-Broadway musical got into trouble and they asked me to come fix it, which I did. Then I had a lot of things to do, Broadway, television, just music—once I conducted an evening for Beverly Sills—and I was music director for Metromedia, Channel 5."

Raposo gets the same assignment sheet that goes to the writers ("a funny little memo that I think comes out of a computer in Princeton"), and begins to think about the music from the list of concepts: "What's today's 'auditory discrimination'? Then you sort of get to a place where you feel the need for a song. Or it's something most easily taught by music, rhythm, anything abstract. You think it would be good to have a song for Susan and Oscar, about how it's good to be nice—'It's Nice to Have Someone to Be Nice To.' Or a writer will say, 'Can you write anything that says this?' I can write out a piece in twenty minutes. Sometimes they give me something at seven tonight for a show tomorrow. . . .

"And I've been in so many theatrical situations that music to me is not the most important thing. I'm always conscious first of the theatrical situation. I'm proud of some of these. Remember Ernie and Bert and the Cookie Monster baking a cake? The clock ticking, the pan sliding in the oven, Ernie saying, 'Make a wish,' the Cookie Monster saying, 'I wish it was a cookie'? I wrote all of that."

Sam Gibbon has been dazzled by Raposo's ability to keep in mind the nonentertainment as well as the nonmusical aspects of the job. "He's endlessly interested," Gibbon says, "in how music can be used to teach, about the placement of key information content in the musical line. I remember once he explained to me that you can't always use the obvious musical stresses. The highest note in the song may be the most difficult for the singer to sing, which means that the singer will destroy the vowel. The fact that Joe's interested in that sort of thing has been just fabulous." Stone, whose open shirt and Bohemian beard conceal his status as *Sesame Street*'s chief worrier, pays Raposo the ultimate compliment by saying that "Writing is the single most difficult commodity to come by." Without Raposo it would be music that was the constant felt need.

In the first year, Raposo used some standard pop material, and he has had some help throughout from Jeff Moss, who took over from Stone as chief writer when Stone moved up to be producer (Gibbon having gone to plan the reading show). Moss is musical, and wrote both music and lyrics for the Princeton *Triangle Show* in 1963; and some of the most popular songs on the show (notably "Rubber Duckie") have started with a "lead sheet" from Moss, to be orchestrated by Raposo. In 1971 Raposo began to get offers of help from outside: "Pete Seeger and Art Garfunkle want to work for the show. And it ought to be like that—if we don't invest the best of our song-writing talent this way, what use is it?

"For myself, I've socked every bit of emotional energy and all I know about this business—which is considerable—into this show; and it's been worth it. I'll be on a street and a kid is singing one of those damned game songs, a little Puerto Rican girl—what a kick! And it's true that the prettiest sound in the whole world is the sound of a tiny child really laughing. I can forgive television everything when I see my three-year-old sitting in front of the set, watching an old black-and-white *Krazy Kat* and just laughing. . . ."

5

Decisions, decisions . . . *Sesame Street* needed a director. Stone had roomed at Williams with Bob Myhrum, now a veteran of daytime soap operas (and off-Broadway). A deal was made by which Myhrum and David Roth would divide their time between *Sesame Street* and *Secret Storm*. People for the street: "We needed exclusivity," Stone recalls. "For an unknown kiddie show for a long period of time. Anyone who signed up would have to guarantee that he would not do commercials." The first to be signed was Loretta Long (Susan), a young black pop singer who had been a schoolteacher and was married to the publicity director of the Apollo Theatre in Harlem, who was a friend of Stone's. No black actor could be found who could play the male role, Susan's husband, a high-school teacher with a positive attitude toward the street, but casual. Finally Connell asked his fellow producer Matt Robinson to step before the camera, and "Gordon" was cast. As the white neighbor,

Stone chose another singer, Bob McGrath, who had made most of his career with Mitch Miller. The only professional actor hired for the show was Will Lee, a character actor with forty years of New York experience, to play the candy-store proprietor, Mr. Hooper.

WNDT studios were needed for WNDT and NET programs, and none of the commercial stations had space to spare. Fortunately, Teletape, one of the larger producers of commercials, had just stretched its resources to buy and remodel the old RKO Eighty-first Street Theatre on Broadway. They were willing to make a deal which would give *Sesame Street* seven months' exclusive use of the studio, so that the set could be built and permanently emplaced; perhaps equally important, they could supply, for as much of the year as needed and no more, the raft of technicians required. (There turned out to be a drawback here: a show aimed at the building of black self-image was given an all-Caucasian crew in 1970–71; pressure got blacks hired the next year.) Meanwhile, outside film studios were commissioned to do bits like the sky-writing airplane (a legitimate slow-motion alphabet, easily tied into the street through the conceit that the characters were looking at it). Several dozen animaion and film houses were given pieces of work relating to the twenty-six letters and the first ten numbers, and Henson began doing some brief bits with Stone scripts for Muppets.

The Man from Alphabet was to have Hollywood production values, so Gibbon and Connell went off to Hollywood to make the first twenty minutes of it. When they came back, in July 1969, they felt ready to go to the psychologists with four sample shows to be tested out on real live children. For *Sesame Street*, after all, was not being funded at $8 million to be entertaining: it was supposed to hold a deeper level of attention, and to teach.

6

Mrs. Cooney ended (probably started) her Carnegie study highly skeptical of the Early Childhood Education professionals. "Far from considering the 'whole child,'" she wrote, "educators were virtually ignoring the intellect of preschool children. They seemed to proceed on the notion that, between birth and five years old, a child's physical and emotional development (rather arbitrarily, it seems to some) should take precedence over his intellectual development. Indeed, we may have been performing a tragic disservice to young children by not sooner recognizing that their emotional, physical and intellectual needs are doubtless interdependent from infancy on." But in the kingdom of the blind, contrary to popular belief, the one-eyed man is strangled for lack of funding. Especially if the U.S. Office of Education was going to pay the bills, Mrs. Cooney would need a card-carrying developmental psychiatrist. Morrisett found her one that everyone could live with happily: Gerald Lesser, professor of education at Harvard.

A casual, blond young man wearing square-cut glasses, Lesser had a usable background in both worlds. He had been working at the Hunter Elementary School in New York, trying to create tests that would measure a wider range of "giftedness" than Hunter had been recruiting (i.e., give Hunter a chance to enroll more Negro and Puerto Rican children than could cross the IQ gate), when Newton Minow "uttered his well-publicized remark about wastelands. Each of the networks began scurrying around to get off the hook by doing something for children. NBC started *Exploring*, and called me to help. I did their child-watching for them—I watched the show with children, then talked with the children about it, then talked with the producers and writers about how it had gone over. Every Saturday morning, as an avocation, for four years. So when Joan began talking with developmental psychologists and talked with me, I could bring more practical experience; and having had the experience I knew it was going to be interesting."

Lesser, however, was not going to give up a Harvard professorship to work every day for *Sesame Street*; the most he would accept was the chair of an advisory committee. None of the other people Mrs. Cooney had met during her exploratory visits seemed right. Fortunately, the project officer responsible for the CTW grant at the Office of Education turned out to have another project under his supervision—a study of children's reactions to television, being conducted by Dr. Edward L. Palmer at the Oregon State Division of Higher Education. Mrs. Cooney called Oregon, then flew Palmer to New York for a round of interviews ending with an enthusiastic job offer. Palmer, a tall, bony redhead, is the Raposo of the academic side: versatile, highly skilled and very smart, able to vamp research as Raposo can vamp a bass line. He also knows what he needs from his staff: Mrs. Cooney likes the thought that the psychologists think up skits for the show, but Palmer says, "I hire with a view to people who do *not* want to be television producers. I want a man who if he has fifteen minutes is thinking about making a research study rather than a TV script." This feeling for cobblers and lasts, of course, minimizes the clash between the academics and the artists.

What Palmer had been doing in Oregon (and prior to that at Florida State University) was developing measures of children's attention to television. The first, central operation was a "distracter" test. In this experiment a slide projector with an automatic carrousel is aimed at a screen set up next to a television set. As the program to be measured plays on the television screen, the carrousel clicks through a series of slides interesting to children (animals, cartoon figures, etc), with an audible click and a new slide every 7½ seconds. An observer sits at the other side of the television screen and simply notes how often the child turns away from the television set to look at the slides. In testing for *Sesame Street* productions, Palmer gave his employers none the better of the deal: the *Sesame Street* episodes were presented in a black-and-white dub from the color videotape, while the slides were all in color.

To establish base-line data, the distracter test was run on a number of films and television shows that had nothing to do with *Sesame Street*. The creative staff learned with some interest that *Captain Kangaroo* did not do well on the test: it was slow-paced and talky for the children, who turned away to see a slide on more than a third of the clicks. An animal movie did very well, and so did *The Monkees*. But the discovery of the testing sessions was a film called *Neighbors*, made by Norman McLaren for the National Film Board of Canada. Up to the last minute (when the message, about loving your neighbor, was presented verbally in a number of languages and the children all turned off), *Neighbors* held almost everybody. Though it was a violent show, with people beating up on each other in various ways, Palmer and Gibbon decided that the content was not in fact what held the children's attention. The secret of *Neighbors'* success was its technique, which is known as "pixilation." Instead of showing a smooth, continuous action, a pixilated film presents a series of individual frames, with the connecting frames omitted, so that characters move very jerkily. The result is like an extreme version of an early Chaplin comedy, and children (including, incidentally, the older children on whom CTW tested material for its new reading show) think it's marvelous.

(So, for wholly practical reasons, does Jon Stone. "In one eight-hour day in the studio," he says, "I can tape maybe eight pieces of pixilation. Then I go to the computer, which makes a continuous tape of each piece. The computer costs me $200 an hour. I have $1,600 computer cost; $400 studio cost; $1,200 production cost, and I have eight three-minute pixilations, which do everything for me that animation can do. That's about $400 each. To get animation, the cheapest animation, is going to cost me $2,500 to $3,000 a minute, twenty times as much.")

The values of pixilation extend beyond its enhancement of live action. One of the techniques employed from early on is called "clay animation," best described through an excerpt from a paper by Gibbon and Palmer (which also gives some of the flavor of the literary style of their collaboration):

Clay is molded in successive stages, each photographed on a single frame of motion picture film. When the film is projected, the clay appears to reform itself into a succession of shapes. In a typical piece of clay animation produced for *Sesame Street*, a small blob separates itself from a larger narrator blob and forms into the letter "E." Next, from the clay "E" are rapidly produced two "G's," and the three letters are aligned to spell "EGG." A clay egg forms behind the word and hatches to produce a baby eagle. The word "EGG" changes to "EAGLE," and the eagle eats the word.

And for each such episode, Raposo writes and performs bouncy, memorable music.

Keeping children's attention, of course, was merely the start of what *Sesame Street* was supposed to do. In measuring the quality of children's reactions to the show, Palmer could not get such easily quantifiable data as those from the distracter test. Instead, he trained his assistants to slot children's activities while watching into one of several categories, ranging from an active response to what was happening on the tube (most children do talk back to a television set, at least once in a while) to turning away and playing. The most common reaction—in a word Palmer now regrets—was "zombie": the child was concentrating on the set, but doing nothing, and whether he was in fact thinking about what was portrayed before him was beyond the wit of observers to ascertain.

One aspect of children's reactions was noted for use, however: while their visual attention might wander from the tube, they remained alert for auditory signals, and something interesting on the sound track quickly pulled them back. One episode from some months later, reported by Palmer and Gibbon, illustrates how this information was used. Trying to make sure children could distinguish between letters they sometimes confuse, the writers developed an episode in which Big Bird, the seven-foot canary (with a man inside, of course) draws an "E" and an "F," and then watches in horror as the bottom line migrates of its own accord from the "E" to the "F," turning the first to an "F," the second to an "E." Children stopped watching when Big Bird, who does not accomplish such tasks with ease, completed his initial drawing. When the migrating line was accompanied by a slide whistle—traditional sound of a pratfall, on stage and screen—the children quickly returned their attention to the tube.

Among the attitudes CTW most devoutly hoped to instill in viewers was the feeling that wordplay is fun. Puns turned out not to work well with the younger children, at whom the show was primarily aimed (and rhyme worked too well—children neglected even the plainest substance of the song, sometimes couldn't even recognize the letter though they could sing a song about it, if the rhymes were really clever). Raposo was therefore given the continuing problem of writing songs around unrhymed and relatively unaccented lyrics; few observers seem conscious of the degree to which the game songs, for example, do without the spur of rhyming tags. But alliteration worked with everybody, and, of course, taught the sounds of the letters. And some of the puns were kept, because older children enjoyed them.

Some of the ideas people liked best did not stand up under Palmer's testing, most notably *The Man from Alphabet*. In fact, though they had shown an episode from it at their first press screening of the show in August 1969, Connell and Gibbon were not surprised by Palmer's results. They had returned from Hollywood delighted with the rushes, but when the film was assembled and shown to them as a unit, they didn't much like it. Connell went to Mrs. Cooney and said it would have to go. "Call it," he told her, "Connell's folly." Jeff Moss, who was recruited to the program about the time this idea was dropped, recalls that "the show changed 25 percent during the month before we went on the air, and another 50 percent in the month after we began." The street was dull, and there was only one quick way to brighten it: whatever the psychologists might say, Muppets were going to live there with the real people. Big Bird and Oscar were created as denizens, and Ernie and Bert moved from the never-never land of puppetry to an apartment in the basement of 123 Sesame Street.

Some other good ideas fell away as the show progressed. One of the things the writers liked best was a recurring episode with two dumb men, Buddy and Jim, who would find simple problems beyond their grasp while a background piano (Raposo again) played silent-movie honky-tonk. (One memorable episode, for example, found the two men unable to set their chairs down on opposite sides of a table so they could face each other and play cards.) "The problem was," Lesser recalls, "that Jim had a rather competent look about him, and the children were confused by his inability to do what anybody could do." In other areas, the producers found they could be more daring: where they had started with seven-year-old children on the street to play with Susan and Gordon and Bob, because they were afraid they couldn't handle younger ones on camera, they moved down the age scale until the children to whom Susan tells stories or Gordon explains the world were roughly the age of the program's target audience. One of the most charming things about the day-to-day production of the program is the way the children actually play on the set between takes, riding the tricycles, tossing the basketball at the basket, just running wildly about on Sesame Street while their mothers watch from a visitors' gallery above. A few of the children are professionals; others are recruited from day-care centers where the *Sesame Street* psychologists are child-watching; and the group turns over every four weeks or so.

In a few cases, ideas turned out to be better than anyone had dreamed. From the beginning, Mrs. Cooney had planned on having celebrities visit the street, to give the show importance. (Early childhood educators were grumpy about that, on the ground that little children didn't know who was a celebrity. *Sancta simplicitas.*) Among the early visitors was the Negro actor James Earl Jones, then playing the role of Jack Johnson in *The Great White Hope.* What the celebrities have usually been asked to do on the show is to recite the alphabet (something that is done on every program, partly because a minor objective of the program was to get children to recite the alphabet and impress their parents, partly because the consultants were afraid that children who missed some shows would not know there were letters other than the ones featured on the programs they had seen). Jones came on and fixed the camera with a terrifying stare, and slowly, ominously, called the letters. On the child's screen, each letter appears briefly at either side of Jones's head before he calls it. He takes no less than ninety seconds to go through the alphabet, his lips very visibly forming each letter just before he speaks. Gibbon and Palmer describe what Palmer has called "the James Earl Jones Effect":

The first time a child sees the Jones performance, he begins almost at once to respond to the implicit invitation to say the alphabet along with the performer. On somewhat later repetitions he begins to name the letter as soon as it appears, before Mr. Jones has named it. Mr. Jones' naming of the letter then confirms or corrects the child's identification of it. With still further repetition, the child begins to anticipate the printed symbol as well. As soon as the preceding letter disappears, the viewer names the next. The effect is significant because it demonstrates the feasibility of stimulating with the one-way medium of television the feedback and reinforcement so instrumental in learning. The instruction is clearly individualized, even though there appears to be but a single, one-way message involved.

Rivaling the Jones bit for Jon Stone is a scene in which the Knicks do simple lay-ups, and as each basketball drops through the net children's voices count "One—two—three—four," etc. "Simplicity is the key to everything we are doing," Stone says, "or it should be. And we violate it so often."

7

How much difference *Sesame Street* will make in the lives of those who watch it is a question nobody can answer at all. The odds against anything new in education are enormous—probably hundreds to one—not because the bureaucracy is inert and stupid (though that doesn't help), but because schooling is a fundamental activity of the society, which means that the patterns it follows tend to be generally acceptable and reasonably effective by comparison with other real possibilities. And the notion that television can deeply affect individual lives, though still plausible, is very far from being proved. And the research for *Sesame Street*, says research director Ed Palmer, has been nothing more than "a set of small experiments in the accountability context."

Primarily, of course, the research done on reactions to the show is designed to improve what goes out. Early on, Palmer had to abandon a good deal of cherished belief in the power of objects themselves to teach without words. "We had round things and music, with no text," he recalls, "and the children didn't see them as round things. So we put a voice track over it—children talking back and forth—'That's a Frisbee; that's a round thing'—and then it worked. Voice-over technique is powerful for us."

The mail arrives and is sorted into four stacks: "Requests," "Thank Yous," "Kids" and "Protests." Palmer looks at the protests, and learns from them. There was, for example, a "game" unit built around Joe Raposo's song "One of these things just doesn't belong here, one of these things just isn't the same" (which is used in the majority of programs, driving everyone, especially Raposo, up the wall). The illustration was a box divided into four squares, three of which held a "W" and one of which held a "3." A distressed father wrote in to point out that because two of the "W's" were written with curving lines, the "W" with sharp edges was also not like the others, and his child had got it wrong. After that, the psychologists worked harder at pretending they were preliterate themselves.

The first and most obvious measurement of the success of the program is obviously its ratings (though the proposal firmly stated that CTW would not "play the ratings game," Nielsen and the executives of the workshop developed quite an elaborate program of possible ratings studies based on Nielsen's need to oversample Negro slums for its standard service; but Nielsen's price was too high). During fall 1970, *Sesame Street* earned Nielsen ratings between 3 and 6, heavy for daytime viewing, by far the strongest ratings for anything broadcast over noncommercial channels at any hour of the day or night. Meanwhile, CTW hired Daniel Yankelovich to do special surveys in Brooklyn's Bedford-Stuyvesant and in five slum areas in Chicago, and got back immense figures, probably reflecting in part the show's strong publicity in these areas and the reluctance of mothers to admit that their children *didn't* watch *Sesame Street,* but almost certainly reflecting also a very heavy incidence of viewing in these neighborhoods.

There are some cities, unfortunately, where the only educational channel is UHF, which many slum homes cannot receive at all and some cannot receive satisfactorily, and here viewing is much lower. (Cleveland is an almost total disaster for *Sesame Street,* because municipal law forbids outside TV antennae on schools and housing projects, and the noncommercial UHF can't get in.) In New York in the first year, Channel 13 would not clear 9 o'clock, which is CTW's preferred air time (mother has packed the older ones off to school and wants to relax awhile with junior otherwise occupied), and *Sesame Street* went to WPIX, Channel 11, which garnered twice as many viewers as Channel 13 achieved broadcasting the same show later in the day. In 1970–71, however, WNET got out of its prior contract with the Board of Education, cleared 9 o'clock, and demanded that *Sesame Street* go exclusively to Channel 13.

Similar demands have been made in the cities where the educational channel is UHF, even when VHF commercial stations offered to carry the show, full-hour, no commercials (indeed, the ABC network offered to carry the whole week's output, five hours in the early morning split between Saturday and Sunday, without commercials). In every case, *Sesame Street* has yielded to pressure, and denied the show to commercial stations. The dog-in-the-manger attitude the Corporation for Public Broadcasting has taken in the *Sesame Street* matter is one of the less attractive aspects of that idealistic service.

The big question, of course, is what the children are learning from the program, and which children are learning what. Given the amounts and the sources of the money being poured into *Sesame Street*, Children's Television Workshop had to be prepared from the first day to demonstrate an educational effectiveness via measurements by an outside body. The Educational Testing Service, as proprietors of the College Board exams, is the outfit with the highest prestige in this rather dubious field, and it has had long relationships with both Carnegie and Ford. It was also, to say the least, *bono animo* to the project. Palmer and Lesser had a number of sessions with test-designers from ETS, and finally agreed on areas of measurement where *Sesame Street* would, in Palmer's words, "go to the wall." These were, inevitably, the items most easily taught and most easily measured—recognition of letters and numbers, knowledge of the names of the parts of the body, and the simplest sorting skills. Though ETS was presumably involved as an outside evaluating agency, CTW participated in the construction of the test instruments, which were made available to the *Sesame Street* psychologists at the beginning. Among the questions asked by the researchers "child-testing" the bits prepared for *Sesame Street* were items from the forthcoming ETS tests; over and above more general educational functions, *Sesame Street* hours were designed to improve scores on the tests ETS would give at the end of the year. Under the circumstances, it is scarcely surprising that the ETS results were positive.

It is important to be precise in what is being said here. All new educational tests are necessarily artificial. The validity of any test can be estimated only after it has been given for some time, and observers can judge whether or not the scores predict anything of importance in the future of the students who achieve them. At best, tests on the first year or two or three of *Sesame Street* could give no better than insufficient information about the value of the program; it is hard to see how the CTW researchers could in all scientific probity have done anything other than plan the tests to show off their work in the best possible light. And what little evidence has turned up about the more complicated, unmeasured goals of the program—or about children's ability to transfer the very simple stuff for which ETS measured into more complicated situations—does argue that the program has indeed made a difference.

As always, the rich get richer—that is, students who do better on the pretests before they have watched the show tend to see the show more often and do better still on the posttests. As Wilbur Schramm and his associates wrote in 1960, "Bright children learn more from television. Other things being equal, they learn more from *any* learning experience than less bright children do." (A statement by the ETS researchers that "such television programs can reduce the distinct educational gap that usually separates advantaged and disadvantaged children" is shockingly false to the real-life situation as illustrated in their own data.) But the problem is not that there is a great gap between advantaged and disadvantaged; the problem is that so high a proportion of unlucky Negro children fall below any possible educational floor, destroying their life chances in the late twentieth century. Equality is not necessary; minimum competence is. "There's a literacy line," Mrs. Cooney says. "Once you're above that line you can participate in American society; below it, you can't."

By an odd chance, the arrival of *Sesame Street*, with its very structured content, coincided with a "progressive" phase in the recurring cycle of educational theorizing. This has been awkward for the creators of the show, all of whom would be enthusiastically with the fashion if they were not involved in constantly testing and working up material. As early as spring 1970 the television critic of *The New Republic* denounced the program for forcing rote learning of things children will pick up anyway when they are old enough to do so, and among those who wrote in to congratulate the critic was John Holt, the most interesting and probably the most influential of the current gurus. (Oddly, Holt later took a precisely opposite tack in the *Atlantic Monthly*, criticizing the show for not being systematic and ambitious enough. Unfortunately, one of the examples he chose to criticize was Big Bird's maneuvering with "OVEL" to produce "LOVE." Holt thought that if left-right progression was to be taught, *Sesame Street* should go whole hog and first present the word as "EVOL." But, of course, "EVOL" would look all right to children with left-right reversal problems, and had to be avoided. People who write about *Sesame Street* have to keep in mind the possibility that Connell or Palmer or Gibbon has already covered the track and gone on ahead.) The escape for the producers has been the insistence that television is a "nonpunitive medium," that because the child doesn't have to *worry* about getting the wrong answer to a *Sesame Street* puzzle he can't suffer the pains usually implied when the words "Rote Learning" are uttered, and cannot be turned off the educational path by feelings of failure.

The most important of the critics of *Sesame Street* has been Monica Sims of the British Broadcasting Corporation, who refused to broadcast the program as part of the children's hours of the world's most influential television service. Under any circumstances, of course, *Sesame Street*'s applicability to the British context is less than a sure thing—what can a Cookie Monster mean in a country where the word is "biscuit"? But Miss Sims, pushed to discomfort by CTW selling propaganda, has gone some distance further. "We're not trying to tie children to the television screen," she says. "If they go away and play halfway through our programs, that's fine. Our educational advisers seven years ago advised us very firmly against trying to teach reading outside the schools. We find it disturbing in *Sesame Street* that there are always right-and-wrong answers. If the child finds he's wrong too often, he may get discouraged. And we think it's a terribly conformist sort of program— the dots always fitting into the right holes. We would leave a little of what we could call creative chaos. And, of course, much of what happens on the street itself, the storytelling, little dramas and such, is extremely amateurish by our standards."

Rebuffed at BBC, the *Sesame Street* salesmen moved over to Britain's commercial network, where they met a number of uniquely British obstacles. The commercial stations are restricted by law to fifty-odd hours a week on the air, except for educational programs. To put on *Sesame Street* "off ration," says Brian Groombridge, educational director for the Independent Television Authority, "we must get approval of the Authority's educational advisers. Our schools committee did look at *Sesame Street,* and split right down the middle. Some were very enthusiastic. Others said it was terrible —entertaining, perhaps, but . . . teaching capital letters! No educational value. Among the anti group were two members who had spent a long time in the States and seen a good deal of the show. Then we have a second hurdle—this is foreign material, and we are governed by regulations on the amount of foreign material we can air. Our unions are very interested in the foreign quota, anxious to see our screens are not flooded with foreign material. The quota would have to be renegotiated."

In spring 1971 Harlech Television, the ITA contractor for Wales, found itself with an air slot for two weeks' worth of *Sesame Street,* and Groombridge got a quick okay for ten hours to be run as an experiment in Wales. CTW paid the costs of transferring the ten tapes to British technical standards. Frank Blackwell of the National Council for Educational Technology made a quick survey on how teachers and parents liked the show, got a very mixed bag of responses to both the structured and free-response sections of his questionnaire, and wrote up the results without coming to anything that could be interpreted as a conclusion.

"We ought to be doing our own alternative to *Sesame Street* or BBC's *Play School*," Groombridge says. "But if I were in a position to take a unilateral decision right now, I would say *Sesame Street* would be enormously valuable in Britain. We screen *Scooby Doo* and *The Banana Splits* and nobody gets philosophical about American influence. But it would stir up the most almighty controversy, a political-educational controversy that would have long-term impact." In fall 1971 ITA compromised the controversy and scheduled *Sesame Street* for one hour a week on Saturday mornings.

Included in Blackwell's collection of comments from teachers, beside a number of skeptical and negative notes, was perhaps the best brief tribute *Sesame Street* has received. An infant-school teacher and parent working in a depressed area of Wales wrote as follows:

> Number presentation—it worked, and many children learnt to count to ten with understanding. Letter presentation—with some reservations from the purists and doubts about unnecessarily alarming and confusing devices—again it worked. Letters became interesting and were found all over the home. General verbal interest—new words, longer words, definitions, especially preposterous, jingles. Methods of consolidating all this from programme to programme were proved. Good presentation of adults and their possible roles. Use of children's voices gave immediacy and often greater clarity.

The massed publicists of the Byoir agency could scarcely do better.

8

To write about other children's shows after *Sesame Street* is a bootless enterprise. *Walt Disney* successfully exploits each week an attitude which survived its creator, plus the best back list in films. *Lassie* runs away from the FCC dognappers, selling Campbell's Soup. *Mister Rogers' Neighborhood* on "public television" is an easygoing, pleasant one-man show, light on production values, better than going out to play but only marginally compelling for most children. *Captain Kangaroo*, the only weekday network show aimed at children, is relaxing and only a little too commercial.

Between 4:30 and 6:00, when the networks do not feed their affiliates, should be children's television time, but the fact is that the stronger stations in most cities do not broadcast anything specifically aimed at children in those hours. A 5 rating for a movie or a situation comedy rerun or Mike Douglas or David Frost will draw greater advertising revenue than an 8 rating for a children's show, because advertisers would rather reach housewives than their fry. (And the problem gets progressively worse with the spread of multi-set households.) In general, the independent station in the market, if there is one, will be the children's channel on weekday afternoons, and what it will carry is reruns of old cartoons and westerns or juvenile comedy (Three Stooges, etc.).

The children's time in American television is Saturday morning. A sociological study of 56 poor families with second-grade children found that in 46 of them the set was on by 10 o'clock of a Saturday morning; and in 41 of them the tube was watched at least six hours on Saturday. From the late 1950s, when the networks first discovered the size of the market available on Saturday morning, to the late 1960s, the great bulk of the hours between 8 o'clock and 1 o'clock were given over to cartoons. A few of them were rather charming—one should note especially the off-beat humor of *George of the Jungle*—but most were pratfall stuff with mouse or moose, or tales of lantern-jawed spacemen and spacewomen, teen-agers, Great Danes, or other specialized heroes. What was wrong with them at bottom was, in the words of Chuck Jones (who was responsible for Warner Brothers' *Bugs Bunny* and *Road Runner* before he became vice president for children's television at ABC), "that the characters changed every whichway, every week. Any program that has lasting value depends on its personalities; and each personality depends on its own discipline."

The ratings war has been more damaging on Saturday morning than anywhere else. Like daytime minutes during the week, Saturday morning minutes are bought by formula, entirely on cost-per-thousand, which will be about $1.50 for most of the year, up to $2.50 near Christmastime, when the toy manufacturers descend on the time salesmen. (During that season, the number of children's programs increases on the independent stations. A study done by Dr. Ralph Jennings for Action for Children's Television showed 10 hours of children's programs, with 54 commercials, broadcast on Channel 5 in New York in the week ended May 4, 1969; and 22½ hours, with 540 commercials, in the week ended November 16, 1969.) And cartoons unquestionably draw audiences. So unsympathetic an observer as the BBC's Monica Sims has written that American cartoons, which the BBC broadcasts for two twenty-minute afternoon periods each week, have "immense popularity with children." But the BBC, perhaps foolishly, never repeats a cartoon strip. "We have thousands of complaints about our canceling *Scooby Doo*," says Miss Sims, "but we never canceled: we just broadcast all seventeen episodes."

The secret of this popularity, as measured by ratings, is the wide age range to which the cartoon form appeals. Network officials have spoken of "ages two to eleven" as the target for Saturday morning, and nothing but cartoons can possibly span that immense age spread. Six showings of a Saturday morning cartoon bring a network almost as much money as two showings of a prime-time half-hour, and the costs to the network are half what must be spent for the prime-time film. Though figures are neither published nor admitted, the consensus of more or less informed observers seems to be that the networks in 1970–71 took in about $75 million for the 8-o'clock-to-1-o'clock stretch on the weekend mornings, and that about $20 million of that total was profit. The exploitation of children's appetites, in other words, brought the networks nearly a quarter of their total profits for the year. That was too much money, and the world was ready to hear about the scandal when five upper-middle-class housewives in Newton, Massachusetts (one of them the wife of an executive of WGBH, Boston's noncommercial station), started Action for Children's Television to press for something better.

NBC led the way in fall 1970 with a program called *Hot Dog*, in which some adult comics with presumed appeal to children offered silly answers—and teams of documentary film-makers found real answers—to the questions children are supposed to ask, like how hot dogs are made. This one appealed to parents (including the ladies of ACT) but not much to children, and it was hanging around a 2 rating (about 1.2 million homes) when it was canceled. For fall 1971 NBC substituted a child-centered adventure show on the *Barrier Reef*, a reprise of a soft-science program called *Mr. Wizard,* and an oddity called *Take a Giant Step,* on which groups of early adolescents were encouraged to be pundits for a day on the big problems of the world. The show was aimed, said George Heinemann, in charge of children's programming for NBC, "at the six-to-twelve-year-old, to help him in making value judgments." It died like a dog, and a good thing, too.

At CBS, the contribution for 1971–72 was a return to an old idea—*You Are There,* pioneered by the news division on radio and in the early days of television, offering the conceit that a great moment in history is being created before your eyes, and employing the services of no less than Walter Cronkite. At ABC, the feature was Chuck Jones's *Curiosity Shop,* a grab bag of props, stimuli, stories and explanation, involving a workshop from which the proprietor is always absent (though a seal is always present, "because," Jones says, "I've always wanted to have a seal"), and the children uncover things more or less for themselves. Each show had a topic; Jones was especially excited by his discovery that some physical anthropologists believe men developed a brain because he had a thumb, and worked up a one-hour show entirely on thumbs. Some of the material for the series came from the National Film Board of Canada. "This program," Jones says reflectively, "is to be competitive, not to be the West Berlin of television—pour all that money in and then say capitalism is better than Communism. But the aim of children's television should be to get children away from television. After seeing our programs about the wind, a child will indeed want to make a kite." Of the more ambitious Saturday morning efforts, this one did best at avoiding sententiousness and holding audience.

Some of the BBC program formats might make successful imports. *It's Jackanory,* for example, is simple storytelling by professional actors, with backdrops which make a setting for the tale—another radio idea easily transferred to television. *Blue Peter* (not an Andy Warhol image, but the name of the flag a sailing ship on an exploring mission used to hoist its second day out of port) presents three late adolescents (two boys and a girl) who do things children would love to do, like drive a car in a rally or help service a real locomotive or perform backstage chores at a Royal Tournament. It is all a little stagy and unconvincing to adult eyes, but the response from children is overwhelming—nearly seventy thousand of them turned up, for example, when British Railways rechristened one of its locomotives "The Blue Peter."

One interesting series, reportage to scale, presented pairs of children (one British, one from some foreign part) who had visited each other's homes and lived with each other's families, and were asked what they thought of it all. The program is ungrammatically entitled *If You Were Me* (Jeff Moss says one of the pleasures of working on *Sesame Street* is that "if a kid shoots a water pistol at a screen you don't have to worry about the network receiving complaints that kids will shoot water pistols"); and it may be possible only in a country where racial bias and distrust of foreigners are accepted as part of reality, because lots of what the kids say is pretty offensive to sensibilities honed by antidefamation leagues and the like. And if nothing but such material were shown on Saturday morning, of course, our parks and streets would resound more loudly with the cries of children investigating the world outside.

But no television on Saturday morning would be an improvement on what we have now. (Action for Children's Television has prepared a "survival kit" for mothers and children in case its activities produce the collapse of Saturday networking.) Advertising specifically directed at children is despicable as an idea, and is not unlikely to be improved in execution. Demanding that the networks and stations supply children's programming without compensation will not produce much to look at, and if we must have children's programming (which we probably must—it would be repressive to require parents to take care of their children at times when they'd rather stay in bed), the best solution may be a government fund, like the National Endowment for the Arts, to sustain the production of children's shows. Assuming an average cost of $80,000 an hour and two reruns for each hour, $10 million would pay the program cost of filling every network's schedule from 8 in the morning to 1 in the afternoon every Saturday. Though one shudders at the thought of who would in fact wind up on the board allocating the money and what most of it would be spent to produce, a certain number of talented people could probably force their way to the trough. Networks as well as outside packagers could be made eligible for production grants, and it isn't hard to imagine incentive systems (like a renewed grant simply for the right to use again in another season particularly prized programs from previous years). There would still be competition, and probably there should still be ratings (by narrow age groups, please, especially commissioned); but there might be some sense that someone real is watching, and there might not be all that pandering.

CHAPTER 7

Sports: The Highest and Best Use

From the point of view of the BBC Television Service, the General Election could not have been called at a worse time. We were deeply involved in the World Cup, an immense operation. The Commonwealth Games were shortly to follow. The Derby, Ascot (in colour for the first time) and a Test Match were on during the week in which the Election would fall. Wimbledon started four days later. A large part of the Service was consequently involved, as soon as the General Election announcement was made, in a multitude of rearrangements. . . . In the event, all was done, and everything was covered. It was a mammoth operation, the phrase justified for once, and a great reassurance. Those of us who had been particularly concerned with our reorganisation during the year drew particular satisfaction from the fact that while all these things were going on, there remained the enterprise to bring to viewers (out of the blue, a last minute extra, as it were) one Sunday night during that hectic month, the last strokes of the American Golf Championship and the sight of Tony Jacklin's victory.

> —Huw Wheldon, Managing Director, Television, in
> his annual report to the
> BBC Board of Governors, 1971

1

The three ABC-TV trucks, each fifty feet long, pull up beside the stadium Friday afternoon for the football game to be telecast the following Monday night, and cable begins snaking out of them. One of the three is just a truck, carrying sturdy miscellanea ranging from golf carts to coffee makers; one is a superbly equipped television control center, the body of the truck attached to the axles through compressed-air machinery rather than through springs to minimize the jar of pot holes, level crossings and the like as the trucks roam the country. The third holds a less elaborate control unit, plus two big Ampex two-inch Video Tape Recorders (VTRs) and two slow-motion machines; the ABC-TV maintenance men who have to keep these items functioning wish someone would put this unit on air suspension, too, but so far it's just on springs, and on Tuesday mornings when the game is over all the pieces are restrained with big rubber bands, masking tape and incantations before the trucks roll off.

Nine cameras come out of these trucks to be spotted around the stadium, and that's a lot of cameras—for Sunday afternoon football, NBC and CBS tend to use only five (because they must by contract cover at least five games each on every Sunday; people complain bitterly and not without reason that the home games of their home team are blacked out on television, but nobody ever considers the size of the logistical effort needed to assure that in every city the away games of the home team are always available). The eight color cameras (there is also one black-and-white for titles) are all Norelco, made in Holland with the Plumbicon camera tube invented there. The most versatile of them, incorporating four breathtaking English-made Taylor-Hobson lenses that cost $10,000 each, is the one that will stand in the box over midfield, and can be used for "wide shots" covering the length of the field, for "tight shots" that will show a fly on a setback's face mask—and also, swiveling around, for close shots of the ABC announcers, Howard Cosell, Frank Gifford and Don Meredith, about five feet away. Showing off his machine, cameraman Steve Nikifor zoomed in to picture the texture of the padding on the goal post in the distance, then placed a pack of cigarettes on a ledge right in front of the camera and got a usable clear picture of that.

While covering the game, Cosell, Gifford and Meredith sit side by side, looking both at the action on the field and at individual monitor screens that tell them what of that action the good old viewer has seen at home. "Of course I can look at both at once," Cosell says with that resentful immodesty that has made him a national character.

"It's a matter of expertise, training and talent." Meredith's problem is a little harder, because he has four monitors before him, one for the picture that is going out on the line to the world, the other three for bits of the scene—pictures from cameras "isolated" on one man or another—that are being preserved for possible replay. Meredith has to do the description—the "voice-over"—when such replay occurs, and most of the time the only warning he will have is the producer's voice saying "Roll VTR-2" or "Slo-mo A" into his earplug just before the picture pops up on the line. Especially when the slow-motion machine is used, there may not be time to tell Meredith anything at all about what is on it. Tape must be rewound, but the slo-mo offers among other miracles random access to stored material; by pushing a cue button when a play begins the operator of the machine guarantees that he can return to that spot in a fraction of a second. "While Frank is setting the stage for the next play," Meredith says, "I glance at the monitors and see where those cameras are when the play starts, and that gives me a slight edge."

Two of the cameras, which can be used for either wide shots of the whole play or close shots of specified participants, stand in "camera baskets," metal cages suspended from the balcony of the stadium over the twenty-yard lines. In any stadium which does not have such baskets, the three networks will chip in to pay the cost of building them. (Neither the stadia nor the ball clubs undertake any extra expense for television, if they can avoid it. In St. Louis in fall 1971 the custodian of the Municipal Stadium tried to hold up ABC for $225 a minute for lighting the park during the camera rehearsal on Sunday night.) Then there will always be a camera on the roof of the stadium behind one of the goal posts, giving an angle down through the posts for extra points and field goals; this camera is also particularly useful in isolating an end and following him as he runs his pattern and catches or does not catch a football thrown (if the television director is lucky) in his direction. In a stadium built high all around, there will be two end-zone cameras; in a baseball park, with bleachers, the fifth high camera will probably be on the other end of the midfield box that holds the three announcers.

ABC likes to have three cameras on the field, if possible. Two of them are mounted on wooden platforms built over golf carts, driven from behind because the cart driver needs to see what the cameraman is doing more than he needs to see where he is going (which he can do only imperfectly). Both of these cameras are on the near sideline, and move so as to be roughly at the line of scrimmage for each play; one can then catch eye-level close-ups or trick shots of the offense, while the other watches the defense. (At some stadia the stands come down close to the sidelines, and management will not permit ABC wagons to block the view of the VIP contingent in the ringside seats; then the cameras are more or less permanently mounted in the corners of the field. At other stadia the management okays but the spectators do not, and the cameramen get pelted with whatever is in the lunch baskets, including beer cans. "Well," says Bill Morris, the unit's chief engineer, a slight man with thinning sandy hair, deep sideburns and a worried expression, "cameramen have strong shoulders.")

Across the way at the far sideline, where the players' benches are, a man with a hand-held camera seeks targets of opportunity. A microphone is built into each of the cameras, but the mike in the hand-held camera is usually kept silent to protect the public from the language players in their excitement (even, indeed, in their moments of calm contemplation) have been known to use. Two other men are on the field to catch sounds: they hold the "shotgun" mikes, long black tubes that can be aimed at the field to catch the quarterback calling signals or a kicker's foot pounding the football. And there is also a man walking around on the far side in an ABC blazer who has no function in reporting the game. He tells the officials, through hand signals, which breaks in the action are being used by the network for the broadcast of commercials, and when the commercials are up and the game can resume.

Each cameraman wears a headset that permits the men in the control units to communicate with him, and permits him to communicate with them through a mike that appears to be no more than a piece of wire bent around in front of the mouth. The mikes before the three announcers feed directly into the signal that goes to the waiting world. For those moments when an announcer feels a need to say something to the truck, he has a telephone which rings in the truck whenever he picks up the receiver. Meredith is the most likely to use this device, to suggest that one of the cameras isolate somebody who seems to be doing something interesting. "I can say, 'Hey, Bob Brown is really working on Joe Jones,'" Meredith reports, using what are, oddly enough, real names; "'see if you can get that on isolation.'" Usually, however, even Meredith will wait until a commercial break, when his mike is live to the speaker in the truck but not to the television audience, which is receiving a more utilitarian message.

Setting all this up takes two days. First the union electricians at the stadium plug the trucks into the local power supply, and the union telephone men from the local telephone company connect the trucks to a local exchange for telephone calls and to the local terminus of the coaxial cable to permit the feed of the picture when the time comes (and also, incidentally, a feed to the truck from the network cable—the local station being blacked out—so the people on the scene can see when the commercial ends). Sometimes ABC technicians will be able to take their miles of wires into the stadium themselves, sometimes a local union will have a contract.

Saturday is occupied entirely with stringing the wire and hooking everything in; on Sunday, with the cameramen absorbing their spare time by looking at other networks' coverage of football games on the monitor screens in the booth the announcers will use, the equipment is tried out to make sure everything works. Never yet has everything worked the first time around, though the usual troubles are minor matters of communication circuits between the headsets and the control console, or between the two trucks. Often enough, though, there are malfunctions that have to be reported back to New York, to ABC's BO&E (Broadcast Operations and Engineering) division—unbalanced output among the three color elements of a camera or "noise" in the signal, expressing itself as streaks on the screen. Then there may be a request for substitute equipment to be rushed to the scene, and calls for later repair work to be done at the ABC technical center in Lodi, New Jersey. Some of the equipment was hand-crafted in the technical center; a blue wooden structure which holds the control units for the slow-motion machines carries the legend, "MADE BY LITTLE ELVES IN THE LODI FOREST."

At dusk on Sunday the lights are turned on in the stadium, and engineer Mike Michaels starts the delicate job of getting the color right—and, maybe even more important, the same—on all the cameras. Michaels and two assistants work in an isolated section at the back of the main control truck, surrounded by floor-to-ceiling stacks of integrated-circuit modules. "Each of the four banks of lights here," Michaels said, looking up Cleveland's Municipal Stadium in a neatly handwritten loose-leaf book of lighting characteristics that he keeps current for all the pro stadia, "has a different mixture of mercury and incandescent lamps. The lighting temperature changes from one part of the field to another—the right end zone here is 'warm.' But this isn't that bad a problem—the bad problems are on the West Coast, where we have to start at six o'clock, and the mix of sunlight and artificial light keeps changing. I have to rebalance the white-on-black for each camera during each commercial." When the game is all at night, Michaels can get it right the first time, during the Sunday rehearsal. The technique employed is simplicity itself. A white sheet is dropped on the green grass in the middle of the field, and the cameras are pointed at it, one after the other. When a color camera gets pure white out of a white sheet, its three color elements are balanced for that lighting.

The control center is in the front two-thirds of the truck, and is divided into two sections by a glass wall. At the rear, entered through a separate door from the outside and elevated to give its occupant a clear view of the screens, is the audio control center, headquarters for Nick Carbonaro, a wiry, fussy, not very communicative man who keeps a broom in the corner and sweeps out the place himself every so often. His is probably the hardest job on the show, because he is responsible for seeing to it that the right mikes and no others supply the sound—and just the right amount of it, too—for each moment on the screen. ABC likes as much crowd sound as can be got consistent with the announcer's ability to get his voice over it, no problem with the nasal Cosell or the penetrating

Meredith but awkward in the early games for Gifford until ABC found a mike with a configuration that emphasized some of the overtones of his voice. The PA system in the stadium must be handled somehow, maybe by cutting the wires to some of the speakers while nobody is looking. Every time a different camera is brought on line, Carbonaro must pull down the mike from the previous camera and bring up the mike from the new one; and he must know when to bring on the shotgun mikes. If sound is needed from the VTR (most replays involve the picture only), Carbonaro must provide it.

If he makes a mistake, he gets yelled at by Chet Forte, *Monday Night Football's* producer-director, a smallish young man with big shoulders and a round face and a hard, almost raucous voice, never a football player himself, very impatient, perfection-minded. He sits in the center one of three chairs at a table the width of the truck, in front of and below Carbonaro, facing an array of television screens that takes up the entire front wall. The two rows of four small black-and-white pictures at the bottom of the wall (Forte sits on three cushions to raise himself high enough to see them over the edge of the table) represent the pictures coming in from the eight cameras inside the stadium. Each has a cardboard label beside the upper left-hand corner, giving the camera a number and its operator a name: where possible, which is not often during the game itself, Roone Arledge, big chief of ABC Sports, likes to have Forte address the cameraman by name, especially if he is to be congratulated for a shot. (This personnel-relations technique can boomerang, however, because each cameraman gets ten or twelve minutes of R&R during the game, and the producer may be congratulating his relief.) Forte spends most of his time controlling what the cameramen do, using the backwards language of the trade—to "pan down" means to *raise* the image on the screen, and vice versa; to "pan right" means to move the picture left, etc. Forte can control all eight cameras, but the production team is set up so that he rarely works with more than four at a time.

The cameras Forte is not controlling are supervised from the other truck by co-director Don Ohlmeyer, under instruction from co-producer Dennis Lewin. Lewin sits at Forte's right hand, and manages communications with the announcers and with Ohlmeyer. He decides, sometimes on suggestions from the booth or from Ohlmeyer, what the cameramen on the field should pick up in isolation, and Ohlmeyer then tells the men where to aim, whether to come in tighter or pull back wider, etc. Control of the cameras is switched from truck to truck through a row of square buttons on each console, which light either a yellow "A" (Forte's truck) on the top half or a green "B" (Ohlmeyer's truck) on the bottom one. Whichever truck hit the button most recently controls the camera to which that button refers. Only cameras controlled on the A truck can be fed directly out to the public; only cameras controlled on the B truck can be fed into the second of the videotape machines (the first of them simply records what is going out on the air) or into either of the slo-mo machines.

The slow-motion machine, a steel box with a glass box on top, about the size of a small steamer trunk, is one of the most elegant solutions to a technical problem television has yet offered. One cannot simply slow down a tape as one might slow down a film, because the signal recorded on tape must be synched to the inevitable 60-cycle AC power supply of the set in the home to give the needed 30 frames per second. So the slo-mo dispenses with tape, using instead a disc slightly larger in diameter than an LP record, and about a quarter of an inch thick, made of so dense a magnetic alloy that it weighs almost thirty pounds. This disc turns at a steady speed of thirty revolutions per second. Thus each revolution of the disc represents one frame of the television picture. Recording and playback heads ride on a track running just above the disc on the radius from its edge to its center. The speed of the playback head as it traverses the spinning disc determines the rate of change of the pictures on the home screen: though the signal keeps pulsing out at the rate of 30 per second, the number of *different* pictures on the signal can be reduced to 15 per second (when the head moves at half its normal speed) or 6 per second (when the head moves at one-fifth its normal speed)—or the frame can be "frozen" by simply leaving the head motionless over what is probably best thought of as a magnetic "groove" in the disc. Or—and this is probably the most frequent use of the disc—the head can move at normal speed and the slo-mo can be used simply as a straight replay device. The disc has room for 900 "grooves"—30 seconds of playing time—and it operates continuously, erasing what was there before.

All this gets tested sometime after 8 o'clock on Sunday night, twenty-five hours before the game, by the full crew of fifty men. Forte and Lewin and Ohlmeyer join the technicians at the trucks, and Forte checks out the cameras: "Dorf, what I want you to do for the top of the show is pan to the end zone low and get the crowd. . . . Jack, you got a two-timer [a device that doubles the focal length of the lens] on that? . . . That's the high end zone? Jesse, I'm going to take you on field goals going away from you and coming toward you, you're my only end-zone camera. I know it's hard. . . . Okay, Steve, you got any problems? I'll look at you in a minute in chromo-key [the device that makes it possible to superimpose part of one picture, like Howard Cosell's head and shoulders, on another picture, like the greensward football field]. . . . Drew, after the field goal, you always stay tight on the kicker. I'll probably take you. . . ."

A number of men have been working all week on aspects of this broadcast. A two-man graphics department, for example, has been pasting white letters on black cardboard to spell the names of all the players on both teams. These pieces of black cardboard now reside in one of two rotary drums controlled by a little gray box with red buttons: by pushing the button an operator controls what name clicks to the exposed position on the drum, ready to be registered by a black-and-white television camera. Beside the drums stand black placards, already lettered for this game (here Cleveland v. Oakland in October 1971), ready to reveal how many passes Daryle Lamonica has thrown and completed, or how many times Leroy Kelly has carried the ball and how many yards he has gained. Others can be added to the collection during the game, as needed. During this game, the directors in the truck became especially interested in an Oakland linebacker named Phil Villapiano, and the graphics men dutifully prepared a special board announcing how many tackles and assists "41 Phil Villapiano" had made, to be "matted"—superimposed—on a picture of the player trotting back to his position between downs.

But the most important planning has been done by the three announcers, especially by Gifford, who has looked at the films of all the recent games played by both of the teams who will meet this Monday. At least one of the three will talk with the opposing coaches, and usually with some of the players. From a long playing and broadcasting career, Gifford knows players on most of the teams ("It's amazing," Arledge says, "how many ex-Giants there are"). Meredith, of course, was the Dallas quarterback; Cosell has been interviewing players on the radio for years; Ohlmeyer, the director in the B truck, has a close friend on the Jets. . . . "We go in like a coach," says Dennis Lewin. "A football game is tendencies, and we try to learn each team's tendencies the way a coach would." A study of the films might show, for example, that a team blitzes its defensive linebackers against second down with long yardage; and Lewin in such a situation would isolate a camera on the meanest-looking of the charging defenders.

Every once in a while a coach may tip off a television producer about something special planned for this game. Arledge's fondest memory is of an Oklahoma-Army game in 1961, when he was still relatively new to sports, and Bud Wilkinson called him in with his announcer, Curt Gowdy. "He said," Arledge recalls, " 'I know you guys like to catch cheerleaders and such things between plays, and I think you ought to know we've got a trick play for today. On third down with one or two to go, we won't line up—just walk up to the line and go. If you don't keep the cameras on the field, you'll miss it.' So we did keep the cameras on the field in third-down situations, and sure enough one time they just walked up and went, all the way. If you know people, you'll get information. Of course, if you breathe it around at all, you'll never get information from anybody again."

All these subjects are discussed at a Monday preproduction meeting of Arledge, Forte, Lewin, Ohlmeyer and the three announcers; and the production team develops its own "game plan." Then Cosell goes out to the field to tape the voice-over for a twelve-minute half-time insert of highlights from the games of the day before, fed by the National Football League from its elaborate film center in Philadelphia, according to requests made by Arledge the night before. (The NFL film center has fifty employees, and processes twenty miles of film every week, much more than any Hollywood studio.) All Cosell knows when he goes out is the list of names from which the excerpts are being drawn, but his absolute memory for the numbers worn by the players on all the teams and for the major episodes of the previous day's games as reported by the wire services enables him to do a perfect extemporaneous commentary the first time he sees the tape. "He really is remarkable," Arledge says. "If you tell him he has three minutes and eighteen seconds to fill, he'll start talking, he'll really say something, it will be in sentences and paragraphs, and it will come to a full close in, say, three minutes and seventeen seconds."

Cosell, who was a lawyer before he was in broadcasting, has always been a little ashamed of being a sports nut (though he always has been a sports nut), and his ambivalence about what he is doing gives him an irritability on the air that no other announcer can offer. Arledge felt this show needed that. "On the weekends," he says, "really there's nothing else for you to do, and if there's a dull game, you watch it anyway. But on Monday night you'll be tempted to look in on Bob Hope or see what's at the movies. I needed something that would keep people tuned in even for weak games, which meant an interest in the people. I hired Howard first. The combination of Howard and Don gave us that old *What's My Line?* Howard was Dorothy Kilgallen and Don was the guy in the white hat."

At 8 o'clock on Monday night (no sooner, so as not to give AT&T any more money than necessary) the truck locks into the master control of the nearest ABC-owned station. Forte runs through the communications system with the booth, the cameras, the other truck, and Morris on instruction pulls pictures from each of the cameras to the outgoing line, represented in the truck by a bigger color monitor screen right above the eight camera monitors, right in front of Forte. The announcers rehearse on camera the opening statements they will make. Forte checks the matting of names onto the freeze-frame pictures, and the VTR delivers the "tease"—the ritualized opening of the program—for his examination. This opening is an accurate picture of the array of monitor screens Forte faces, but it is misleading in one minor respect: it shows the program going on the air as the sweep-second hand of the clock touches the hour, though in fact the program does not feed onto the line until 9:00:20, permitting the stations to sell a final, valuable adjacency. At 8:59:20, Arledge leans forward from a post behind the three men at the console and speaks into the mike to the announcer and cameramen: "Okay, guys, let's have a great one tonight." And the show is on the air.

In the opening minutes there are no instant replays or isolated cameras, and the wide cameras show the lines and backs of the opposing teams, while Gifford identifies them and the graphics unit supplies names to be matted. Forte is in complete control: "Give me a quarterback, 5. . . . That's it. . . . Take 5 . . . matt it . . . lose the matt . . . dissolve to 4. . . ." There are various ways of changing the picture on the line from the output of one camera to that of another, the most striking being the "wipe," with the edge of one picture moving across the screen and eventually squeezing out the other. This can be done through "special effects" buttons, with the line of the new picture moving up or down or across, or a circle or square of the new picture appearing within the old and expanding to fill the screen, or on various diagonals (these are known as "windshield wipes"). To go to replay, by Arledge's decision, *Monday Night Football* always uses a "curtain wipe," which seems to pull the old picture apart from a slit that forms in the center, like an opening theatre curtain, to reveal the new picture apparently masked behind it. Morris pushes one button to bring in the VTR or the slomo, pushes another to start the curtain wipe, pulls a lever to control the speed with which the curtain parts, and everything happens too smoothly to be remarked upon. When the replay segment is completed, Morris with equal smoothness closes the curtain and returns the scene to "live action." There is an injury on the field: "Get that injury, 8," says Forte. "Take 8. . . . Who is that, Frank? Say who it is. Matt it. . . . Lose the matt. . . . Punch 1. . . ."

In the B truck, Ohlmeyer's voice instructing the cameramen competes with the sound of Forte's voice on the loudspeaker, so Ohlmeyer always knows what instructions Forte is giving the cameramen being controlled from the A truck. Ohlmeyer stands while working, sometimes putting his left foot on the seat of the chair he isn't using. As he is never actually sending anything onto the air (his cameras feed into the replay devices), his voice carries less urgency: "See if you can get 41 on defense. . . . Okay, this time give me 82 and 81. . . . Danny, if the play goes toward you, go with it; if one of the backs flares toward you, go with him. . . . Feed 8 to VTR-2. . . . Seven, you take the two backs. Chet, we've got that offensive holding in the line, on B." Ohlmeyer's technical man, John Fredericks, says, "Don, we've lost 1," meaning that Morris in the A unit has punched that button. "Chet," Ohlmeyer says, "can I have 3? . . ." Forte's voice rises, irritably, on the speaker: "Give me that defensive line, 5. . . . I want to see faces. . . ." Ohlmeyer says, "Eight, take the near-side receiver. . ,. Chet, we have that run on VTR-2. . . ."

After a scoreless first quarter replete with fumbles and intercepted passes and missed field goal attempts, the Cleveland running backs score two quick touchdowns. Both on the air and to his colleagues in the booth during a commercial break, Cosell says something about the Cleveland offensive line chewing up the Oakland defensive line (Cosell would not dream of denying to his fellow announcers the enlightenment he gives the people). Lewin during a commercial break reminds him cheerfully that he picked Oakland for the Super Bowl. Forte is a little less cheerful. "I guess we have to root for Oakland," he says, thinking of audience flows. But Oakland needs little rooting. A quick march brings a touchdown, and then a long return of a pass interception (dutifully replayed) leads to a field goal by "ageless George Blanda" just before the half. During the half, Arledge allows himself a little fun, having a camera pick out the box of Art Modell, owner of the Cleveland Browns, and telling Gifford to identify the man next to Modell as Bob Wood, president of CBS-TV. "That'll get us all fired," he says with some satisfaction. Forte adds, "Especially since Jim Duffy [president of ABC-TV] was at the game last week, and we didn't show him." The game resumes. Cleveland misses two touchdown opportunities and takes field goals to make the score 20-10, and then the roof falls in on the home team and in ten minutes of the fourth quarter Oakland scores 24 points to win, 34-20.

"Should have been a very exciting game," Meredith says by the truck when it is all over. "But somehow it wasn't." Morris is supervising the packing up of the gear, which is to be put on the road to Dallas Tuesday morning. "In Dallas," he says, "we're going to have the Goodyear blimp. That'll give us a tenth camera." Forte and Meredith, who swing, go off swinging; what Arledge does, nobody knows; the technicians work, then sleep the sleep of the just and catch a plane for New York. "This is a good job," says one of the cameramen. "You know you're going to get home for the same two nights every week."

2

Sporting events have been from the beginning the most popular and artistically the most successful use of television, and every technical improvement has helped sports coverage more than anything else. The European Broadcast Union was put together for the first time in 1953 to bring to the Continent the Coronation of Queen Elizabeth II, but it did not become a high-priority item in the planning of the various national television services until the World Cup soccer matches in 1954. By far the heaviest use of satellite transmissions to date was the World Cup of 1970, broadcast live from Mexico late at night to an increasingly bleary-eyed Europe. Both BBC and ITA carried these matches, after bidding the price to an astronomical level, because neither could afford to be without them; in France both networks of the state system carried the matches, apparently on the grounds that nobody would want to see anything else anyway. In America, Sunday afternoon was scornfully referred to as the "cultural ghetto," where the networks hid shows of artistic and intellectual distinction, at hours when nobody ever watched television anyway. But the Super Bowl games in that time slot draw larger audiences than any nighttime entertainment show—24 million homes, at least 70 million people, in 1971—and that unquestionably understates the audience, because it doesn't include people in hotels and college dormitories and bars, who assuredly are watching nothing else. In 1971 the three networks among them spent about $130 million on sports—about as much as they spent on news and public affairs.

It is—always has been—fashionable to deplore this overwhelming interest in sporting events on the tube, but the fact is that the audience for sports is the best-educated, richest and generally liveliest of all the audiences television draws. Sports on television is the best match of, to coin a phrase, the medium and the message. As everyone has said from the beginning, television's glory is its ability to convey the feeling that the viewer is in attendance at a real event. But reality is usually not all that interesting, because its time scale is wrong; significant moments arise too rarely. So we set up an artificial reality—a Senate hearing, a quiz show, an athletic contest. The first wears out its welcome, because it lacks a known end point, a denouement; the second corrodes in the corrupt air, or palls because the skills displayed lose interest with repetition. But a sporting event is real by the simplest definition (it is happening while you see it), it has a known time frame and visibly progresses toward a conclusion that will be a result, and the skills on display are human yet often astonishing. "It's the perfect combination," says Pete Rozelle, president of the National Football League; "you watch a news story developing while you're being entertained." Add to this a degree of personal interest in the outcome (because it's the home team, or because you have a bet on it), and the sum is an irresistible reason to sit in front of the box.

There was even more sports on the tube in the beginning than there is today. "As equipment improved in the early Fifties," William O. Johnson, Jr. of *Sports Illustrated* has written, "one could find as many as five boxing matches, eight football games, perhaps eight or ten hours of wrestling, plus several nights of Roller Derby roughhousing in a single week of programming."

Both baseball and college football were telecast by local arrangement between the ball club or college and the station. The result in both cases was shrinking attendance. In 1951 the National Collegiate Athletic Association intervened to limit the quantity of college football on the tube, and eventually NCAA worked out an arrangement whereby a single college football package, usually a different game in each region each week, would be sold to a network. The NCAA contract guarantees the network that there will be no television of any game involving an NCAA college anywhere in the country within two hours of the network broadcast. Two colleges, however, have been able to syndicate later broadcasts of films of their games— Notre Dame and Grambling. Ethnic power.

In 1971, ABC had the NCAA contract at a price of $12 million for the season, and was losing money on it. "The advantage of the college package," says Herb Granath, who sells it for ABC, "is that it's exclusive on Saturday. The disadvantage is that Saturday has less heavy viewing." In televising college football, Arledge follows principles very different from those he uses on the pro game. "College football," he says, "is a spectacle, with bands and cheerleaders and coeds and spirit. The idea is not to bring the game to the viewer, but to bring the viewer to the game. In pro football, you're *dissecting* the game, showing the viewer something he wouldn't otherwise know about." Pete Rozelle, who started professional life as a PR man for the Los Angeles Rams, has moments of worrying that the television producers do this sort of thing too well. "We're being criticized," he reports, "that the person in the stands doesn't get as much information as the viewer at home. We're improving the PA to help on that. And there's talk about a future with big screens in the stands so people at the game can see instant replays, too."

Baseball was never able to organize itself enough to take advantage of the national aspect of television coverage. The individual ball clubs were unwilling to give up their own local deals and revenues, and each has its own pattern of local coverage, ranging from nearly all the games (home and away) in New York to nearly none of the games in San Francisco and Los Angeles—indeed, Horace Stoneham and Walter O'Malley took the Giants and Dodgers to California to get away from the television-saturated ambiance of New York. The Baseball Commissioner's office sells a national package including a Game of the Week, the All-Star Game at night in midsummer and the World Series, but the Saturday afternoon Game of the Week has been a feature of dubious value. One reason, as television executives have pointed out in the course of haggling with the owners, is that all the year's 1,944 major league games are equal, and a Game of the Week is not an event the way the pro football games are. An even worse problem, though, is that the local team is often on the air on another channel in competition with the nationally televised game. A New Yorker often has a choice of the Yankees on Channel 11 or the Mets on Channel 9 or a Game of the Week involving two sets of foreigners on Channel 4. On those Saturday afternoons Channel 4 is lucky to get a 1 rating for the nationally televised game. The clubs divide up the Game of the Week revenue equally, but only three of them—San Francisco, Kansas City and Milwaukee—get more money from their piece of the Commissioner's contract than they get from their local deals.

The conventional wisdom says that baseball is not a good television game; Marshall McLuhan offers some pages of hokum on the subject. Roone Arledge, who failed to do anything much with Game of the Week when ABC had it in 1965, feels that the real problem is that "baseball is a *geometric* game, and if you shoot wide enough to get the geometry of the game all the people turn into ants." But Tom Dawson, who went from the presidency of the CBS network to be consultant on television to the Baseball Commissioner, points out that the All-Star Game and the World Series (on weekend afternoons or, in the 1971 novelty that will be repeated and expanded in 1972, on a weeknight) draw an enormous audience. The nighttime World Series game in 1971 was the highest-rated show of the fall season, and the Sunday afternoon seventh game outdrew NFL football on CBS by five to one.

"People like these games," Dawson says, "because they're shot with ten cameras. The local telecast of the local game is shot—regretfully—with three-camera setups. The director becomes obsessed with the duel between the pitcher and the batter, and ignores everything else in the park. This is a thinking man's game, with a lot of strategy—you can watch the fielders shift, see the coach reposition his infield—but you don't get that on television. With handheld cameras, pickups of what the manager says to the pitcher, interviews with the pitcher warming up in the bull pen—what's *he* thinking about?—it's going to get better."

The biggest production jobs in television sports are the golf tournaments, which can command as many as twenty cameras, placed high above the course on cranes and beside the tees and greens. "You have to have at least two units," says Carl Lindemann, vice president for sports at NBC News, "one for the eighteenth and seventeenth, one for the sixteenth and fifteenth. You're doing instantaneous live editing. Sometimes you can be lucky, as ABC was in its famous split screen at Baltusrol, with Palmer and Nicklaus putting on separate greens simultaneously. But you can also be unlucky. There are thousands of yards of landscape. And when you miss a crucial shot, you hear about it." The Tournament of Champions at Akron in September 1971—a relatively small tournament in production, with only fourteen cameras employed—cost NBC $216,000. (Production costs for *Monday Night Football* are only about $100,-000.) "Or so they tell me," Lindemann says. "We are by far the largest users of those mobile units, and your costs are a function of how fast they want to write off those facilities. I have to pay rate card down the line."

Basketball has been an important television game from the beginning, but mostly on a regional basis. It has been the mainstay of the independent sports packagers, most notably Richard Bailey's Sports Network, Inc., which in 1968 was sold to Howard Hughes and became the Hughes Sports Network. Bailey, a matter-of-fact businessman in a world of personalities, was network program coordinator for ABC-TV in the early 1950s, when he got the idea for a separate sports operation that could feed attractions into any station in any city, regardless of its network affiliations. "I offered the idea to ABC," he recalls, "but they didn't know what I was talking about, thank goodness."

Since 1955 Bailey has had a contract with the Big Ten to deliver its basketball games around the Midwest, and since 1963 he has handled regional basketball telecasts for the Pacific Coast Conference. (Any regional game with national appeal, like the USC-UCLA game of 1969, he may offer nationally; for that one, he cleared time on stations covering 90 percent of the country, and got a Nielsen rating of 13.) HSN has outbid the networks for several major golf tournaments, and in fact outbid ABC for the rights to *Monday Night Football* ($9 million-plus against $8⅜ million), but NFL decided to make its deal with the network, ostensibly because ABC could clear more stations in top markets, actually because all three networks were terrified of what Hughes would do to their Monday night line-ups if Bailey were free to offer football to everyone's affiliates.

In 1970, ABC made a major commitment to the National Basketball Association, and since that year ABC has telecast NBA tournament games in prime time, with improving audiences. (Several New York advertising agencies have been touting basketball as "the urban game" destined to sweep the 1970s, which makes the minutes easier to sell; and it's a four-camera sport, with low production costs.) CBS meanwhile made a deal with the rival American Basketball Association, which desperately needed television coverage and apparently gave away the first two years' rights for nothing.

Pat Weaver says that hockey should be the best television sport of all: "You can mike it very close, get all the sounds of the roughest contact game there is, and you don't have to worry about anything because all the cursing is in French." But in fact neither NBC nor CBS has been able to do much with hockey in the United States (the game dominates Canadian television: the government network takes half the games and the private network the other half, because neither could survive without it). Though only forty-eight stations had cleared time for its regular season National Hockey League games on Sunday afternoons, CBS in spring 1971 gave hockey the full treatment, with prime-time pre-emptions for the Stanley Cup, but the audience was not forthcoming. Whether this effort will be made again is a much more difficult and important decision than anybody not in the television business can imagine.

CBS has already had one total failure in sports—a two-year, very expensive effort to introduce soccer, the major sport of the rest of the world, to an American market never receptive to it before. William MacPhail, vice president for sports at CBS and the only man so situated to have come to television from the sports world (his father invented night baseball while general manager in Cincinnati in the 1930s), blames the failure of televised soccer on the failure of the fans to attend the games themselves: "If you play in Yankee Stadium before eighteen hundred people, there's no way an announcer can do a good job." Carl Lindemann, an MIT graduate who started as a cameraman for Kate Smith and produced shows and ran the network's daytime programming before coming to sports, says reflectively about MacPhail's soccer gamble, "I'm so pleased Billy took that one."

CBS had been by some margin the leader in sports, with the National Football League, the Triple Crown in horse racing and the Masters, still probably the country's premiere golf tournament (sponsored every year, as a clever salesman might expect, by Cadillac). In 1971, however, there was very little profit in all this. "We're not doing as well as we have done," MacPhail said. The one thing he saw on the horizon that might make a difference was tennis: CBS had a five-year franchise at Forest Hills. "It's just beginning," MacPhail said. "We'll promote it a lot on the network. The move toward open tennis has been very important." And his associate Ronald Bain added, "There's the age-old draw to the public: money."

Again, CBS was hoping to duplicate a European experience, for in Britain the Wimbledon tournament is one of the biggest events of the television year: one or the other of the BBC's networks carries tennis all day long, and BBC-2 presents a sixty- or ninety-minute summary of the action that night. The technical production is the most elegant and imaginative in televised sports, superbly timed to catch everything from the ramrod line of the server hitting his serve to the facial expression of the man at net hitting a clever half-volley. NBC has bought the Wimbledon rights, but was unwilling to pay BBC's price—apparently well into six figures—for the use of its coverage; and the American network did a much less interesting job with its own crew. MacPhail planned to put six cameras around a single court at Forest Hills for CBS, and do it up brown. But the box-wallahs of tournament tennis got into a fight with the upstart organization that controlled the contracts of the most important professional players, early September of 1971 saw floods of rain, and the venture was such a disaster that one of the advertisers on the show publicly complained about having been had. MacPhail will probably try this one again; televised sports is a business for a patient man.

Wrestling survives in the form of a weekly film (complete with managers rushing into the ring to assault referees) packaged by KTLA in Los Angeles and sold around the country to independent stations. Even the Roller Derby still appears on television once in a while and—astonishingly, like wrestling—sells out big arenas. With the demise of the clubs, boxing has become a sometime thing, the rare highly promotable attraction cablecast to big screens in theatres, lesser events thrown into the great bouillabaisse of ABC's Saturday afternoon *Wide World of Sports*.

This two-hour magazine of mostly minor sports is Roone Arledge's invention, resting on his early insight that the real costs of sports programming were going to be the rights, and that even obscure events, properly packaged, would draw more than enough audience to pay all the bills if the big item of expense was the production costs. "We go after quality," says Bill MacPhail of CBS, expressing a gentleman's horror at what Arledge has done. "We won't carry these barrel-jumping contests and demolition derbies." Arledge himself speaks cheerfully of the genre of "arm wrestling contests," and mentions as the nadir of the show the afternoon when the ABC softball team made an appearance. Events on *Wide World* are not necessarily live (especially not the fights: Arledge will happily put on the films of a fight remarked upon in the sports pages, though it happened several days before). But sometimes the production skips back and forth between live auto races, live swimming meets, live bowling tournaments and live ski jumps, catching highlights from each, with lots of "natural" breaks for commercials.

Arledge is a rather unprepossessing senior executive, a puffy man, puffs of curly blond hair, puffy smile, a puffed button of a nose, a puffed body. He wears cowboy boots that zip up the side and shirts with very wide stripes. Unlike Howard Cosell (than whom he is probably smarter, though it would suit neither of them to say so), Arledge was never immensely interested in sports, and may not be to this day. He came out of Columbia College in the early 1950s to a job at NBC—Carl Lindemann remembers that Arledge was a studio manager in NBC's Sixty-seventh Street theatre when Lindemann was coordinator on the Arlene Francis show—and moved over to public affairs and children's shows. "I remember," he says, "every year, I lit the White House Christmas tree. It shocks me that they can do it without me; did they, this year?" In 1960, for reasons never explained to him, ABC came and asked him to produce NCAA football, which they had just acquired, and to the sorrow of his serious-minded friends he said yes. A money-spender, Arledge fancied up the NCAA coverage, adding pictures of girls and local color, until he was promoted, taken out of the direct production end and asked to think up new ideas for things ABC could do in sports.

"Wide World," he says reminiscently, "came within five minutes of not going on. We were to start it on Saturday with Le Mans, but the deal was that they'd go only if it was one-half sold, and a week before it was only one-quarter sold. Then L&M dropped NCAA football, which was then very hot, and the sales department offered a quarter of NCAA football to anyone who would buy one-quarter of Wide World. Brown & Williamson offered to take one-eighth of Wide World, and Ollie Treyz said no. Then, that Friday afternoon, Reynolds called and said, 'That show you're talking about—what's-its-name. We'll take a quarter of it.'"

Arledge reports that the wildest event in Wide World's history was a race up the Eiffel Tower, celebrating the seventy-fifth anniversary of its construction, with the cameras pointing down, the heads of the intrepid climbers silhouetted against the Seine and the bateaux mouches increasingly far below. These pictures could not be taken from a fixed vantage point on top. "We had cameramen with hand-held cameras, going up the tower ahead of the climbers,"

Arledge recalls, "backwards. Our guys were talking about the bravery of these contestants, and the cameramen were staying ahead of them, carrying all their equipment, doing it backwards."

3

What television brought to sports, of course, was money. The most remarkable example of the importance of money was the efflorescence of the American Football League after NBC in January 1964 bought the rights to five years' worth of Sunday afternoon games for $42 million, an annual rate five times what ABC had been paying for the same rights in previous years. (CBS had just bought up NFL football again, at a price of $14 million a year.) With this money, AFL owners went out into the flesh markets, bought up enough important talent to make their league a ponderable competitor to the established National Football League, and bid up the bonuses for graduating seminarians to the point where the NFL felt it was the better part of wisdom to amalgamate the two groups into one noncompetitive syndicate.

The great bulk of this money has gone to the players. Stars have become authentically wealthy men, even by the highest capitalist standards, through combinations of bonuses, salaries, fees for appearances on commercials and for the rights to the use of their names and photographs in nontelevision promotions of various products. But the workaday linemen and shortstops and not-quite-great golf pros have benefited, too: at least during the years of athletic prowess, sports are now a reasonable living for fairly large numbers of men, which was never true before. And the ownership of major franchises, especially football franchises, has become finger-licking good in terms of capital gains opportunities.

Much concern is expressed on sports pages and elsewhere about the "domination" of sports by television, but it is hard to see exactly what the critics mean. All professional sports have always been tailored in one way or another for the excitement and convenience of their paying audience—Yankee Stadium was built with a short right field so Babe Ruth would hit more home runs; well before television, football rules were changed to permit unlimited substitution so players with crowd-pleasing specialized skills could be used despite weaknesses in what had previously been considered "fundamentals" of the game; professional basketball lengthened a forty-minute game to forty-eight minutes so a single event could be stretched through a whole evening and the gate could be split by two rather than four teams. Against this background, it seems something less than blasphemy to lengthen the natural breaks in an interrupted game like football to give advertisers commercial minutes. All games have a component of artificiality: that's what makes them games. As Arledge puts it, "If nobody made lines on the field, who would care—or know—how far a jumper could jump?" Television probably measures the skills of football players to a finer scale, but it has not made the game any more artificial than it was.

Games played in long stretches of continuous action have been hard to adapt to the commercial needs of television. When CBS was fighting the good fight on soccer, it was commonly alleged (and never denied) that referees called for corner kicks, which stop the play briefly while everybody gets reorganized, to leave time for commercials. Soccer, then, was distorted for American television; but soccer didn't make it. Hockey offers similar problems, and when NBC did the Stanley Cup in 1966, it solved them by running the game on a tape-delay basis—that is, the commercial would be slotted in at a break before a face-off, and would still be running when play resumed. The announcers would continue as though they were on the air, but in fact words and pictures would be put on tape, and would go out to the public a few seconds (ultimately a few minutes) after the actual events on the ice. "But," Carl Lindemann says, "you run into trouble with listeners who are also following on radio, and you get out of phase."

Tennis is probably the sport that has been most violently changed by television: the sudden-death tie-breaker that brings a set to an abrupt end was invented primarily to guarantee broadcasters an event of finite length. But the scoring system of tennis had never been one of the triumphs of man's rational analysis, and there were probably as many tournament players for the new system as against it, quite apart from the putative demands of commercial television.

The prime influence of television's money seems to have been to raise the tone in most sports. Though there is, God knows, plenty of rough stuff in most of the more popular games, the men playing them are much less likely to be roughnecks than they were a dozen years ago. In a funny way, the growth of career lines in professional football and basketball has legitimized what was once the scandal of big-time college athletics. Fifteen years ago, the athlete recruited to a college with a major sports program was simply being exploited for the interests of others, with a junk heap ahead; now he is engaged in vocational preparation that may greatly improve his life chances and those of his children. Moreover, the psychic income can be considerable: nobody today would echo Babe Ruth's astonishment that one could be paid for playing ball, but the great majority of those who play any sport well enough to become professionals enjoy their work to a degree unusual in more ordinary occupations.

Sports producers and announcers are often shocked by the depth of emotion and commitment roused in audiences by the games they telecast, and some of them come to feel guilty about their part in it. As a rule, of necessity, they are shills for the game: local announcers tend to work for the ball clubs rather than for the stations, and even the national announcers (except for ABC on Monday night) hold their jobs at the sufferance of the teams or leagues. This is their problem, and a real one, in terms of self-image and status, but theirs is not the only trade where it is difficult to be both a celebrity and a man. William O. Johnson quotes Howard Cosell as saying that "In sports today a truly good journalist must know the black movement, the labor movement, the law—not just who the hell stunted on a red dog or blitzed on a zig-out or batted .267 in the middle of May. No, for God's sake, we have an obligation, a moral responsibility to do more." Johnson adds, "No one really would disagree with what Howard Cosell said," but that's not so. This page is here as witness that it's not so.

A public interested in the political opinions and marital troubles and salary negotiations of movie stars doubtless takes a similar interest in similar matters relating to athletes, and it is journalistically necessary—not just valid—to deal with these concerns. But they are not inserted into the middle of the movie, and they cannot be inserted intelligently into the television coverage of a game, because they don't have anything to do with the game. There is an obligation on sports reporters to tell the public that, say, Duane Thomas of the Dallas Cowboys has called its management racist and its coach a "plastic man." But to retail this information as part of the description of Thomas carrying the ball on an end sweep would make even Howard Cosell look a little foolish. What draws the audience is the game, and the public is entitled to its game, presented as completely and as interestingly as art can arrange. The rest need not be silence; clearly, sports reporters should have the right to tell the public on news and comment shows what they have learned about any game, without worrying that they will jeopardize their jobs as announcers. But confusing the role of the "sportscaster" with the role of the "journalist" is in practice not only unwise but impossible.

Apart from the rewards they offer to talents not otherwise rewarded in an increasingly meritocratic society (which is not a trivial benefit), sports probably do not have any very important redeeming social value. Bruce C. Ogilvie and Thomas A. Tutko of San Jose State have argued convincingly that sports do not build character where character did not exist before, but that the selection process in competition brings to the top "those who already are mentally fit, resilient and strong." God forbid the nation as a whole should develop a *jockische Weltanschauung*—but surely there is more good than harm in having ordinary people, especially the young, draw their heroes from the ranks of those who can be described in such terms.

What sports offer the society is the sight of men and women who have schooled their skills and can thus display human talent and grace and resourcefulness in unusual measure. On one level (obviously not the only level), Joe Namath throwing a football or Ken Rosewall hitting a backhand or Juan Carlos crossing a hurdle gives pleasures very similar to those given by Vladimir Horowitz playing the piano or Edward Villela dancing or Leontyne Price singing.

That people find great satisfaction in such spectacles and hunger for more is no discredit to them, especially in our time, when so many publicly prominent figures are merely faces painted by publicity on what are really faceless men.

Television is well used, not debased, when it satisfies this hunger in its uniquely ubiquitous fashion. There may be costs as well as benefits; there usually are. Nationwide—world-wide—exposure to the highest orders of talent through electronic media may indeed have serious negative effects on society by diminishing the respect and admiration and rewards accorded to second-order but still admirable (and much more widespread) talent—just as the spread of literacy and the availability of the world's finest storytelling talent on the printed page choked off oral traditions in all modern societies. But the notion that all accomplishment must be looked at with reference to "solving" social "problems" grows less from social concern than from envy and from the normal but discreditable fear of any form of greatness.

CHAPTER 8

The Nightly Network News

We have certain difficult problems in television. The pictures which become available to us from various agency services, and through our own cameramen, do not always reflect what a reader of *The Times* would think of as the most important news of the day. It is in the nature of pictures to reflect action. It is very difficult for them to reflect thought or policy.

—CHARLES CURRAN, Director-General, BBC

I think television news *is* an illustrated headline service which can function best when it is regarded by its viewers as an important yet fast adjunct to the newspapers. When I read statistics that show sixty percent of Americans get all or most of their news from television, I shudder. I know what we have to leave out.

—AV WESTIN, executive producer,
ABC Evening News

In Africa . . . we are dependent to a very large extent on foreign sources for our foreign news. . . . We have suffered through a situation in Nigeria in the last three years which has opened the eyes of those of us in broadcasting to the dangers of this situation, because you accept a film coverage of the events in Vietnam or the events in the Middle East as the truth, because this is a majority medium, as we say. And suddenly the next day, you get a film coverage by the same foreign newsfilm company of an event in Nigeria, and you begin to ask yourself whether the infallibility of this foreign newsfilm company is not something that you doubt. We are beginning to wonder whether we have been giving a distorted view of the world to our viewers all the time.

—CHRISTOPHER KOLADE, Director of Programmes,
Nigerian Broadcasting Company

I remember I was on a newspaper in Tennessee, and Frank Clements was running for Governor and my paper opposed him. I was covering his campaign, and I would go with him to those ball fields—two hundred, three hundred people. He'd come near the end of his speech, and he'd lean over and say, "There's a young man here, perfectly nice young man" —and he'd sort of peer around—"there he is, Wallace Westfeldt, and he's been sent here to write lies about me." Then he'd see me on the way out and say, "Hope you don't mind, Wally." I'd say, "Not at all, Governor." You get used to it.

—WALLACE O. WESTFELDT, executive producer,
NBC Nightly News

1

Walter Cronkite does not in fact do his day's work at the table where he is seen at work, five nights a week, by twenty-odd million Americans. He comes to that table at about 5:45 in the afternoon, Eastern Standard Time, drapes his jacket over the back of the chair, loosens his tie and opens his collar, sets his pipe in his teeth, places his stopwatch into a cubbyhole built onto the top of the table beside the invisible microphone, takes the sheaves of paper interleaved with carbons which constitute the working script, and bends over them clutching a pencil. Cronkite does not write his own stuff (the show has a staff of three writers whose entire job is to write for Cronkite), but he edits it minutely, employing what may be a unique gift for hearing precisely what a phrase or even a word on a page will sound like in his own voice. When he has finished editing a discrete section of what he will say on today's *CBS Evening News,* he mutters through it soundlessly, stopwatch in his right hand, putting the watch down to pick up a pencil and note the timings on a separate sheet. Edited pages are ripped apart from their carbons and distributed, one copy to the teleprompter above the TV camera that is slaved to Cronkite. Cronkite uses the teleprompter, though he also keeps a script in his hand, mostly to review in the breaks. The men in the control room down the hall want to know exactly what Cronkite will say, so they can roll the film or the tapes (including the commercials) exactly on the button ("ten . . . nine . . . eight . . ."). A roll cue is not like a stage cue for an actor, who can start saying his line immediately: getting the tape on screen (film is usually put on tape before broadcast) takes seven seconds.

Though the script is written and edited by 6:15 or so, the show is not closed. Under the world map painted on the wall to Cronkite's left stand the AP, UPI and Reuters teletypes, batting out in caps the news of the world: men rip the paper from the machines every few minutes and carry their trophies to three or four men working around the rim of Cronkite's desk, who may pass items into the slot. As air time nears, Cronkite is very much the boss of the show; he is not just talent or anchor man, but (the title by his own choice) Managing Editor. Not infrequently, he will edit his script for the last time viva voce, on camera, losing "roll cues," driving to distraction the directors in the booth down the hall.

Cronkite's strength with the American people—with the departure of Ed Sullivan he is the doyen of television personalities—rests on the near-universal perception that he is what another culture calls a *Mensch*. When he was removed from the central position before the camera in the CBS coverage of the 1964 Democratic Convention (because Huntley-Brinkley had won the ratings laurel at the Republican Convention), he did not wince or cry aloud, and presently he was back on top, simply because he was the best in the business. He has never been just a man in front of a camera. Night editor of his college paper at the University of Texas (where he was also an actor in the Curtain Club, until the demands of the two conflicted), he made his first contact with network broadcasting as a UP correspondent in London during the war, when Ed Murrow used him as a stringer, and he went full time with CBS in 1950. Hugh Baillie, late president of the United Press, used to say that Cronkite was one of very few men he regretted having lost through his organization's refusal (it was not an inability) to pay competitive salaries. Though Cronkite does not feel that newspaper experience is an essential for the television newsman—"After all, Ed Murrow, the greatest of them all, never worked for a newspaper"—he cherishes his background with the UP: "A man should know about press service, because he has to work with it; it's a handicap not to know how a press service works."

But for all his intelligence and background and tireless tenacity (much of Cronkite's hold on the public derives quite simply from shared experiences through the night, at conventions, elections, space explorations), Cronkite's basic function for CBS News is that of a performer. Only in America is the announcer a significant factor in the broadcast news operation; and even here, Cronkite says, "station managements think of anchor men as front men for the station, not as newsmen." On BBC in England, l'ORTF in France, RAI in Italy, the German regional stations, the man (often woman) who reads the news is a talking model, who may also be used during the course of the evening, on camera, to tell the viewer about tonight's entertainment schedule. (The English say "linkman"; the French, "présentateur." In the early days of television, the BBC linkmen wore tuxedos.) Because they are only presenters, most of them tend to have a fixed expression, which on the girls is almost always a grin. In America, offense is caused when a commerical for some kissable toothpaste follows a news film about, say, an earthquake in Peru; but this abomination is no worse than the typical English sequence in which a news film about, say, starving Bengali refugees is followed by a very toothy girl saying in beautifully modulated tones, "Now, let's go to this afternoon's garden party at Buckingham Palace. . . ."

Though the stress on the anchor man seems wrongheaded to nearly all American commentators—the charm of a man telling of stories others have seen is scarcely a valid criterion for judging the quality of a news service—there is also an unmade case for the proposition that the perceived glamour of an announcer has saved American television news from total dependence on the accident of available film. Where the inescapable man on camera is a nonentity, all the stress in a television news broadcast falls on the stories which the news department has been able to place before its cameras. Though nobody ever seems to have gathered authoritative statistics, casual observation indicates that European television news (except perhaps the Americanized Independent Television News in England) makes much less of still photography—the shot projected onto a screen beside the anchor man's head—and the presenter reads many fewer items than an anchor man does in America. Important stories lose form under the European pressure to have some—any—film; and, in Germany especially, television news becomes a sequence of fat middle-aged men taking seats around a conference table.

If Cronkite feels a story is worth a full minute, which is a lot of time on a network news show, he can have it written to a minute's length, even though the producers have nothing of real visual interest to go with it. The schedule for a *CBS Evening News* show lists perhaps half a dozen pieces of film or tape, and between them, taking up perhaps a third of the broadcast time, the word "Cronkite." Because the audience does not feel cheated when it is merely watching the anchor man, an American news director has more freedom of motion than his European counterparts.

On ABC, Av Westin, executive producer of that network's evening news, has worked up with his graphics director Ben Blank a library of slides ("light boxes," in the local argot) which advertise the evening's more important stories at the start of the show and serve as visual filler while Howard K. Smith or Harry Reasoner is reading the news. The library is steadily enlarged by a staff of five full-time artists, headed by Jerry Andrea, a young man who "can do art in any style for any story," and who turns out every afternoon four to six striking charcoal sketches to be superimposed on color backgrounds for that night's broadcast. The artwork (prepared with a silhouette of Reasoner to the right or Smith to the left, to see how it will look in use) is not delivered to the studios until 5:45 in the afternoon. "There's a small panic," Ben Blank says, "every day."

Still, the executive producer of an American news show, too, must build his half-hour on an armature of available film. Every morning, by around 10 o'clock, the staff supplies a list of films already in the house, plus the stories which the network's reporters and cameramen expect to cover that day. There is always some prejudice against running tonight anything rejected for use last night, but in fact many of the most adventurous efforts of a news department will have to bide their time. While he was producing the Huntley-Brinkley show, for example, Reuven Frank (now president of NBC News) sat for two weeks on a filmed report of the civil war in

Yemen, very exotic with soldiers on camels, and far from trivial on the world scene to those who had a sophisticated interest in the world scene. Some $30,000 of NBC News money had been spent to send crews to Yemen, bribe the right people and get the film to New York. But the story needed at least six minutes, because the audience had to be told where Yemen was and what issues were involved, and Frank could not use it until a day when the shortage of "hard" news opened that large a hole in his schedule. A story on a bad drought in the West, or on attempts to put no-fault automobile insurance through state legislatures, or on the financial crisis in the Catholic schools—each a subject of considerable importance but "soft" news, not what-happened-today—may have to wait many days before a hard-news hook forces an opening for it or a dull afternoon leaves space. Most such "mini-documentaries" will not appear on the network evening show at all, but will end their days at CBS on its low-audience hour-long *Morning News* (7–8 A.M. Eastern, 6–7 A.M. Central, Time), at the other networks on a special late-afternoon feed to the local stations for use, if desired, on local news shows. But any network news show that does not have a bank of such films is clearly not doing its job.

The day's assignment sheet (called "troop movements" at ABC News) will get a more careful look. Each of the networks has fifteen to twenty bureaus scattered around the world, and bureau chiefs are usually left free to decide for themselves what is most likely to be airworthy in their bailiwick on any given day. New York may want eminent visitors covered (i.e., Senator Barry Goldwater at the Paris Air Show), or something, somehow, uncovered on the correspondent's end of an American story (French involvement in the heroin traffic). Under the executive producer at each network are two or three just plain producers who take turns working through what CBS calls the "Outlook" for each night's news show. All the bureaus report in every morning by teletype (and many of them call in telephone reports for use on radio: one of the reasons the networks established separate subsidiaries for news operations was the desire to consolidate an organization to serve both forms of broadcasting).

A big story breaking abroad may call for the use of a communications satellite, to get the film on the air tonight. COMSAT sells time to networks for news purposes only on a minimum purchase of ten minutes, which in 1971 cost $2,490 from London, $2,520 from Tokyo or $2,280 from Hong Kong. As few if any stories require as much as three minutes of feed, the network news departments consult with each other about the possibility of splitting the ten minutes and the costs; this collusion has been approved by the Justice Department. Because they share satellite costs routinely (except when one of them believes its story is a beat) but do not share domestic AT&T circuits, the networks in 1971 spent as much in line charges to cover a West Coast story as they would spend in satellite costs to cover a European story. AT&T color-television circuits are sold only in full-hour units, at a cost of $1.30 per mile per hour, which means that a two-minute snippet of, say, a Reagan press conference will cost each network the full-hour price of more than $2,500 for transport alone.

All three networks now originate a piece of their evening news in Washington, and the wires between the two production units go much of the afternoon. The other domestic bureaus come together with New York for a conference call around midday, with each office reporting on stories of possible network dimension in its region. A breaking story may be covered for the network on a holding basis by its local affiliate, but local managements are no longer eager for such work. "They used to have one fifteen-minute news program to do," says NBC's Frank, "but now they have three programs a day, one of them an hour, and they don't have the time. They'll give you a story on a sudden news break, a natural disaster or a police thing, but after the second day you're expected to send your own people."

Frank is talking about the feeling of local executives, of course: the local reporters are more than happy to go on the national network and display their talents in old home towns. Moreover, they will be paid extra for their appearance on the screen. The form and on occasion the content of television news have been influenced by the AFTRA contract, which requires a special payment ($50 to $150) to every union member who appears on the screen or is heard in a sponsored network television show. On European television a news story is usually presented on film with a reporter's voice but not his picture; in America the habit is to show the fellow standing before, say, the smoking ruins of Ah Ch'yt Hospital in Vietnam, saying, "This is Sumner Beech, for AlPhaBet News." In recent years, the network news divisions have signed more and more of their people to annual salary contracts, eliminating the AFTRA requirement and a minor but not trivial source of pressure on a reporter to get the wildest story he can, and get himself on the tube.

The day's assignment sheet reflects the fact that most "news" is predictable. The Senate has scheduled a vote on a bill to preserve the *petite marmite;* the wife of the Secretary of Labor will christen a new atomic submarine, the first time such a personage has performed such a function; Japanese students have announced they will protest a Texas Instrument executive's visit to the Emperor by lying down in the Ginza in the middle of rush hour and bellowing like longhorns. A reporter and a camera crew will be covering each of these. Railroad service is resuming on some struck lines, and several bureaus have cameras out in the yards and the stations ("Strikes me," says the CBS "Outlook" ditto—this one is real—"like a dog-bites-man story, but it's there"). By 1 o'clock in the afternoon the executive producer of the evening news can be pretty sure he knows what film will be available to him; by 3 he can make up a relatively firm schedule. Westin of ABC writes it out lefty, apparently backward, on a pad of yellow paper; Westfeldt of NBC dictates it to a production assistant, a young lady with shorthand.

"Television," the Englishman John Whale writes, "can do very little with events of which it has no foreknowledge: although the clumsiness of its equipment diminishes every year, television can still be the slowest news-gatherer to get to work. A team of people must be assembled: power-supply, exposure, focus, sound-level must all be adjusted." Or, in the words of Fred Friendly, "Reporting the news on television is like writing with a one-ton pencil." But Reuven Frank likes to point out that this is a newspaperman's view, and newspapermen forget that a story does not reach the public immediately upon leaving the city editor's desk: television gets its picture to the viewer a great deal faster than the newspaper gets its prose to the reader.

The typical crew is four men strong: reporter, cameraman, audio man, electrician. Usually the reporter tells the cameraman what he needs in "long shots" to set the scene, and what he wants in "close shots" for his interviews, but it is not unknown for the cameraman to be the more influential of the two and to tell the reporter what *he* needs for a piece. Among the skills of a producer is a knowledge of which reporters ought to be working with which cameramen. The rest is in the hands of the film editor, who works hard.

ABC feeds its evening news to its affiliates for the first time at 6 (5, Central Time); the other two networks do a first show at 6:30. If that first show is "clean," the subsequent feeds are usually taken from tapes, though everybody sticks around to be available for "updates." A White House story can break at any moment before broadcast and get on the air, because the White House is fully equipped both technically and in terms of personnel to get its news right on the wire. (Indeed, anything from Washington can get out very fast: the tape machines and telecine chains in Washington are actuated from the control consoles in New York, and for broadcast purposes there is literally no difference between a piece of tape in Washington and one in New York.) But in most places a story that breaks after 3 cannot be got on the national air, except as a script read by the anchor man—there simply isn't time enough to get the crew out and the film back and developed, cut and edited to reasonable length and coherence, synchronized to a sound track, on the telecine chain and over the specially ordered cable or microwave back to New York. One of the reasons demonstrations are covered by television is that the organizers of a demonstration can time what they do for the convenience of the camera, and give plenty of notice. Similarly, Joe McCarthy would time his big announcements and press conferences to catch the deadlines of the late-afternoon papers in the East.

The first decision about how many minutes to give a filmed or taped story must often be made before any material is in the house, and it creates irreversible effects. "For us," Les Midgley said while he ran the Cronkite show, "time represents space. If we put down four minutes, everybody knows that's a big story. But if it's one and a half minutes, everybody working on it—writer, editor, film man—knows it's a small story." In any event, the length of any story arriving late is to a large extent controlled by the length of the film it displaces from the schedule. Editing film is the high skill of a news producing team, and even if the reporter is in the studio to do a new voice-over, something that runs 140 seconds (84 feet) may be hard to cut back to 110 seconds (66 feet) without starting over again from scratch on the 400 feet that came in from the field; and there isn't time for much of that.

Between 4 and 5—sometimes as late as 5:30—the executive producer has run for him through a wire to a TV set in his office, or steps out to a screening room to see, all the film or tape the show is expected to carry this night. If he feels something is missing or excessive, the item may be recut; if it's "soft" news and he decides he doesn't like it, something may be substituted from the bank. Meanwhile, if there is to be some closing comment on the show, the newsman who will deliver it is away in his office writing it. At around 5 o'clock, he reports in on how long he will be speaking, which is more or less his decision, subject to general policy: in the 1971–72 NBC format, Brinkley may go six minutes, but both Reasoner and Sevareid are expected to stay under two minutes most of the time. At none of the networks does anybody try to tell a "commentator" or "analyst" what to say; at ABC, indeed, it is not unusual for Av Westin to find out what Reasoner or Smith is saying only as they deliver their words to the public.

News broadcasts require no rehearsing. The camera positions and the lighting are the same every day, and the anchor man's audio level is known. Technicians get the film or tape in sequence for the prerecorded sections, and the script is wound on a roller to feed along just over the camera's eye, so the man on the desk can read his words while looking right at you. (At NBC there is no teleprompter, and John Chancellor, who has written his own script, visibly reads it.) White letters are stuck to black cardboard to provide the "supers" that will appear at the foot of the screen to identify speakers. (In France, supers are electronically generated in the control room by a man at a keyboard, and the assistant director who is supposed to approve them before they appear on the screen has

other things to do, so difficult names are sometimes misspelled.)
Because the show is interrupted not only by commercials but also
by film or tape inserts, the anchor man and the director will have
multiple opportunities to communicate with each other, and problems can be straightened out on the spot. All such communication
is necessarily electronic: the men in the control room are sealed off
from the world, and their only view into the studio is through the
cameras; and the anchor man will get feedback on how things seem
to be going through a headset or a telephone while films or commercials are on the screen.

At NBC, the producing staff goes down one flight in a waiting
Rockefeller Center elevator and does the show in a studio dedicated
to the news division. At ABC, those who have business in or near
the studio will dart through a rabbit warren of basement corridors
in the converted horse stables just off Central Park West which serve
as their offices, to the former dining room of the Hotel Des Artistes,
now hung with lights and paved with cables to serve as the point
of origination of the *Evening News*. At CBS, the news department
is housed on the far West Side of Manhattan, in the studio building
which fills a former dairy warehouse, and the smaller of two
news rooms is dedicated to the Cronkite show. Two television
cameras are trundled through the door and placed where they always go (the number of possible angles is limited at best by the
clutter of desks and doors). Cronkite himself has emerged from
his book-strewn office, separated off from the newsroom by a glass
partition, and is in his chair. As he checks his final script against
the last line-up—one page out of order guarantees disaster—the
second hand on the clock enters the last minute before air time. A
make-up girl quickly brushes a powder on his face, and holds a
mirror which he uses while combing his hair. With fifteen seconds
to go, he buttons his collar and pulls his tie tight, and reaches around
the chair to don his coat. There is often only a second to go when
he swings his swivel chair to expose his profile to the camera that
will catch him against the background of the map on the wall as a
voice-over says, "Direct from our newsroom in New York, the *CBS
Evening News* with Walter Cronkite . . ."

2

Broadcast news improved out of all recognition in the 1960s. In the early days of television, news was an unloved stepchild, partly because broadcasters perceived (correctly) that radio was a superior medium for the transmission of news, partly because doing this job at all well would be a major expense. In 1945, NBC hired a newsreel man who worked with a camera stolen from the Office of War Information, and in 1948 the network turned the job over to the Fox-Movietone newsreel. CBS used Hearst-MGM and Tele-News. Murrow's *See It Now* unit was still using Hearst personnel on loan to the unit in 1954, when the McCarthy broadcast brought down on Murrow's head the unexpurgated wrath of the Hearst organization, including personal attacks on CBS reporter Don Hollenbeck, who had excitedly acclaimed the Murrow broadcast on the 11 o'clock local news in New York which followed directly after *See It Now*. Hollenbeck committed suicide a few months later, and Fred Friendly demanded that whatever the cost CBS sever relations with Hearst, which it did.

But the fifteen-minute nightly news that all the networks broadcast in the 1950s and early 1960s was essentially a radio service with occasional films. In the absence of videotape, film from out of town had to be flown in (network news motorcycles racing from the New York airports were a familiar sight) or sent twice over the wire to New York at extravagant line charges (for the two separate feeds to the stations: the whole show had to be done twice). In the absence of satellites, and of high-capacity undersea cable, film from abroad was always at least a day late. Live remote coverage was possible, but it required much advance planning and very bright, very hot lights. When the propagandists for television acclaimed the medium's capacity to communicate reality, what they were talking about was the studio-originated press conference or interview show (Presidential press conferences were not televised live before Kennedy), the televised Senate hearing (Kefauver catching crooks; McCarthy reveling in nastiness), and the documentary, the nonfiction film produced at Hollywood ratios of ten feet shot for every foot used, with Ed Murrow telling the viewer what it was all about. News qua news was fifteen rather perfunctory minutes at 7:30, then (when CBS destroyed NBC's early-evening ratings by programming entertainment at 7:30 and moving the news back) at 7:15.

The technology of the news show in a studio was primitive and inevitably amateur. "When I first started in 1951," documentary producer Perry Wolff recalls, "Sig Mickelson [director of the undermanned TV news department] said, 'What do you want to do?' I said, 'I want to be the guy who says, "Take one, take two."'' He said, 'That's a director; why don't you take over the morning news tomorrow?'"

Moreover, television had no tradition whatever of interrupting programs to handle even the most important external events. In the days when network time was sold in hour or half-hour pieces and advertisers provided their own programs, a pre-emption cost the network not only the lost time charges but also payments to the producing company that would otherwise have supplied the program for the time slot. When President Eisenhower went before cameras to discuss what the Seventh Fleet would be doing about Quemoy and Matsu off the China coast, NBC and CBS delayed broadcasting his statement until after the end of the prime-time entertainment schedule. As late as 1961, when President Kennedy wanted half an hour at 8 o'clock on a Sunday night to explain to the citizens of Mississippi and the rest of the country what he was doing to ensure James Meredith's right to attend that state's university, the networks haggled about whether they should give up this highest-audience time slot; and eventually Kennedy compromised on 10 o'clock. The time Kennedy had originally requested would have been before sundown in Mississippi; the time he got was after darkness had fallen and the rioting had begun. Not until the Presidency of Lyndon Johnson, who had close personal ties to Frank Stanton, did it become automatic that the nation's Chief Executive would receive television time of his own choosing for his own purposes.

Up to 1955, CBS gave Murrow's *See It Now* a once-a-week 10:30 slot. Among the last of these weekly nighttime shows were two half-hours devoted to the growing evidence that cigarette smoking caused lung cancer—doubly ironic in retrospect, first because Murrow himself, a visibly heavy smoker, was to die of lung cancer; second because FCC Commissioner Nicholas Johnson, attacking "The Silent Screen" in *TV Guide* in 1969, was to place among the centerpieces of his argument the snide question, "Would it surprise you to learn that the broadcasting industry has been less than eager to tell you about the health hazards of cigarette smoking?" (In his answer to Commissioner Johnson, incidentally, CBS News president Richard Salant listed four documentaries on this subject between 1962 and 1969, but forgot the Murrow shows.) For 1955–56, Murrow had seven irregularly scheduled evening hours; by 1956–57 he had been pushed into a Sunday afternoon time slot. In large stretches of the country the show was not available at all, because the CBS affiliates refused to clear time for it. Perhaps the most striking failure to clear in the 1950s, however, was the restricted network that carried the first-ever televised interview with Nikita Khrushchev, a CBS beat in 1957. Only 105 stations carried the interview on a Sunday afternoon; that same evening, 220 took the CBS feed of *The Ed Sullivan Show*.

By the 1958–59 season, news and news-related shows occupied minimal fractions of the network nighttime schedules. Then the roof fell in on the quiz shows, and in return for some degree of protection against an outraged public and Congress, the Eisenhower Administration exacted a specific promise from the three networks. Chairman John C. Doerfer called down to Washington to a special off-the-record meeting the Messrs. Robert Kintner, Stanton and Goldenson, and told them the FCC would require them to produce each week at least one hour public-affairs show that did not conflict with any similar program at the same time. When the objection was raised that such collusion would violate the antitrust laws, Doerfer pulled from his desk a written opinion from the Justice Department that network cooperation in so virtuous a cause was entirely legal.

The resulting burst of inexpensive and ill-prepared public-affairs programs was important mostly in terms of personnel: because ABC could not possibly do this volume of work itself, Drew Associates, a film-making group associated with Time, Inc., got the chance to illustrate the possibilities of cinéma vérité on television; David Brinkley tested his bright wings on NBC; Howard K. Smith and David Schoenbrun received much more exposure on CBS. The key matter, however, was quantity rather than quality: the news divisions had to be beefed up considerably to carry the weight. And the public-affairs shows did not begin to pay their way: typically, they had to be sold at prices that covered little more than the cost of the air time alone. Corporate executives noted that ratings and sales were better on the evening news; if the fifteen-minute format could be doubled, the larger staffs could be more profitably employed. In 1963, within two weeks of each other, both CBS and NBC went to a half-hour news program.

By now the technology was moving. AT&T's first-approximation Telstar satellites were in the air, permitting occasional long-distance transmission as the bird flew into the line of sight (Brinkley from Paris told American viewers in 1962, live on camera in spectacular illustration of what the new gods had wrought, that not much was happening in Europe). Ampex was turning out the first videotape machines. Both film and television cameras had greatly improved, reducing the need for artificial lighting, always an aggressive interference by the medium with subject being covered.

Further improvements in television cameras, however, have not produced the benefits one would expect in network news; the problem is unions. By 1971 it was possible to equip a cameraman with a very lightweight television camera and a back-pack transmitter; the result could be got on the air much faster (because film would not have to be processed, a job which takes nearly an hour in the case of color film), and the intrusiveness of the equipment could be even further diminished. But television cameramen are members of the National Association of Broadcast Employees and Technicians (NABET) or the International Brotherhood of Electrical Workers (IBEW), and film cameramen are members of the International Alliance of Theatrical Stage Employees (IATSE). The substitution of tape for film in normal news coverage is simply impossible under existing NABET contracts, and it seems unlikely that this situation can be changed without a strike that would affect all television production for some months. Another union problem reduces the chance for "magazine" features in the context of the news shows, because the Directors Guild of America has insisted that directors be assigned to any camera crew that is going to film anything more complicated than a talking head on anything not a hard-news instant story. To the news divisions, as NBC labor relations counsel Richard N. Goldstein has written, "the very idea that footage to be included in a news documentary was 'directed by' or 'staged by' was abhorrent." Even the Writers Guild of America makes trouble: a strike was once threatened over Fred Friendly's insistence on deleting a "written by" credit for the man who had prepared the introductory announcement to a long interview show with former President Eisenhower; Friendly argued that the credit line implied that Eisenhower's statements had been written for him. He lost the argument.

Up to late 1970, when the economy soured, the ratings war between the CBS and NBC evening newses produced ever larger staffs for the news divisions; ABC stayed out of the fight until 1968, then became ponderable competition—running about 60 percent of the average CBS audience level in 1971. Now all three national network news divisions are very big operations. Among newspapers, only the *New York Times* and the Los Angeles *Times* support so many people in the news business.

CBS News in 1971 had more than eight hundred employees, and was budgeted at $47 million; and president Salant estimates that about 80 percent of the budget "feeds into" the Cronkite show. NBC News is much bigger, because the division runs *Today* and the local news shows on the network's own stations as well as the national service, but president Reuven Frank says "there is no budget. To bring accounting procedures into a news division as though it were a newspaper doesn't make sense. We do a lot of things the network couldn't possibly *not* do, and it's silly to consider those a loss." ABC News is considerably smaller, but at $35 million it is many peanuts.

Covering the news is an inherently wasteful business: to have a man on the scene when something happens, you must also have him there through the long days and nights when all is well. The key decisions in a news service are the financial decisions, determining the ambitiousness of the service and the odds that somebody will be there to get a story when it happens. One fine night in spring 1971 the rumor ran around New York that Fidel Castro was about to visit Chile, and the next morning all three networks had people and equipment on planes flying to Santiago. But the tip was wrong— Castro had stayed in Havana. This incident of the heavy expenditure on the visit that didn't happen rouses all sorts of questions, which the class is invited to discuss after the close of the lecture and the departure of the lecturer.

Ben Bagdikian and the RAND Corporation have estimated that on "large metropolitan dailies (over 300,000 population)" the gatekeeper, the man who decides which stories run and which go into the wastebasket, sees ten times as much copy as the paper actually prints. Each of the three network news shows receives every week about 100,000 feet of film, which is just under fifty hours' worth, and uses perhaps 2 percent of it. No one man can see it all and do anything else for a living. In addition, there are the same wire service machines that pour news into the newspapers, and reports from fifty to one hundred correspondents and stringers all over the world. And the news producers assiduously read the newspapers, too.

What can be put on the air in half an hour is a very minor fraction of what a newspaper carries. Richard Salant, president of CBS News, once had the complete spoken text of a half-hour Cronkite program set in *New York Times* type, and slotted it into a dummy of the paper. It occupied less than four of the eight columns of the front page *alone*. "We have forty-eight permanent correspondents," Salant says, "bureaus in New York, Washington, Chicago, Atlanta, Los Angeles, London, Paris, Bonn, Rome, Moscow, Beirut, Tel Aviv, Cairo, Saigon, Hong Kong and Tokyo. But if you look at the results of this great world-wide news operation, it's a fraud." A rival news executive says Salant's example is the fraud, because there is only one *New York Times*, and the network news looks much better by comparison with, say, the Denver *Post* or, indeed, the New York *Post*.

Oddly enough, the need for drastic compression makes the gate-keeper's job—the creation of the line-up for each night's show—easier rather than harder: there are only a handful of stories that can possibly qualify for inclusion on the program. As Av Westin puts it, "A television news broadcast is based on elimination rather than on inclusion." A little personality used to be possible: Reuven Frank while producer at NBC once led off with the fact that *By Love Possessed* had *not* won the Pulitzer Prize for fiction. Today there is much more feeling (with luck it will go away) that judgment must be tightly controlled by events.

Abroad, only natural disasters, revolutions or other changes of government and disputes between nations are likely to qualify; at home, Presidential elections or announcements, legislation or Congressional hearings, other political wurrawurra, strikes, airplane or train accidents, big crimes, big trials, big bankruptcies, civil disorders will be presented, provided the film is interesting. In August 1971 one of the nuttier bands of black separatists lost a gun battle in Jackson, Mississippi; NBC's Wallace Westfeldt, who had the third day of the Nixon freeze to handle, looked at the footage, which was all postaction (inevitably), and said, "Nah. Just a routine shoot-out." Something like four-fifths of the stories covered are the same on all three networks on an average evening, though the order of presentation and the time for each story will probably be different. (ABC usually runs shorter pieces; NBC, longer ones.) The intellectual cross-fertilization is considerable: because each network feeds to its out-of-town affiliates before it broadcasts in New York, a man can look at what his rivals are doing right after he finishes his own work. At CBS the viewing is done by an assemblage (including Cronkite) packed into the executive producer's small office, hooting and grunting as the NBC News snakes across the screen.

These are the day-to-day operating procedures of a national news service, whether the locale is the United States or Western Europe. (In Europe, the conference call, scheduled every day at 10:15, links not the bureaus but the news directors of the various Western

European broadcasting systems, who describe the stories they can make available, if desired, on Eurovision; the language of the conference call is English.) Some customs are different. European news shows may include live interviews at the studio, which almost never happens in America on network news. Individual pieces may be larger. During the Nigerian Civil War, Britain's commercial ITN gave 12 of its 26½ minutes to an interview with Major General Yakubu Gowon, the Nigerian President, and features lasting as long as eight minutes are not uncommon on *News at Ten*. Bavarian Television needs three days' notice to do a live remote news coverage, but the entire cinematographic resources of the organization are at the service of the news division, which can call cameramen from entertainment productions, as needed.

The first French network makes the final line-up for its 7:45 news at 4:30, and anything that happens after that will have great difficulty getting on the air. The second network, on the other hand, holds a formal conference at 10:30 in the morning to discuss what should go on its nightly *24 Heures*, which consists of a magazine half-hour at 7:30 and a news half-hour at 8; but the line-up keeps changing right up to air time—indeed, the *présentateur* often gets the timings for his announcements, or for any live interview he will conduct, as he goes on camera; and the eye he keeps on the clock will be as important as the eye he keeps on the script. The executive producer himself does the final editing of the script, in the fifteen minutes before the show goes on.

A 10:30 meeting of the *24 Heures* staff in summer 1971 was entirely recognizable to an American visitor. The acting executive producer was Jacques-Olivier Chattard, an extraordinarily handsome young man, an early Mike Wallace with French style, who was normally the foreign editor for the service; and the group crowded into his narrow office consisted of thirteen men and two women, each with a separate beat or producing function. The meeting opened with a sharp discussion of a poorly edited tape that had gone on the air the night before, then proceeded to an announcement by Chattard of the evening's magazine layout—a feature on the forthcoming festival in Aix-en-Provence, a discussion of a new book on Coco

Chanel, with the pretty authoress filmed in a verdant setting in the Bois, and a long takeout on the next day's featured horse race at Longchamps, including live interviews at the studio, filmed interviews at the race track and films of recent races.

For the news section, the political analyst proposed another lead on the proposed Nixon China visit, which had been announced only the day before, but Chattard thought it had been covered as hard news quite thoroughly the previous night. Chattard suggested that M. Kissinger was reputed to have a *"petite amie"* in Paris, and that she had offered films of M. Nixon's adviser both in his home in Washington, which she had visited six months before, and at stages of his recent trip to Paris. Someone in the room had seen the film: "You learn when he left the embassy, when he crossed the bridge, when he ate dinner with a blonde, when he got on the plane speaking English and flew off. From Washington, you know when he ate a beefsteak, when he ate bacon. . . . It doesn't make a film." Someone else suggested that perhaps the lady could be put on camera to talk about her friend. Chattard offered to look at the film, from which, in fact, he subsequently pulled a very funny segment showing the President's Adviser on National Security Affairs struggling unsuccessfully to get the combination lock into the position that would open the safe in his living room where he keeps top-secret papers.

From England came a note that former Prime Minister Harold Wilson would be addressing a Labour Party conference on the Common Market; the speech would be delivered only twenty minutes or so before *24 Heures* went on the air. Chattard made a note to have his London office film preliminaries of the conference, and to get the speech from Eurovision live for taping while the magazine section was on the air, to be used after the break to the news section. The day's prize press conference was being given by King Hassan of Morocco, who had just weathered an attempted *coup d'état*. Chattard nodded, then asked, "Do we have pictures from Belfast of the Catholic protests? Very interesting story."

He turned to his domestic nonpolitical correspondent and asked what was available on "the affair of the lady insurance salesman and the gendarme in Lyon?" (The lady had shot the gendarme, and 24 *Heures* interviewed him in the hospital.) An economics expert had four minutes and fifty seconds of film on the problems of growing and selling peaches—interviews at stores and orchards, at the weighing stations where the wholesalers bid for truck lots, statements from various authorities that the costs of distribution alone were greater than the prices to which peaches had fallen in the Paris fruit stores. Chattard was happy to have it, and scheduled it; but he asked also for something on an aspect of a new Paris financial-political scandal, a real-estate speculating firm called Garantie Foncier, which had just gone noisily and perhaps criminally broke.

French national news also includes sports coverage, and it was now the turn of a lean, long-haired young man in a rough tweed jacket, who had spent all the meeting up to now reading that morning's edition of *L'Équipe*. He mentioned a touching incident in Bordeaux in connection with an injury suffered by a leading cyclist in the Tour de France, the bicycle race that would end in Paris the next day, and suggested live coverage of a few moments of the French Open Golf Championship, or perhaps something on an American-African track meet. "But we have no pictures," Chattard said; and his political analyst commented, *sotto voce*, "It would just be two *noirs*, anyway." No, there were really only two strong sports stories for this evening—one, a filmed report from England on that afternoon's Silverstone auto race; the other, the annual feature on the *Lanterne Rouge*, the man who was running last of the sixty-odd cyclists still in the Tour de France. "Red Lantern" was an extremely bewildering term of art to a foreign visitor, and it was explained—like the freight train, the red lantern that hangs out of the last car . . .

Chattard now had enough for his line-up, and everybody went back to work. During the course of the day, he would talk to these subeditors again, and to his bureau people in Europe (and perhaps in Washington: the first network has a New York office; the second network has an office in the capital). Two in-boxes sat on Chattard's desk, one labeled "FLASH," the other "DÉPÊCHES," and young men kept running in flimsy from the news tickers. His visitor returned that evening at 6:45, as invited, and asked if anything had happened that day, and Chattard said, in English, "Little things."

Then he pressed the lever of the intercom and asked into the news room: "Have you followed the operation at St. Germain des Prés today? . . . What operation? Drugs. Here it says, one hundred and eighty arrests." Two minutes later, a young man walked rapidly into the room and took the flimsy from Chattard's hand and read it. "May be a fantasy," Chattard said. "Verify." Then he pushed another button on the intercom, and finally found someone on his staff who already knew the story. "When did it happen?"

"This morning."

"How many arrested?"

"Six."

"It says here, one hundred and eighty."

"That was the number stopped. Only six were held."

"Thank you," said Chattard, and went down the hall to a screening room to look for the first time at the cut film on Aix-en-Provence. He was not happy with the ending: "Perhaps I am not intelligent enough, but I don't understand what it is trying to say . . ." and another snippet was cut out in the twenty minutes before broadcast time. "That's our only film tonight," he said to his visitor as he left the screening room. "All the rest is *sur* Ampex."

There were differences between all this and American procedure. The news judgment reflected in the meeting and on the program was somewhat lighter-weight than one would find at an American network news division. The total rupture between police and journalists that left Chattard scrambling on the drug story would be almost unimaginable on either side in the United States. Though Chattard himself was a fully professional figure, there was an air of slapdash quite impossible in America, and quite visible in moments of amateurishness on the screen. Most important of all, Chattard was not entirely the boss of his show—before he could authorize the telecast of his excerpts from the Kissinger film, he needed approval from the news director of the network, Mme. Jacqueline Baudrier—and she may have needed approval from higher up. On the desk of Pierre Desgroupes, the news director of the first French network, there is a sign reading "SILENCE. LE BOSS BOSSE," and Desgroupes is immensely proud of the progress that has been made in freeing French television from its former role as a mouthpiece for the government in power. But if French television news need no longer say what the Ministry of Information wants it to say, it still may not say what the government forbids. Desgroupes is not, in American terms, really "Le Boss."

At the Amercan networks, by contrast, nobody above the level of the executive producer will know on the ordinary night what is going on the air. "News judgment," says ABC's Elmer Lower, "must be that of the people on the scene." Richard Salant says, "The only way I can exercise my responsibility is in a postaudit. I don't know any American business with the kind of total delegation of authority that we have in broadcast news. I take responsibility, because I picked the people, but I have nothing to say about the program." Neither corporate manipulation nor New Left conspiracy controls what goes out over the networks on the evening news; what is represented is, simply (and for those involved it *is* simple: news is an unreflective business), the professional judgment of the men who make the program.

This does not, unfortunately, answer all the questions. The time horizons of intelligent analysis and decision-making are much longer than those a newsman can employ in making news judgments. Ithiel Pool of MIT once asked nervously about "the situation that results when good judgments piled up day after day somehow produce an unbalanced diet." We live, after all, in an Age of Aquarius, which the economist Kenneth Boulding has defined as "a time when everyone is all wet."

3

Among the powers given to the heads of the news divisions at all three networks is pre-emption of the program service to cover a breaking story. ("There isn't much top management could do about it," says ABC's Elmer Lower, both large hands planted firmly, palms down, on his desk. "The wire runs through here, and we flick the switch ourselves.") But this power normally extends only to the unexpected occurrence: time for an "instant special" to give background on a news story, or for a Presidential press conference or announcement, must be cleared through the president of the network or the broadcast group. On November 13, 1969, however—each news president acting (he says) on his own motion, without any request from the speechifier or his boss, without requesting approval from anyone upstairs—the three network news divisions pre-empted their own 7 o'clock feed of their evening news. Some time after 5 that afternoon, a teletype message went out to every affiliate, telling him that if he wanted his network's *Evening News* he would have to carry it at 6:30 EST: at 7 the networks would be feeding a speech that the Vice President of the United States planned to deliver in Des Moines, Iowa.

"I was having lunch with my bosses across town," Salant says reminiscently, "when the call came from Bill Small in Washington: he'd seen an advance text of what Agnew was going to say. They left the decision to me, and I suppose I decided to carry it following the *Times* tradition—when you're attacked you have an obligation." At NBC the pieces of paper filtered up more slowly, and Reuven Frank did not see the text until 4 o'clock, but the minute he read it he made up his mind. "There were technical problems," he says. "There wouldn't be time to process film. None of the Des Moines commercial stations was planning to carry it; we had to take a feed from the local educational channel. We didn't know when we went on that all three nets were carrying it." There is good reason to believe that Agnew himself was surprised, for his speech entered its peroration with the words "Whether what I've said to you tonight will be heard or seen at all by the nation is not my decision, it's not your decision, it's their decision."

Agnew's speech took off from the comments that had been made by network correspondents (and, on one network, by a guest: Averell Harriman) immediately following a television talk on Vietnam by President Nixon. But he soon moved on to the larger focus of the evening news:

How is this network news determined? A small group of men, numbering perhaps no more than a dozen anchor men, commentators, and executive producers, settle upon the twenty minutes or so of film and commentary that's to reach the public. . . . They decide what 40 to 50 million Americans will learn of the day's events in the nation and the world. We cannot measure this power and influence by the traditional democratic standards, for these men can create national issues overnight. . . . They can elevate men from obscurity to national prominence within a week. . . . For millions of Americans the network reporter who covers a continuing issue—like the ABM or civil rights—becomes, in effect, the presiding judge in a national trial by jury. . . .

Now what do Americans know of the men who wield this power? Of the men who produce and direct the network news, the nation knows practically nothing. Of the commentators, most Americans know little other than that they reflect an urbane and assured presence seemingly well informed on every important matter. We do know that to a man these commentators and producers live and work in the geographical and intellectual confines of Washington, D.C., or New York City. . . . We can deduce that these men read the same newspapers. They draw their political and social views from the same sources. Worse, they talk constantly to one another, thereby providing artificial reinforcement to their shared viewpoints. . . .

The American people would rightly not tolerate this concentration of power in the government. Is it not fair and relevant to question its concentration in the hands of a tiny, enclosed fraternity of privileged men elected by no one and enjoying a monopoly sanctioned and licensed by government? . . . As with other American institutions, perhaps it is time that the networks were made more responsive to the views of the nation and more responsible to the people they serve.

Agnew eschewed any thoughts of censorship, but he stressed several times that broadcasting was a government-licensed medium. The FCC had been dramatically successful with "regulation by lifted eyebrow"; Agnew had lifted a shillelagh. The networks at first thought there would be a great rush of public support for their service against bullying by a man neither New York nor Washington took seriously. In fact, the public reaction to the speech was mostly favorable.

"In Honolulu," the Du Pont-Columbia *Survey of Broadcast Journalism* reported, "KHVH broadcast a fighting editorial that began, 'Do you want the government to choose your news for you?' and ended, 'Intimidation is implicit in this situation, and the current administration seems willing to take advantage of it. This is a situation that Americans should not tolerate, not at the hands of any administration, be it national or municipal. We want your support. Let us know.'

"Five days later the station followed up with an editorial beginning: 'We have just found out that we don't know our audience . . . and our audience doesn't know us—that's a shock for any medium.'"

An ABC poll soon after the speech found that 88 percent of the public knew about it; that 51 percent agreed with Agnew that television news was biased, while only 33 percent disagreed.

Much of the support for Agnew probably derived from the superb quality of the speech itself, which is one of the most accomplished pieces of sustained political rhetoric written in the United States in this century. Some, no doubt, came also as revenge on the news services for having carried information people did not want to know. News professionals, quoting some education-school wisdom about Babylonians who killed messengers bringing bad news, put all the blame there, though their experience should have told them that at moments of really bad news—a Cuban missile crisis, a Kennedy assassination, a riot, an earthquake, a Tet offensive—people hang by their radios and television sets, and are grateful.

The newsmen were especially bitter at Agnew (as they have never been against somewhat similar criticisms from the left), because his characterization of their personal views was more or less accurate—a nonpolitical British researcher had noted a year before in a paper for the Columbia School of Journalism that "the sympathies of decision-makers in TV news are overwhelmingly Democratic or Liberal Republican."

All but a handful of young twerps among the television journalists spend much emotional and intellectual energy making sure their personal views do not influence their professional decisions. Moreover, they are not linked together in a cabal: they compete ferociously. On the decision-making level they don't by any means "talk constantly to each other": Westfeldt of NBC and Midgley of CBS had never met in their lives when Agnew spoke. Americans could not perhaps go quite so far as Stephen Murphy of ITA in England (now the censor of all live theatre in London), who said that "the typical situation in this country is where a man produces a program that is biased against the party in which he as a private person believes." But all the network television news producers were certain they had kept all the bias they could identify in themselves from influencing their news decisions, and to get Agnew's speech as their reward was the last straw.

Among the practicing TV newsmen, only Howard K. Smith publicly announced any sympathy for what Agnew had said, which was interesting because personally he was probably the farthest left of all those who survived the 1950s. (It was Smith who was the reporter behind the CBS camera when Bull Connor loosed his dogs on the civil rights marchers, and who quoted Burke's line that "The only thing necessary for the triumph of evil is for good men to do nothing"; and Smith, again, who put Alger Hiss on the network air in a show kissing good-bye to Richard Nixon after his defeat in the California gubernatorial race in 1962.) But Fred Friendly, who had resigned from the presidency of CBS News and become television consultant at the Ford Foundation, thought the networks themselves had to take some of the blame for public loss of confidence. "When Ed took on McCarthy," he said some time after the Agnew fuss, "we had a hundred thousand letters supporting us the next day, because people trusted Murrow. Today, the networks just won't get the support, because there's so much arrogance that comes out of that tube."

The people who supported Agnew—as Friendly, of course, did not —would probably agree with the word "arrogant," though they would mean something else by it. The problem is felt rather than analyzed, and it has not been clearly articulated. What happened during the 1960s was that in the general inflation of self-importance that characterized the decade people in the news business—especially in the television news business—lost much of their grip on the difference between news and reality. This difference has been most conveniently stated by an anonymous Canadian, who observed that it is not news when the 7:05 from Vancouver lands safely at Toronto Airport. In the real world where people live, the airplanes do arrive safely and one goes out to greet one's family; in calendar 1970, after all, there was not a single passenger fatality on a scheduled airliner in the United States. But the news world is alerted only when the plane crashes.

Of course, planes *do* crash; 1971 was not so fortunate as 1970. Nobody of any sense would seriously argue that newspapers and radio and television newscasts should not play up the occasional airplane accident, though such stories doubtless depress air travelers. That's news. Murder on the street is news. The collapse of a dance floor is news—in fact, the story of this tragedy in Grenoble drove virtually all other news off French television for several nights. But statistical reality is that the plane arrives, people walk the streets without getting murdered and dance the evening away without falling through the floor. Societal as distinguished from individual reality is always statistical. What society asks from newsmen is that they maintain their sense that they deal in a construct called news, not in the always statistical reality of life. This is hard on the newsmen, because someone who deals with reality is clearly a more important fellow than someone who deals just in news.

David Nicholas, who runs the night news show for British commercial television, tells visitors about a popular saying that "If it hasn't been on *News at Ten*, it hasn't happened." Such stuff is common in the United States, too, but of course it isn't so. Television news presents a very small fraction of the news that happened and is reported as news in the local paper. And the flow of events that absorb people's lives is not news at all; even when they die, in Auden's phrase, they suffer deaths "unmentioned in *The Times*." It is, perhaps, unwise to tell them that what happens to them hasn't happened.

Television has the further problem that it *appears* to be presenting reality. Its "uniqueness," Frank Stanton said in 1959, "is in its power to let people have that intimate sense of meeting the great figures of the world and actually seeing many major events as they happen. . . . Great events of all kinds do not have to be filtered through the appraising accounts of reporters and editors. They can be witnessed by the people themselves, who can make their own judgments." This is complicatedly wrong. At best, television shows a picture composed by a cameraman. The CBS News "Guidelines," despite Stanton, warn producers that "The important thing is to convey to the viewer that he is seeing only the impression of an event, not an event itself." In his book *They Became What They Beheld,* the anthropologist Edmund Carpenter makes much of the fact that "in TV studios, idle employees watch programs on monitor sets, though the live shows are just as close"—but what is on the television set, not the action on the stage, is the "truth" of an entertainment show; and the news show is different only in degree. Pierre Schaeffer of the research center of the French broadcasting system (who is himself, incidentally, way to the left of anyone in American broadcasting or government) puts the matter drastically:

"Cinema offers itself as a production that starts from a simulation. Radio and television seem to the confused like direct branches of reality, merely relayed by diffusion, not created by production. It is forgotten that these images are carried and multiplied in space, not as objects or as authentic happenings, but as shadows, as transformations of reality fully as great as those of the cinema."

For news coverage, these pictures must be composed very quickly, often under extraordinarily difficult conditions. Reporting to the British Parliament, the Pilkington Commission commented that "triviality is a natural vice of broadcasting"; the viewing eye reacts differently to a picture that inevitably includes aspects of reality normally sloughed away by live perception. Unlike the newspaper or magazine photo editor, the television film editor does not mask or cut out or air-brush away the irrelevant area of the picture.

In a bitter speech in spring 1971 about the Nixon Administration's efforts to control the network news broadcasts—and there have been many more such efforts than the public or even the news divisions know about—Walter Cronkite said that "Radio and television journalists have spent thirty-five years convincing the public that broadcast news is *not* a part of the entertainment industry. It is a shame that some would endanger that reputation now." But of course broadcast news is part of the entertainment industry, and nobody knows it better than Cronkite, who draws down every week an entertainer's salary. Nor is entertainment necessarily unreal: surely the artist as well as the newsman seeks to convey something about reality.

Wilbur Schramm has tried to draw a dividing line between fantasy-seeking and reality-seeking television, but there is a continuum here, running from the deliberate inanity of *Beverly Hillbillies* to, say, an American Marine setting fire to a Vietnamese hut with a flick of his cigarette lighter. At the fantasy extreme, producers and writers and actors can make their own versions of their own reality; all they need is an audience willing to go along. At the reality extreme, in news film, the producers and editors and reporters and cameramen must seek to see the world as others would see it were they present. Cronkite's definition of the "bad news" his enemies would suppress—"aberrant behavior and dissent from establishment norms"—is somewhere some distance up the continuum: real enough, but with a substantial component of theatre.

There is no escaping the tendency of "news" to seek the most colorful and interesting stuff that can be brought into camera range. Agnew's scorn for television's apparent legitimization of George Lincoln Rockwell and Stokely Carmichael is perfectly reasonable as far as it goes, but one step further lies the fact that the American people quickly grew bored with Rockwell and Carmichael, who thus quickly became no longer "news." Indeed, the true complaint lies not with the complacent forced to hear of others' discontents, but with the concerned forced to hear significant criticism demeaned in the mouths of trivial critics. Television burns up issues as it burns up comedians, but Women's Lib might not have rolled over so fast (and died, of course—what did you think was meant, you sexist?) if more sober spokeswomen had been chosen by television to present its unanswerable arguments about the unjust division between the sexes of the burden of change in late-twentieth-century society.

Finally, there is the daily deception inherent in the fact that news has no memory: the reporter makes today's story as snappy and important as can be without worrying too much about its coherence with past stories. In November we are told that Lake Erie is dead as a doornail, nothing can live in it; in March we learn that among the economic problems resulting from mercury pollution is the loss of income of some thousands of Ohio commercial fishermen, who will not be permitted to sell the tons of mercury-bearing bass they have been pulling out of Lake Erie. Monday's expert worries about the need to beef up all our institutions to handle a hundred million more Americans by the end of the century; Tuesday's story about education mentions in passing that the American birth rate declined steeply and steadily in the 1960s, so that enrollment in the elementary schools is down. The viewer's memory, of course, is even worse than the news producer's, but eventually there arises a malaise, a feeling that the world as presented by television news doesn't hang together, that somebody is conning us.

All this is recognized under the wrong rubrics: news executives speak sadly about their inability to cover "trends" or "long-range stories." But nothing much can be done about the time horizons of news, because news is by definition a construct of what happened today. Yesterday's newspaper is used to wrap fish, and yesterday's news broadcast does not exist at all. In a better world, television news personnel and the rest of us would be more energetic, more analytical, more imaginative, more perceptive, more accurate. Failing that better world—toward which, of course, Telemachus, we should all seek and strive and find and not yield—the best way to avoid distrust is an open recognition of the limitations of news. Important, instructive, useful, entertaining (at least in Stephenson's sense that communication gives communications pleasure), news is an indispensable aspect of a broadcasting service. But it does not, cannot, present "reality."

"I know, I know," said Richard Salant. "If I were a tyrannical boss, I would forbid Walter to end the evening news by saying, 'That's the way it is.' But . . ."

CHAPTER 9

Right Before Your Eyes: The Political Nexus

Because of television, history will never again be quite the same. By putting the viewer on the scene at the moment news is made or shortly thereafter, television is transforming history from something we read about into something that happens to us, involves us and becomes a permanent part of us through our participation.
> —1968 Annual Report
> to the Shareholders of CBS

The United States, like other countries, was subject to the fevers of jingoism long before the rise of the current rapid mass media of communication. These media, however, make the situation even more precarious, so that contemporary society sometimes appears like a ferry boat with a great many passengers who rush madly first to one side, tipping the boat until it nearly keels over, and then to the other side, tipping it the other way. Radio and television have reduced the distances in space and time which once could buffer individuals so as to delay the impact of news and its interpretation. And such instantaneous knowledge in the absence of instantaneous remedy may perhaps have increased our sense of helplessness, although it is doubtful whether we are in fact more helpless than when we knew less of what was going on, or learned of it more slowly.
> —David Riesman

I suspect that the nerve of America, and its unity and wholeness and happiness, have been broken by the mass media—especially television—as much as by anything else.
> —Maurice Wiggin, TV critic,
> London *Sunday Times*

1

The first really important event covered live on television was the pair of political conventions of 1948. The Democratic Convention was especially significant, presenting as it did the convulsions of a doomed party cursed with the need to nominate an incumbent President who could not win (who was Harry Truman, anyway?). The left wing was already breaking away to a new Progressive Party and the Wallace Presidential candidacy, and the deep thinkers of 1948 felt it absolutely essential to prevent a Southern breakaway that would strip the party of its right wing. Truman and Hubert Humphrey, then Mayor of Minneapolis, did not agree, and Humphrey forced through the convention a civil rights plank on which the Southern Democrats felt they could not possibly stand. With the television cameras watching, they departed the hall, depositing on a table by the door as they left their badges as convention delegates. The television cameras trained on the aisle and the table registered their departure and the mounting pile of badges. "I remember sitting in the NBC booth," says William Ray, now head of the FCC Division of Complaint and Compliance but then the news director for NBC's Chicago station, "and watching that heap of badges grow on the screen, and saying to myself, 'This is the most sensational thing in the world.'"

William S. White, among others, has argued that this episode, presented by television in American living rooms, won the election for Truman: "At Philadelphia, TV—with all its matchless capacity for flat, surface disclosure and never mind its disabilities as to interpretative disclosure—had unforgettably shown one climactic moment. . . . The Negroes in the Eastern part of the country had seen that picture of marching and angry men. . . . Mr. Truman had, indeed, 'stuck out his neck' for them and this they knew in a deeper sense than any number of printed words could have conveyed. . . . It is possible to make a case that the medium of TV literally saved his candidacy. I, for one, have always believed this to be the plain truth of it."

The story will do service as a paradigm. The episode of the badges on the table had been staged for the cameras, and once the demonstration was accomplished most of the delegates returned to the table, picked up their tags and resumed their seats. And its influence simply cannot have been what White recalled. There were fewer than half a million sets in the country in summer 1948, and convention coverage was in the Northeast quadrant only. Painfully few Negroes even there had access to television. And Truman did not carry the important states in which the convention had been telecast (New York, Pennsylvania, New Jersey and Maryland all went for Dewey). But all these negatives are not necessarily important. What the Renaissance Italians believed about ancient Greece was more important to the modern world than the truth about ancient Greece; and what politicians and academics believed about the political impact of television may have been more important than the impact itself.

This screw can be turned once more by noting that though the incident with the badges had been staged it had not been in any way false to the feelings of the participants. Like the antiwar demonstrators of twenty years later, the Southern delegates were seeking a way to get their rage and anguish before a public that might not listen to simple declarative sentences. There is supposed to be something new and wrong with this, but really it's all perfectly proper and historically sanctioned. Think of Jack Ketch, the public hangings and beheadings and the hoopla surrounding them—or of Coxey's Army, or of the NAACP march against lynching in 1919— or of the Great White Fleet sailing idiotically around the world for the greater glory of America and Teddy Roosevelt. "Investigative" Congressional hearings are entirely staged and always have been (though most committees still bar television cameras from the premises); the committee hears witnesses in secret and then selects those whose testimony, already taken, it wishes to present live-action in public. Elections themselves with the hullabaloo surrounding them are staged in the manner of sporting events, and the theatrical elements of a national election have political importance—indeed, one of the ponderable objections to the fancy prediction analyses now performed by all the networks is that by reducing the drama of election night they may well be diminishing the apparent importance of the elections themselves.

Losers as well as winners are entitled to stage ceremonial functions, and circumstance requires them to be more imaginative if anyone is to pay attention. The Negro leadership of the early 1960s was especially talented in this direction, and the climactic event of the March on Washington was surely one of the most powerful and moving dramatic presentations in the country's history—especially the speech by Martin Luther King, whose presence defined the vague term "charismatic personality": nobody who had the good fortune to be in a room with him when he flicked the switch will ever forget the experience.

There are risks in the use of demonstrations to make political points, because individuals in exciting crowd situations can easily lose any self-control they may happen to have. But because they are held during the day, when television cameras can use natural light and people of criminal instincts tend not to show them in public, mass political demonstrations are usually peaceful. How much good they do is, of course, a separate problem, beyond the purview of these pages—though it should be noted that they are at bottom the tactic of a minority in a democracy where majorities rule. Unlike the guerrilla theatre of the hippies, which merely seeks to annoy grownups and usually does, the mass political demonstration is an effort at persuasion, using theatrical devices to call attention to an argument that is believed capable of gaining majority support but for various reasons would not otherwise receive the necessary level of attention. Any demonstration shown on television may be presumed to have been staged: no spontaneous demonstration (except perhaps a riot following an athletic event) would have enough human fuel to burn until television apparatus arrives.

In a sense, the greatest of modern demonstrations was staged by the American government itself in the panoply of the moon shots, though here the problem confronting the networks was not the coverage of the events but the filling in of the time between them. "It takes a hell of a videotape library to cover a moon shot," says Elmer Lower of ABC, "a terrible big job of backlogging the stuff and putting it in the library." Competition among the networks to get the most vivid "simulation" of space travel cost them literally millions of dollars a year, and hundreds of thousands were wasted on the "stake-outs" of the homes of the astronauts and even their parents' homes, in the worst privacy-invading traditions of American journalism. But the result was the slow education of most of the population in what had been pretty arcane science—plus an experience given only once in history, simultaneously to hundreds of millions of people around the world.

Television cameras accompanied man to the moon because the networks insisted, against strong opposition from the National Aeronautics and Space Administration, that the American people had the right to know all the details of what was being done with their tax money. NASA wanted a technological, not a human, demonstration; the agency was scared, and with reason. The first television documentary on the subject, Ed Murrow's *Biography of a Missile,* had shown a failure. On my own first visit to Cape Canaveral, in 1962, I was told that the initials "IRBM" stood for Indian River Banana Missile, because Indian River was where the things fell down when they misfired. Everyone involved was deeply conscious that when a man went up in one of these things his life was in serious danger, and NASA did not wish to risk the future of its projects on the life or death of an astronaut whom television had made not only a hero but a companion in the living room.

"They wanted it to be very impersonal," says Robert Wussler, chief of the CBS special events unit, who has organized that network's coverage of every man-in-space venture. "They didn't even want to tell us the name of the man who was going into the capsule. There were always three possible astronauts, and it wasn't easy to find any of them. We were staking out barbershops and churches— it was a game: they would always take men from two different Protestant sects, plus one Catholic, so we had to stake out lots of churches."

Nevertheless, in the end NASA came around, and cooperated on full, live coverage every time a missile went up with a man aboard.

It was a triumph of courage forced by the press. No other country in history would ever have permitted such public access to the possible *experience* of failure in a major governmental endeavor. It is hard to know which is more remarkable—the openness of a society in which live coverage of manned space flight was made possible or the ignorant ingratitude that took such openness for granted.

2

But the conventions are still the biggest, the most costly and maybe the most important effort television makes. The battle for delegates between Senator Taft and General Eisenhower at the 1952 convention was the first incident in American history that really large numbers of Americans had watched as it happened, seeing the delegations polled on the key votes on the report of the credentials committee. Headline stuff was in the home. Bill Leonard, now waxing stout as a desk-bound vice president of CBS News, whose *Eye on New York* was for years the most admired local show in the country, remembers standing outside William Blair's house in Chicago and watching Adlai Stevenson, the newly nominated Democratic candidate, come onto the balcony to accept the cheers of his friends. "I said into the mike," Leonard recalls, " 'These are the pictures you'll be seeing in your newspaper tomorrow.' That was a *tremendous* thing." And from the networks' point of view, to descend rapidly from the sublime, the early conventions were commercially valuable: Betty Furness became a national figure, and Westinghouse became a much more significant factor in the home-appliance business.

The coverage of the 1972 conventions will cost the three networks little if any less than $22 million, of which little if any more than $7 million will come back in payments from advertisers. In 1968, when about the same amount of money was spent but more could be bought for it, there were 1,700 people working on network payrolls to produce the four evenings of program. CBS alone deployed

111 television cameras in 1968—60 at the Republican Convention in
Miami Beach, 51 at the Democratic Convention in Chicago, which
was harder to cover because a telephone installers' strike had
prevented the laying of the cable the networks wanted to have
available to them outside the convention hall itself. For 1972, CBS
will use only 50 cameras—27 for the Democrats in Miami Beach,
23 for the Republicans in San Diego.

The equipment employed starts with the trucks that are used
for sporting events, and many of the technical people are those a
visitor will find at the ball parks in the fall. (ABC has informed the
Republicans that it will be unable to field its technical first team
in San Diego, because the Republican Convention comes only a
week before the Olympics, and the senior technicians are needed in
Munich.) And, indeed, covering a convention has some of the
feeling of covering a football game, with a number of cameras to be
pointed at this or that, and the output of one at a time selected for
transmission. Wussler of CBS—a man with thinning brown hair
above a round face, white shirt striped with red flowers, pink
checked suit, very stylish—wouldn't know about football games (he
has been with the CBS special events unit since emerging from a job
in the stockroom in the 1950s); to him, a convention is basically just
another special events show. "I produce space, conventions, pri-
maries, election nights, assassinations," he says; "there are great
similarities: they're all live things, with multiple program sources,
a need to share the quarterbacking with an anchor man, and an
order from the sales department to get so-and-so many minutes of
commercial on the line every hour."

Technically, the convention is staged not only for television, but
by television. For each convention, one of the three networks (aided
by the Mutual Broadcasting System for audio services) runs "the
pool," a basic coverage of what goes on in the hall, especially on
the podium. The choice is made by lot; for 1972, ABC drew No. 1
for the Democratic Convention; NBC, for the Republican Convention. The pool designs and builds the convention podium and a
platform about thirty feet in front of the podium where television,
film and still cameras will be mounted. The pool determines the

total power requirements the local electrical company will have to supply to the convention hall; it designs and installs lighting for the podium and (subject to argument) for the rest of the hall; it lays out and installs a new public-address system for the building, all the mikes that serve the podium and the delegations, and the control unit that will enable the party leadership to decide which mikes should and should not be live at each moment (nobody without a live mike can possibly make himself heard in a convention). The control unit is in fact manned by a network engineer, who takes his orders from the party chairman. In 1972, for the first time, the pool rather than party officials will actually count the votes on all roll calls.

To build all this equipment for a Democratic Convention starting July 10, the pool manager at ABC demanded access to the Miami Beach auditorium on May 25 (for work on changes in the lighting), and exclusive occupancy of the hall from June 1. Pool engineers also vet the air-conditioning systems of the halls, and for 1972 this problem was worrisome. "San Diego," says Walter Pfister, Wussler's opposite number in special events at ABC, "has a good enough chiller, but insufficient circulation; the way things look, it's going to be the biggest smoke-filled room in history." TV cameras have trouble getting clear pictures through a haze of tobacco smoke.

In addition to two cameras on the platform pointed at the podium, the pool for the Democratic Convention has a camera on the podium pointed out at the delegates and audience, plus two at the sides of the hall, in the gallery and level with the rostrum, to catch what goes on beside the seats of the mighty. The pool director treats the input from these five cameras as full television coverage of the convention, choosing one at a time and blending in a sound track of what is publicly said at the hall. This pool feed will be, indeed, the coverage that goes abroad through the facilities of the European Broadcast Union, for which the pool makes all arrangements (including accommodations for ninety-five television journalists of different nationalities); each foreign reporter finds a way to add his own voice to the pool picture. And it will be the central element in the coverage by the three American networks, because, after all, the most important things at the convention are the official things that happen on the podium and in the miked statements from the delegations. At the least, each network will continuously tape the feed from the pool while broadcasting something else.

But the networks have a lot of something else. George Murray, who produces conventions for NBC (subject to some direction from NBC News president Reuven Frank, who still sits in the control booth as he did while just a news producer), says that plans even in the austerity year of 1972 call for sixteen cameras in and around each convention hall, with three producers in three separate control rooms to handle all the images. Murray and Frank are the producers in "air control," making the final choice of what gets broadcast, but they see only what has been passed on as plausible by the producers for "hall control" (coming from the cameras in the building) and "outside control" (coming from cameras outside the building, the hotel suite headquarters of the candidates, etc.). NBC used to build a separate studio complex beside the convention hall, placing its air control between the two subsidiary control rooms, so producers could talk back and forth. In 1972, with money tight, the network will take two mobile control units normally used for sports, knock out the end walls, and build a small unit for air control between them. There is also a glassed-in VIP booth behind air control, where NBC and RCA *grandi ufficiali* can entertain the great men of politics, business and labor with a backstage look at convention coverage.

Unlike the football game, where directors can delimit in advance the possible activity their cameras must cover, the convention can be managed only if a stream of information flows to the producers to tell them where the cameras should be pointed or carried. Correspondents wander around the floor and in the halls, a button in the ear keeping them up to date by giving them the sound track of the television transmission currently going over the air from their network. When a producer wants to speak to a correspondent, he has an engineer flick a switch that interrupts this flow of program material to that individual (hence the name of the system: IFB: "interrupted feedback"), and puts the producer's voice in the button. Meanwhile, the correspondent carries a mike he can use to communicate with the control room: "I have Kennedy's valet up here in the balcony. . . . I'm with the Wisconsin delegation and they're raising hell. . . . I have Mayor Daley out here by the south gate. . . ." He speaks to the producer responsible for him—in hall control or outside control—who turns around and asks Murray or Frank, "Do we want . . . ?" If something more important is currently being broadcast, the correspondent and his cameraman may be told to do a piece for storage in one of the ten tape machines NBC has at the convention, and possible later use. But the bias is always toward the live ("When we're live, everything tingles," says Gordon Manning of CBS); and Murray is more likely to ask the correspondent to keep Mayor Daley there for five minutes, if he can. . . .

The heart of the operation at NBC and CBS is a booth each builds high over the floor (steelwork and wood lathe and acoustic tiling: $75,000 to $100,000). Here the anchor man or men try to look both at the convention floor and at the monitor screen that tells them what's going out to the public. There are two cameras in the booth, permitting interviews to originate there (there are also two cameras in a studio near the control room, permitting floor correspondents to invite people in for less hectic interviews than a convention aisle or a hallway permits). One of the switches in audio control permits the anchor man to receive through the button in *his* ear whatever of importance a floor correspondent may have to say, and the anchor man can speak through an intercom to anybody downstairs. His mike is not always live to the public; he has to push a switch and "request air" whenever he wants to say anything for broadcast. Normally, the request is obeyed automatically by the audio man, but sometimes audio control may query air control; "and sometimes," says Murray, "you may not want two minutes of Huntley, so you'll say no. Then you call him and say, 'Chet, you can't talk now,' and then ask, 'What did you want to say?' "

Meanwhile, cameramen and correspondents are staked out in the hotels near the headquarters of the more significant candidates, picking up interviews with denizens and visitors; and these, too, feed into "outside control." The maelstrom of images in the control booths is made more complicated by the fact that there is normally some action on the podium, which is what the party would like the networks to be carrying. (In 1972 both parties plan lots of films, to regale delegates and the folks back home.) "I try to hold the full nominating speech for every candidate," Frank says, "but then you get the five seconding speeches with the full Warner Brothers cast— one black, one Jew, one city, one farm, one woman—and I'll cut away to other things. I remember Eisenhower himself bitched about that in '56—they'd gone to great trouble to get a real live black, except that you didn't say 'black' in those days, and we never showed him."

"People ask me," Wussler says, "how can I format anything as complicated as a convention, how can I plan ahead? But I never plan more than five or six minutes in advance. They pay me for my judgment, and for my reaction time."

ABC works in a very different way. Lacking the station clearances the senior networks routinely receive, ABC was never able to sell convention coverage at anything like the same price-per-minute, and lost more money while offering a lesser service. In 1968 the network dropped the traditional "gavel-to-gavel" coverage of the conventions, and went to a schedule that kept ordinary summer programming on the air until 9:30 EST, to be followed by an hour and a half of combined summary-of-the-day-so-far and live coverage of the action at the moment. "It doesn't work," says Wussler. "You can't do a summary of a live event that's going on while you do your summary." Wally Pfister of ABC disagrees violently: "We went to ninety minutes because we had financial trouble," he says, "but when you run a tighter operation, you do things better. The guys at the other networks now wish they could do what we do." Elmer Lower, who covered conventions for both the other networks before becoming president of ABC News, says that the worst problem of his format is just looking at all the film that gets shot and processed. "In 1968," he says sadly, "we had some real good stuff nobody got around to seeing until after the convention was over."

Making a ninety-minute show out of a night's convention events requires a collection of unanticipated choices. In 1968, for example,

ABC cut the Republican keynote speech down to fifteen minutes, because it occurred before the network went on the air and that seemed all it was worth; but the network gave the Democrats all twenty-eight minutes of *their* keynote speech, because it occurred when ABC was live from Chicago. Both parties were unhappy: the Republicans because the importance of their keynote speech was downgraded, the Democrats because they felt they had been made to look windy next to the Republicans.

Unlike the other anchor teams, Smith and Reasoner on ABC will work from a studio in the bowels of the hall, and the engineers will use a chromokey technique to show them in silhouette against a background of the activity in the hall, taken simultaneously from another camera—much as Howard Cosell appears against the background of the field in the pregame episode of *Monday Night Football*. "The other networks could do that, too," Pfister says, "but when you spend all that money to build a booth actually in the hall you trap yourself, you want to show the anchor man looking out his window through field glasses. Actually, because our anchor men face all the monitor screens, they really see more of what's happening in the hall than anybody can see from a booth."

ABC in 1968 found it impossible to do everything the producers considered necessary before the 11 o'clock (EST) deadline, when the affiliates wanted their air back for their profitable local news programs; on the average, the network didn't sign off until 11:45. And the heavily "formated" ABC show, including the fascist v. fag nastiness between Gore Vidal and William F. Buckley, Jr., made the conventions into a rather different circus from the one the parties were staging. But the audience to the three networks in the 9:30–11 period in 1968 seems to have been larger than it was in 1964, and ABC takes credit for drawing a bigger crowd to the conventions through the better lead-in of early-evening entertainment.

"You can make only a fair case for complete coverage of these conventions on a hard-news basis," says Bill Leonard of CBS. "That three networks should spend a whole week to produce that limited information about the processes of a democracy . . . well, it's a little much. But the fact is that this was the first thing, back in '52, and it was a kind of miracle we could do it. And it's still a good time to put your whole organization in action, under pressure, in tension. It's good for morale. It's a great place to discover how good your people are. The lines cross all the pyramidal things of a great organization, you find a young reporter or a young editor you might not otherwise know about. It's a hell of a price to pay for that, but . . ."

3

It is as near to certain as anything can be in the political world that there will be an absolute explosion of complaint after the televising of the Democratic National Convention. For the reform movement in the Democratic Party has proceeded from the plausible political science argument that the way to take care of the people who demonstrated in the streets in Chicago in 1968 is by coopting them, giving them a voice in the councils of the party. (Plausible but not wholly convincing, by the way: without defending the boss-run convention, it might be noted that there is a strong case to be made for the proposition that a Presidential nominee should be chosen primarily by people who know the candidates, rather than by people to whom they are merely names, faces and positions on issues.) In any event, the reformers' attitude is pretelevision, and the result of the reforms in delegate selection has been a guarantee that Miami Beach will see a zoo of creatures from various university towns and passionate movements, ranting and cussing on the floor of the convention itself.

An election chooses a government, not a political party; and without impossible self-censorship by the network news departments, the Democrats who make the most splash on camera at Miami Beach will be boys and girls nobody in his right mind would wish to see entrusted with the tasks of governance. As one Democratic leader said, "We lost in 1968 because the American people saw the Chicago police clubbing demonstrators outside the hall; so in 1972 we're going to show the American people our sergeants-at-arms clubbing our own delegates in the aisles." Some leading Democrats did not seem at all conscious that there was a problem, but party chairman Lawrence O'Brien and the Arrangements Committee took a shot at keeping the networks off the floor entirely, barring not only the hand-held "creepy-peepy" cameras and the floodlights, but also all broadcasting correspondents with microphones.

CBS didn't particularly care about permission to use hand cameras. "I can get better pictures from the baskets," Wussler says, referring to the small platforms hung from the balconies. "If I've got a guy on the floor, the signal is wireless, which means it's unreliable." All the producers say that there isn't that much interesting on the floor—a correspondent on camera is likely to offer a "situationer" in which he answers questions rather than an interview in which he asks them, and for this purpose a man can move near a camera in a basket. Still, Pfister wanted access: "Without the camera on the spot, you don't get the involvement; you can get intimacy with a lens from a remote camera, but you can't get excitement. And that business of the photofloods drawing a crowd—that's backwards. It isn't that our lights attract, it's that we go where things are happening."

On the general principle that anywhere a newspaperman can go a television camera should also be permitted, the three networks fought the issue through the Arrangements Committee. "If they're going to keep all the press off the floor," Bill Leonard says, "that's all right—it's their convention. But once they let guys with pencils in there, they have to let us in there, too." Still, the network producers are fully conscious of the dangers. NBC's Frank has suggested unscrewing from the cameras the red bulbs that indicate the producer is taking the picture from that camera, to lessen the danger that media freaks will know where to posture. Though both NBC and CBS expect to be live nearly all the time, and will not go to a "tape-delay" procedure to guarantee control over what goes on the air, the producers feel they can react fast enough to avoid scandal. "If all the 27 percent of the delegates who must be under thirty by the rules of the McGovern Committee start doing something disruptive," Wussler says, "that's news, and I'm going to cover it. But if there are only, say, fifty delegates being disruptive, that's theatre and I won't show it." Fifty delegates and one stink bomb are news, though.

The dangers should be seen in concrete rather than general illustration. One Gay Liberationist in semidrag, wearing a Democratic delegate's badge and waving a sign reading, say, "MUSKIE IS A MOTHERFUCKER," might convince even a black welfare mother that the country needs four more years of Republican rule. Fights over the credentials of delegates, more or less guaranteed by the complexity of the McGovern Rules, always bring out the worst in conventions, and students must be a special concern, because the gut issue of amnesty has been raised. The October 7, 1971, issue of the Columbia *Spectator* ran a feature article on student voting, beginning with the words, " 'Yeah, I'm gonna vote, 'cause I think Nixon sucks,' stated Peter Wise '74. . . ." Such a young man might easily be a delegate.

In the end, the Arrangements Committee exacted one concession from the networks: they would not be permitted to take their own floodlights into the hall, and would have to rely on the ambient light of the floor. Should there be bad trouble, the Convention chairman would be able to dim the lights (as they must be dimmed for films), and black out the picture on the home screen. But such an action would be a political gesture of major significance all by itself, and a great danger to the Democrats.

Meanwhile, the remnants of the Yippie colonies that threatened the Democratic Convention in Chicago in 1968 have announced that they expect to demonstrate *outside* the Republican Convention in San Diego in 1972. Given the prospects of television in every home showing the Democrats struggling to maintain order on the convention floor and the Republicans calmly ignoring an outside hubbub of nuts, President Nixon must feel he can safely order new doormats or whatever else it is the White House needs these days, with his initials suitably embroidered on them.

4

It is in this worst-case focus that one should look at the significance of what the FCC has come to call its Fairness Doctrine. For despite the handful of foolish virgins who follow St. Ursula to the abattoirs of New York, the Democratic Party really has no more identification with the counterculture than the Republican Party has; yet by the time September 1972 arrives television reportage may have labeled the Democrats as hopelessly unfit to govern. There is a class of events so *talkable* that anyone who sees them soon convinces the people who did not see them that really they were witnesses, too. The creation of such events, usually by disrupting other people's events, is called "handling the media." It is no good to say, as everyone at the news divisions does, that the media being handled are just "the messengers bearing bad news"; they are more like detectives uncovering false evidence that has been planted on the person of an innocent man. Like the detective, they can't be blamed: they're just doing their duty. Television is supposed to be fair—especially in an electoral context. What are we talking about?

The Fairness Doctrine is a queer bird in the American legal aviary. It is not in the Communications Act and it appears in a later amendment only by indirection. What has sustained the Commission more than anything else is public confusion between fairness in general and fairness among candidates running for office. Here the law clearly does empower the FCC to move: Section 315 requires a station that gives time to one candidate to give "equal time" to every other candidate for the same office. Thus "debates" between major-party Presidential candidates have been possible only in 1960, when Congress suspended Section 315 for a year to make them possible. (Normally, there are a dozen or so unknown but duly qualified Presidential candidates, all of whom would otherwise have to be given "equal time" under Section 315.) Free "equal time," however, applies only when the candidate who has already been on the air was given his time for free: a station can *sell* time to anyone without incurring obligations to do anything but sell time to his rivals. As broadcasting became the quickest and best way to become known to the body of voters, the budgets for political advertising rose into the tens of millions of dollars.

But elections are much less than all of political life. Herbert Hoover went on the radio ninety-five times during his Presidency without provoking great complaint from the Democrats, apparently because he was as ineffectual over the air as he was in the White House. Franklin Roosevelt was something else again. "During his first ten months in office," according to Edward W. Chester of the University of Texas, "FDR spoke over the radio 20 times, Mrs. Roosevelt 17 times, and Roosevelt's cabinet 107 times." Congress became highly upset; in 1934 Senator Arthur Vandenberg publicly complained about Roosevelt's domination of the air, and that year, again according to Chester, the two radio networks gave free time to Senators and Representatives on 350 occasions—almost every day. This was done partly as a courtesy and partly in fear (the Communications Act was in the works in 1934), and without any intervention from a government agency.

To some extent, these brief political broadcasts substituted for news, which the networks of those days didn't have: by contract with the wire services (which were protecting their newspaper constituency), they had agreed in 1934 to broadcast no more than ten minutes of news bulletins a day—and those ten only after the news involved had appeared in the papers. Barred from hard news, sponsors offered news commentators, some of whom took strong positions on controversial issues. Here there is some record of FCC intervention. Chester writes of Boake Carter, who "charged that the Roosevelt Administration was attempting to generate a pro-war atmosphere after the Japanese had sunk the American gunboat *Panay*. . . . The Chairman of the Federal Communications Commission informed the Washington managers of the various broadcasting companies that news programs should be impartial; Carter's program clearly did not meet this standard." General Foods let Carter's option drop, and presently he was off the air. But the FCC did not at that time believe it had the power to make stations carry replies —all it could do was to send out little warning notes that the station was asking for trouble in connection with its next application for a license renewal.

The great worry of the later 1930s was Father Charles E. Coughlin, who had a bit of a weakness for, among others, Adolf Hitler. There was a good deal of scurrilous anti-Semitism on Father Coughlin's programs (for which he bought the time himself, soliciting money from his followers on the air to pay for the broadcasting and much else). In 1939, in large part to curb Coughlin, the National Association of Broadcasters amended its rules to require member stations not to sell time for the presentation of controversial views. Meanwhile, Roosevelt had noted with increasing annoyance the demand for radio licenses by newspaper proprietors. Roosevelt had 80 percent of the nation's press against him in election years, and he did not wish to face the same negative reactions on radio. In 1941, apparently at the President's desire, the FCC ordered a Boston station not to editorialize: "A truly free radio cannot be used to advocate the causes of the licensee. It cannot be used to support the candidates of his friends. It cannot be devoted to the support of principles he happens to regard most favorably. In brief, the broadcaster cannot be an advocate."

This "Mayflower Doctrine" survived eight years, the last two of them in that familiar FCC limbo where parties at issue wait for a decision. Then, in 1949, the Commission reversed itself (or seemed to do so: in fact, only two of seven Commissioners voted for the change in rules as promulgated; two abstained, two opposed, and one leaned far enough toward the change to be counted on that side). "The Commission is not persuaded," said the "Report on Editorializing by Broadcast Licensees," a triumph of atrocious prose, "that a station's willingness to stand up and be counted on these particular issues upon which the licensee has a definite position may not actually be helpful in providing and maintaining a climate of fairness and equal opportunity for the expression of contrary views." Amen, brother, amen. If a station did editorialize, of course, it would be obliged to afford proponents of different positions some time on the air to reply. And the Fairness Doctrine was born.

First crack out of the box came a complaint against New York radio station WLIB for broadcasting editorials in favor of a Fair Employment Act. "The broadcast by the station of a relatively large number of programs relating to this matter over a period of three days," the Commission opined, "indicates an awareness of its importance and raises the assumption that at least one of the purposes of the broadcasts was to influence public opinion." Under these circumstances, the Commission ruled, WLIB should have given time to those who were in favor of discrimination in employment practices, to make their reply.

This was still "should have" decision-making: the FCC did not claim the power to order a station to do anything. Complaints against fairness, and reports of what had been done about the complaints, would go into the file to be considered when renewal applications were received: "regulation by lifted eyebrow." Not until 1963 did the Commission raise the other eyebrow, and assume the power to require a station to give time, pronto, to persons aggrieved by something a station had broadcast. At first, this power was applied only to enforce the "personal attack" section of the doctrine, to ensure a right of quick reply by an individual who had been personally maligned.

For reasons never entirely clear to anyone, the broadcasting industry decided to fight the Commission's assumption of these powers in a case where a station had clearly misbehaved. The attack involved had been made over radio station WGCB in the town of Red Lion, Pennsylvania (pop.: 5,594); the aggressor was the right-wing revivalist preacher Billy James Hargis; the victim was Fred J. Cook, a New York reporter who had written an unflattering book about Barry Goldwater. Rev. Hargis had told the listeners to WGCB that Cook had been fired from the New York *World-Telegram* for fabricating charges against city officials and had then worked for a "Communist-dominated" magazine (*The Nation*). Cook requested time to reply and was refused. The FCC ordered the Red Lion Broadcasting Company, owner of the station, to give Cook time; the station went to court to appeal the order; and the National Association of Broadcasters supported the station, "as though," says NBC general counsel Corydon B. Dunham, rather bitterly, "this stuff was *broadcasting*." The case wended its way slowly, as cases will, through the federal court system, and in June 1969 the Supreme Court unanimously upheld the Commission. Indeed, the opinion by Justice Byron White went some distance beyond anything the Commission had ever said.

Justice White rested his argument on the fact that anybody,

with a little help from his friends, can start a newspaper—but only someone with a government-issued license can start a broadcasting station. In these circumstances, he wrote, "It is idle to posit an unabridgeable First Amendment right to broadcast comparable to the right of every individual to speak, write or publish." There is thus an obligation on broadcasters to convey to the public all varieties of opinion: "The licensee . . . has no constitutional right to monopolize a radio frequency to the exclusion of his fellow citizens." Not only *may* the FCC require broadcasters to present all sides of controversial issues; it *must* do so: "The right of the public to receive suitable access to social, political, esthetic, moral, and other ideas and experiences . . . may not constitutionally be abridged either by Congress or by the FCC."

Exactly what all this may mean was and is rather mysterious. Richard Jencks, then president of the CBS Broadcast Group, said that a few weeks after *Red Lion* was decided he received a letter from a man in Pennsylvania who announced himself in disagreement with some of what Walter Cronkite had been saying on the *Evening News*, and politely requested CBS to honor his constitutional rights and give him time to speak his piece on the Cronkite show. The FCC, which would have to apply the *Red Lion* principle to practical affairs, waited until May 1970, almost a year after the decision, to announce an inquiry into possible revisions of the Fairness Doctrine; and in early 1972 the inquiry is still in progress.

Fairness has been expanded mostly in an unexpected direction, to give a right of reply to product advertising that raises what might be considered a controversial issue—first with regard to cigarettes, by the FCC itself, and then with regard to "cleaner" gasoline for automobiles (after the FCC had refused to act) by the Court of Appeals for the District of Columbia. The same court has also ruled that stations may be compelled to accept paid advertising for ideological positions, even though each such ad would trigger a number of free right-of-reply minutes; but this decision has been taken by the FCC as a sort of advance statement by an interested party to be fed into the forthcoming hearings on revisions of the doctrine.

The worst moments the FCC has had in trying to apply the Fairness Doctrine grew out of the national agony following the Cambodian invasion in 1970, which President Nixon announced and sought to justify in a prime-time speech. Democrats seeking to buy network time to reply were able to do so only on NBC, the other two networks still adhering to the 1939 National Association of Broadcasters ban on such sales. ABC, however, gave time to a previously announced speech by Democratic National Chairman Lawrence O'Brien; and CBS announced that to balance the President's ease of access to air time it would start a series of programs, one every three months or so, in which the party not occupying the White House would get time to present its views on the State of the Nation. But O'Brien used his CBS half-hour for so aggressively partisan a presentation that the Republican National Committee demanded time to reply, and the FCC ordered CBS to accede to the demand, at which point CBS withdrew its long-range proposal. (The D.C. Court of Appeals in November 1971 overruled this FCC decision, but CBS did not reinstate the proposal.) Meanwhile, the Commission had ruled that because of President Nixon's extraordinary use of television to promote his Southeast Asian policies— no fewer than five prime-time speeches in less than a year—some reply time would have to be given for a Democratic statement on this specific issue, and the nation heard an exposition of opposing views from Senator Mike Mansfield.

Neither Mansfield nor O'Brien—nor the Senators who bought time on NBC—got much audience by Nixon standards (indeed, neither did the President himself a few months later, when he tried the experiment of appearing on only one network rather than preempting entertainment programming on all three). But even if audience size were the same, the impact of a Presidential announcement is different in kind, not just degree, from the impact of an appearance by an opposing political figure. For a Presidential appearance is—or can be made to seem—an event; the observer is in some sense watching history.

The problem is extraordinarily difficult. Kennedy alerting the country to the presence of Russian missiles in Cuba, Johnson announcing the Gulf of Tonkin incident and the first bombing of North Vietnam, Nixon asserting the need to invade Cambodia or to freeze the price level—such spectacles are important (and exciting) in a way no statement by an opposition Senator could ever be.* Even a Presidential press conference is, or can be, an event—as Charles de Gaulle discovered and John F. Kennedy perfected. (President Eisenhower's press conferences had been televised only some hours after they occurred: the White House reserved the right to look at and edit the film before releasing it. Roosevelt, of course, never permitted direct quotation in print of what he had said at a press conference, let alone the broadcasting of it.) Opposition politicians can use television effectively—the most imaginative such use being that of the Iraqui revolutionaries who in 1962 dragged the mutilated body of Premier Abdul Karim Kassem into the television studios and before the cameras to show the country that the revolution had indeed occurred. But usually the government in power is going to have the edge.

And, incidentally, should have the edge, because the government and nobody else is going to be held responsible for what happens. Everyone interested in the recent history of the American economy can cite chapter and verse of Lyndon Johnson's deliberate under-estimate of the costs of the Vietnam war, and his failure to go to Congress to seek more tax revenues to pay the bills. But relatively few remember that Congress delayed the tax surcharge for a year after Johnson requested it—and only a handful of people with freakish memories can recall that among the reasons for that delay was the adamant opposition of the *New York Times*, which on economic matters swings some weight in Washington, and which

---

* This observer happened to see Kennedy's Cuban missile crisis speech in the bar of the old Hotel Willard in Washington; at the next table were four members of Congress who had not known anything of what they were about to hear, and they got themselves sodden drunk within half an hour of the close of the speech. The next morning I had to go to a meeting on peaceful matters at the Executive Office Building, and Jerome Wiesner, the President's Science Adviser, came to the meeting. I said, "Jesus, Jerry, am I glad to see you." He said, "Why?" I said, "I thought you'd be busy doing things like calculating how to decontaminate cities where everything has suddenly got radioactive." Wiesner said, "Nah. Nothing's going to happen."

later felt entirely free to denounce Johnson's irresponsibility in delaying tax increases. If television tends to help a government rather than its critics—because only the government commands the resources to make events people wish to see—the bias is proper. The meaning of representative government is that the public periodically judges the results of policy but only rarely, in crisis, seeks to judge the policy itself. Neither the FCC nor the courts nor their Fairness Doctrine can change the meaning of political institutions.

Come election time, moreover, there is some reason to believe that the opposition gets revenge. The English reporter John Whale wrote in 1969, "So long after the beginning of the television age, no politician has yet been elected to high office chiefly because of television, in America or anywhere else," and that's still true. But the marginal efficiency of television advertising for a new candidate, as for a new product, greatly exceeds anything that can be done by advertising for a known political figure. De Gaulle himself was forced into a run-off election for the Presidency of France by a man almost nobody had heard of until French television dutifully gave him time as a recognized candidate. Though Harold Wilson was universally regarded as a more adroit performer than Edward Heath in British television studios, Heath as a much less familiar face seems to have benefited more from television in the 1970 election. In the United States, incumbent Congressmen (who cannot be effectively challenged on television because television coverage areas overlap too many districts) are much more likely to win re-election than incumbent Senators and governors (whose rivals can use the tube).

5

Robert MacNeil, now part of a public-affairs project for the Public Broadcasting Service, once wrote nervously about "the power of a medium such as television to make *appearance* seem to be *reality*." The example given to illustrate that statement is the plausible television appearance of Senator Clair Engle of California, who was dying of cancer when he announced his intention to stand for re-election, on camera, "in a carefully edited film lasting only forty-two seconds." MacNeil also cites general disappointment with the performance of Endicott Peabody as Governor of Massachusetts, and quotes Peabody's television adviser Joseph Napolitan: "Peabody was a big, handsome guy. Immediately after the election he started holding press conferences and people said he seemed entirely different than in the campaign. The trouble was he was getting tough questions and giving bumbling answers. Not the clean, crisp image of the prepared spots. Because maybe you waited eighteen times in filming the spots to get just the right thing . . . They had elected him on one basis and he seemed entirely different."

But television did not invent illusion. Roosevelt was very nearly as sick as Engle when he was re-elected in 1944, and McKinley in his "front-porch" campaign of 1896 successfully created an "image" for himself that television could not have improved upon. The chance to create a disturbance in millions of living rooms at once came with television and is a new political phenomenon the system must somehow digest, but "image-building" has been the focus of politics since the Oracle settled at Delphi. Television hasn't changed that at all.

Four important, not connected points should be made about political advertising on television before we can proceed to more edifying matters:

1. The one-minute spot came about not because of the Machiavellian manipulations of Madison Avenue but because too many voters resented the pre-emption of their favorite programs for political broadcasts. Even the loss of the last five minutes of a show was more sacrifice than many Americans were willing to make to receive political messages in election season. Paul Klein's argument that television decision-makers opt for the "Least Objectionable Program" is more true in politics than in entertainment. One minute was by far the least objectionable kind of politics for many Americans.

There is nothing to be said in favor of the one-minute spot as a means of communication in politics, but there is also no justification for saying that it has degraded American politics, or "oversimplified the issues" to a degree other ways of campaigning did not. Nothing in recent American politics is any more simple-minded than Honest Abe the Rail Splitter or The Happy Warrior or The Hero of San Juan Hill. At the almost dirgelike tempo used for the song at the time, it takes just under a minute to sing:

> Oh, what has caused this great commotion—motion—motion—
> All the Country through?
> It is the ball that's rolling on
> For Tippecanoe and Tyler, too
> Yes, Tippecanoe and Tyler, too,
> And with them we'll beat little Van—Van—Van—Van—
> Oh, he's a used-up man.
> Yes, with them we'll beat little Van.
>
> Oh, let them talk about hard cider—cider—cider—
> And log cabins, too.
> It will only help to speed the ball
> For Tippecanoe and Tyler, too
> Yes, Tippecanoe and Tyler, too.
> And with them we'll beat little Van—Van—Van—Van—
> Oh, he's a used-up man.
> Yes, with them we'll beat little Van

The few seconds remaining would be just enough to say, "Presented in your interest by the Independent Whig Committee for the Election of General William Henry Harrison and John Tyler."

2. The main purpose of political commercials on television—as of all campaigning—is not to convert the heathen but to strengthen the convictions of those who are leaning your way, and get them to the polls. What is still the most sophisticated analysis of a forthcoming election ever done for a candidate was performed in 1960 for John F. Kennedy by Simulmatics, a New York social-science-cum-computers firm owned and operated by half a dozen of the Ivy League's brightest lights in political science, sociology and psychology. The conclusion in August was that Kennedy had already lost the votes he would lose on the Catholic issue, but had not yet gained the votes he could gain by exploiting his religion. The strategy based on this analysis produced an expenditure of about $2 million to show in cities with heavy Catholic populations a one-

minute excerpt from Kennedy's confrontation with the Protestant ministers in Houston, in which he assured them that his being a Catholic would not affect his decision-making as President. It seems beyond question that these selectively placed spots were important in Kennedy's success. But bulk purchasing of advertising has been much less significant. In 1968, according to the political scientist Vic Fingerhut (who worked on the Humphrey staff), "there was a huge eight-million-vote shift to Humphrey—simultaneously with the most lopsided GOP spending advantage on record." Because there aren't enough Republicans in the country to win big, the scheming and the immense expenditure portrayed in Joe McGinnis' funny book *The Selling of the President* seem in fact to have *reduced* Nixon's margin of victory over Humphrey.

3. Even at minimal figures for advertising, however, television has made campaigning so expensive ($60 million in the non-Presidential year of 1970) that the parties must virtually sell themselves to the big contributors to finance a national election. The industry associations, farmers' alliances and unions have become much too influential. The damage done is less severe than reformers like Ralph Nader and John Gardner like to say, because up to a point the "public interest" is in fact the sum-and-difference of group interests; but an important part of political life lies beyond that point. And beyond that point in contemporary American politics one can say, as Gertrude Stein said of Los Angeles, "There isn't any *there* there." New laws restricting total expenses on campaigns, to take effect in 1972, may or may not diminish the influence of the big contributors.

4. Since 1968 an odd and ultimately unstable asymmetry has grown up between the Fairness Doctrine applied to controversial subject matter and the established rules of political advertising on television. Historically, except at a few stations (the most important being WGN-TV in Chicago, which will not sell time for political announcements in less than five-minute pieces), candidates have been able to buy from broadcasters whatever they can afford to buy. Those who can afford less get less; those who can afford none get none. Nobody can defend a system which awards free time

to reply to spokesmen for "different sides of issues" (judges and bureaucrats alike tend to picture "issues" as pieces of paper with just two sides), but requires every candidate for office to pay for whatever he gets. The hope that the multiple channels of cable systems will resolve this problem is more than usually childish, because the aim of this business is not to set up a shop where your partisans can find you but to find people and remind them that you're their guy. One thing is sure: the question can't be intelligently answered in the courts.

And for problems like the television coverage of the 1972 Democratic Convention, the only answer may be the inherited good sense of Americans—fool me once, shame on thee; fool me twice, shame on me. But the networks should be prepared to face the fact that television, not the silly kids, will be blamed for trying to fool people—and perhaps not unjustly, if the news divisions fail to develop policies to guide their producers and anchor men in distinguishing between theatre and news.

CHAPTER 10

Five Thousand Words
with Pictures

This is what TV is for.
—Letter from a viewer to
EDWARD R. MURROW, 1952

1

In December 1950, with American forces in pell-mell retreat from
the Chinese border of Korea, Edward R. Murrow and CBS Radio
launched an hour-long weekly program called *Hear It Now*, presenting a long string of prerecorded statements and comments by
participants in the events of that week, linked by a live commentary
from Murrow. The next summer, AT&T opened the first coaxial
cable between the two coasts, making possible nationwide simultaneous transmission of television pictures. By November 1951
Murrow and his producer Fred Friendly had moved to television
("Good evening," Murrow said to start the first program. "This is an
old team trying to learn a new trade"; and on the two monitor
screens beside him appeared pictures of the Atlantic and Pacific
oceans). *See It Now*, at half an hour, was to run regularly in prime
time through the spring of 1955, then occasionally (mostly not in
prime time) through the spring of 1959.

The program rested on the personality of Murrow and the
budget for film, which was considerably higher than any continuing
show (other than the nightly hard news of recent years) has ever

enjoyed. Ten feet of film were shot routinely for every foot used on *See It Now;* some weeks, the shooting ratio was twenty to one. The show opened with Murrow on camera, usually smoking a cigarette, telling the audience what they were going to see, and closed with a statement by Murrow about what had just been shown. Never a newspaperman, Murrow brought to televised journalism the instincts of a great secondary-school teacher (his early career had been, in fact, at the Institute for International Education). Explaining the obvious gracefully for the benefit of slower learners—the audience to a documentary is and has always been skewed to a somewhat *lower* educational level than most television, for the reasons of data-transmission speed noted in Chapter 2—he could also deliver thought-provoking goodies for the brighter members of the class. And always there was the decency of the American countryman—born in North Carolina, raised in the state of Washington, looking to the practical results of what gets done and especially to the impact on the people involved. Television news documentaries did not have to develop as they did: nothing in the general intellectual climate of the 1950s would have kept them from the romanticism of the Norman Corwin radio documentaries or the Robert Flaherty film documentaries. That they became straightforward expositions was the doing of Ed Murrow and Fred Friendly.

When CBS offered a week-long festival of its documentaries at Lincoln Center in December 1971, only seven of the Murrow programs were included in the forty-five shown, which is something less than their true proportion of the nearly one thousand CBS documentaries aired in the twenty years since Murrow's first. Most of them have probably worn poorly: they were news shows, often planned only a week or two ahead. Their great moments, of course, came in 1953–54, when Murrow led the fight to free the country from the grip of McCarthyism, presenting reports on the man himself; on a scandal at one of his hearings (when Mrs. Annie Lee Moss, a poor Negro messenger in the Pentagon, was badgered because some professional informant had mentioned somebody named Annie Lee Moss as a participant in a Communist meeting; it was, of course, a different lady); on the attempted wrecking of the career of an Air Force officer because members of his family had leftist affiliations; on the refusal of a civic auditorium in Indianapolis to rent space for a meeting of the American Civil Liberties Union. One of these programs remedied the injustice described: the Air Force Lieutenant was reinstated. The others tended to be convincing to those who already agreed, and public opinion polls after McCarthy's televised reply to Murrow indicated that more people thought McCarthy had come out ahead in the exchange. But the foolish attempt to tar Murrow with the wide brush of "anti-anti-Communist" was not the least damaging of the failures that McCarthy suffered during the year when Dwight Eisenhower decided he really did not have to put up with this man any more.

NBC's main effort in the early years was a series on the naval aspects of World War II, with a Broadway musical score in the background. There were also a number of specifically educational features in popular science and in art; the art films, starting with a splendidly dramatic show on Van Gogh, revealing the values that could be gained by a camera moving slowly over the details of a painting, were often unusually interesting despite musical scores of the grossest vulgarity. But NBC did not seriously try to rival CBS in public-affairs programs until the later 1950s and the presidency of Robert Kintner, a gravel-voiced bull of a man, big head, grizzled crew-cut, no neck at all, the opposite of everybody's image of a broadcasting smoothie, who had come from a newspaper background (he had been Joseph Alsop's partner in column-writing in the years before World War II). Even then, except for the brief period when John Doerfer's FCC made heroes of all the networks, NBC did not establish a regular weekly time slot for public affairs. There were more "specials" on NBC anyway; some of the specials would be nonfiction. It is probably fair to say that at all times the NBC documentaries tended to be more heavily flavored with entertainment values than those of CBS—even the one that got the network in trouble, a step-by-step examination of the efforts of some West Berliners to tunnel under the Wall, had obvious elements of a movie thriller.

Among the most highly anticipated values of television was the possibility of portraits in depth of the great men of the age, but these turned out to work satisfactorily only when the individuals involved were of a fundamentally theatrical temperament—J. Robert Oppenheimer, Eric Hoffer, Igor Stravinsky were much more convincing on screen than Pablo Casals or Dwight Eisenhower. Probably the most unexpected star (a stroke of Fred Friendly's genius) was Walter Lippmann, who was called upon mostly for opinion, but infused all with a personality built on a style of recollection.

The most important documentaries were undoubtedly those that picked up from and extended the Murrow tradition, detailing situations that either were or should have been news. They stood to the nightly news show as the magazine article to the daily paper, and like magazine articles, they tended to work best when the subject was one that could be handled intelligently in five thousand words. Stories that were inherently ambiguous, like the plight of the suburban teen-ager (*Sixteen in Webster Groves*) tended to be the best of all, while shows like *The Great American Novel* (a bittersweet exercise supervised by a producer who had once intended to write same) were able to make a statement of more than passing interest about the materials from which fiction is created. Narrowly focused subject matter was always a help—the best of the "investigative" documentaries was probably the CBS *Biography of a Bookie Joint,* because the story, once found and cornered, was simply there to tell. Among the best of the this-is-how-it-happens documentaries was Drew Associates' *Primary* on ABC, with its close-up photography and miking of Kennedy and Humphrey on the campaign trail in one state. Similar techniques gave strength to Frederick Wiseman's pieces *High School* and *Hospital,* shown on public television.

War, of course, is its own narrow focus, and much that was done on the fighting in Vietnam was admirable for the courage of the men who covered the story and for their art in organizing the material. But both those who thought the war a mistake from the beginning and those who came to that conclusion only much later (there is no third category) would agree that television coverage did little to set the context of the fighting.

Inevitably, the documentaries are at their weakest when context is required: they are magazine articles, not books. Like the muckraking articles of the turn of the century, they can set political machinery in motion, and in the light of history they may acquire magnificence, but they are always drastically incomplete at the time of utterance. And they almost all have that splendid if stupid American attitude that anything that is *wrong* (even old age and dying) is a *scandal* and *something* must be done about it. When the matter turns out to be a little more complicated than that, as it usually does, shouts of betrayal and conspiracy rise all 'round.

Twice, for example—in the CBS *Harvest of Shame* in the early 1960s and in NBC's *Migrant* in 1970—network news departments have looked at the horrors of migrant agricultural labor in America. But the cold conflict of interest between the poor in the cities and the poor on the migrant trail (which means that one cannot really arouse the liberal politicians, whose constituency is urban) was something nobody could possibly have deduced from what was on the screens. Similarly, nobody could have guessed from the two-part CBS report on health care that doctors were disappearing from rural America and city slums in part because of Medicare. (Once the doctor's economic situation is so structured that all he needs to make a living is a supply of elderly patients, he has no reason not to live in the most comfortable and congenial surroundings he can find.) Simplification of a complicated world is a talent of reporting as well as of advocacy. But documentary producers like newsmen want to make their stories both better and more obviously important. And with 50,000 feet of film to cut to 2,500 feet, it's easy for a man with a mission to make unfortunate mistakes.

2

Few episodes in the tangled history of news broadcasting have been so generally discreditable to so high a proportion of the participants as the fuss attendant on the CBS documentary *The Selling of the Pentagon* in spring 1971. The subject—the abuse of funds and discretion in the public-relations programs of the Department of Defense—was moderately daring; and given the Nixon Administration's reaction to Joe McGinnis' book *The Selling of the President,* the title guaranteed negative reactions at 1600 Pennsylvania Ave. But most people, even some in the military, would be willing to agree that pushing weaponry and war is an activity scarcely more defensible than pushing drugs. The key scenes in the documentary, the introduction of children and businessmen to the pleasures of putting little fingers on big triggers, were legitimately horrifying, and there were details here and there in the show that were not part of the fund of common knowledge. Still, said Perry Wolff, the CBS News executive producer whose unit was responsible for the program, "This wasn't supposed to be our fast ball for this season. It was just one show in the continuing *CBS News Hour.*"

In fact, what appeared on the air may have been a slower ball than the first cut that had been prepared for viewing by CBS News senior executives. The problem was in part that CBS itself was involved in some of the activities criticized in *The Selling of the Pentagon.* (Wolff, in fact, had been the producer of the CBS series on *Air Power* back in the 1950s.) One of the sillier bursts of anti-Communism in Defense Department promotion films had been uttered by Walter Cronkite as narrator, and the Cronkite film, though dating back to 1962, was one of the most popular things in the Pentagon catalogue, with more than a thousand showings at Kiwanis affairs and the like in fiscal 1970. Cronkite is a multimillion-dollar property for CBS, and while he was willing to be criticized, there were limits. Wolff screened the show first for his immediate superior, Bill Leonard, then for Salant, executive producer Burton Benjamin and Cronkite, and there followed a snow of memos. The one from Salant included twenty-two "review points," things in the show that ought to be made to conform to policy, among them the introduction to the Cronkite film. The re-editing process stretched out, and the program missed its first air date.

Among the pieces of machinery set in motion when the show did air, on February 23, was the bank of videotape duplicating machines at the Pentagon. CBS estimates that the Defense Department made thirty copies of the show, and distributed them about to members of the military who might be able to find demonstrable fault. Within a week, the Pentagon had a brief attacking the show in the hands of every major newspaper, and in print in the *Air Force Journal*. As an attack on the show, the Pentagon effort was easy to brush off, because there really wasn't anything seriously wrong with the show. But in investigating how CBS News had put the show together, the DoD analysts had come up with several embarrassing errors, two of which should be sufficiently disturbing to trouble people who liked the show and hate the Pentagon.

One was the presentation of excerpts from a speech by Colonel John MacNeil at a military "seminar" sponsored by a local business group in May 1970 in Peoria, a city much benefited by defense contracts. The episode began with a statement by CBS correspondent Roger Mudd: "The Army has a regulation stating, 'Personnel should not speak on the foreign policy implications of U.S. involvement in Vietnam.'" Colonel MacNeil was then shown saying the following six sentences, apparently one right after the other:

Now we're coming to the heart of the problem—Vietnam. Now the Chinese have clearly and repeatedly stated that Thailand is next on their list after Vietnam. If South Vietnam becomes Communist it will be difficult for Laos to exist. The same goes for Cambodia, and the other countries of Southeast Asia. I think if the Communists were to win in South Vietnam, the record in the North—what happened in Tet of '68—makes it clear there would be a bloodbath in store for a lot of the population in the South. The United States is still going to remain an Asian power.

The first of these sentences was from page 55 of Colonel Mac-Neil's text; the second, from page 36; the third and fourth (quotes from Prince Souvanna Phouma of Laos, though the CBS excerpt did not indicate that), from page 48; the fifth, from page 73; the sixth, from page 88. They were made to appear continuous by a standard film-editing technique, in which shots of the speaker are alternated with shots of the listening crowd.

Now, there is simply no question that this sort of thing is illegitimate. There is also not much question that CBS could have taken a continuous paragraph from the speech that would have served its purposes *almost* as well. The minority report of the House Interstate and Foreign Commerce Committee, which investigated this story, cites such a paragraph, and says, "It is quite apparent that the excerpt from Colonel MacNeil's speech did not unfairly impute to him a position he did not take." From a public point of view, producer Peter Davis and/or his film editor stand convicted of improving a good story just a little, which deserves a mild rebuke and an injunction to go and sin no more. Even the most sober and responsible reporters (*mea culpa! O, mea maxima culpa!*) do this sort of thing every once in a while. From Colonel MacNeil's point of view, the situation may be somewhat different, for CBS did show him deliberately and consciously violating a Defense Department directive, while the full text of his speech could be held to leave him at least a loincloth of self-image that he was obeying the rules. He is suing CBS for $6 million, and presumably a court and jury will decide whether or not CBS violated his rights by stripping him of his self-protecting belief.

The other matter was considerably more serious as a question of reporting practice (though quite trivial in terms of the message of the program). Roger Mudd had interviewed at length Daniel Z. Henkin, Assistant Secretary of Defense for Public Affairs. They ran down some of the divisions reporting to his department, ending with the "Directorate of Community Relations," which Henkin explained as a service that arranged meetings, supplied speakers, etc. Mudd then asked, "But aside from your meetings in which you disseminate information, what about your public displays of military equipment at state fairs and shopping centers? What purpose does that serve?"

Henkin replied, "Well, I think it serves the purpose of informing the public about their armed forces. It also has the ancillary benefit, I would hope, of stimulating interest in recruiting as we move or try to move to zero draft calls and increased reliance on volunteers for our armed forces. I think it is very important that the American youth have an opportunity to learn about the armed forces."

On the air, this sequence appeared as follows:

MUDD: What about your public displays of military equipment at state fairs and shopping centers? What purpose does that serve?

HENKIN: Well, I think it serves the purpose of informing the public about their armed forces. I believe the American public has the right to request information about the armed forces; to have speakers come before them, to ask questions and to understand the need for our armed forces, why we ask for the funds that we ask for, how we spend these funds, what we are doing about such problems as drugs —and we do have a drug problem in the armed forces; what are we doing about the racial problem—and we do have a racial problem. I think the public has a valid right to ask us these questions.

Henkin's references to recruiting as a purpose in displaying military equipment at state fairs, a reasonable enough reply, had been deleted; and in its place CBS had inserted, from a subsequent section of the interview, some statements from Henkin's answer to a question Mudd had asked him about "the instant availability of military speakers at Kiwanis and Rotary and so forth." This question was not presented on the program; and the result was to make Henkin seem like a weaseler and a fool.

Mudd's next question as the show ran was: "Well, is that sort of information about the drug problem you have and the racial problem you have and the budget problems you have, is that the sort of information that gets passed out at state fairs by sergeants who are standing next to rockets?"

Henkin's actual reply was: "No, I didn't—wouldn't limit that to sergeants standing next to any kind of exhibits. I knew—I thought we were discussing speeches and all."

On the program, Henkin's reply was presented as, "No, I wouldn't limit that to sergeants standing next to any kind of exhibit. Now, there are those who contend that this is propaganda. I do not agree with this."

The second sentence of the reply had been lifted from the context of an answer to an earlier question. (It was added to the rough cut of the show in response to Bill Leonard's complaint that Davis had not given Henkin an adequate exit line.) Henkin looked confused, because he had thought he was answering a question about speakers at meetings; and the look of confusion, of course, is not unlike a look of guilt.

This episode clearly reveals a desire by the producers of the program that the man in charge of the Pentagon selling apparatus shall look bad on the home screen. (To the contrary, incidentally, Henkin appears from the record as one of very few heroes in this story. He told the House committee that what had just happened to him was unique in his long experience with CBS. Asked whether he thought CBS would "do any violence or damage to the First Amendment" by supplying to the committee its notes and records and film "outtakes" not used on the show, he said, "I do want to be candid with this committee and anyone else who may read its record, sir, and I must say that as a newsman on such a matter as this—I would, of course, first want to consult with my counsel—but my inclinations would be not to provide my notes or source material." This drew from Congressman J. J. Pickle of Texas a mock-sympathetic aside about his understanding of why Henkin might wish to "hedge just a little bit for the 'brothers.'") Nobody in or out of the news business should condone the manipulation of the filmed interview with Henkin. John Tisdall, chief assistant to the editor for news and current affairs at the BBC, says that "anyone here who was discovered to have presented as the answer to a question an answer that in fact had been given to another question would be deprived of his authority to exercise discretion in the production of programs."

How much of this sort of thing there is on American television nobody really knows. Producers and film editors are disgusted by the whole controversy, because in the overwhelming majority of the cases where they wield their scissors the purpose is to make some tongue-tied clown sound like a fluent statesman. But some Congressmen and some others who are regularly interviewed on film for documentary programs do believe that deception is commonplace. There have been problems, for example, about the technique of "reverses." Where possible, naturally, television news divisions like to send out only one camera, which is trained on the man answering the questions. After the interview is over, the reporter is separately photographed asking his questions, so both parties to the conversation can be shown on screen. There undoubtedly have been instances in which reporters have rephrased their questions subsequent to the interview, to give a more favorable impression of their work and perhaps a less favorable impression of their respondents' replies. Other editing devices create an apparent dialogue between two respondents at different interviews, who are shown contradicting each other's statements; and one side or the other can easily be shown to be "winning" this artificial argument—a specially vicious tactic, because both sides can be said to have had equal time to present their case.

Other things happen, too. In 1968 the FCC considered a complaint against WTTG-TV in Washington, based on a televised discussion between former Kennedy press secretary Pierre Salinger and Johnson adviser John P. Roche. The discussion had occurred before a studio audience, which was permitted to ask its own questions. Roche had to leave before the end of the taping, and a question which was asked after his departure was spliced into an earlier part of the discussion, where it appeared to have been directed at him. His failure to answer it had made him look foolish. The Commission applied no specific penalty to the station, but its opinion on the case concluded with the comment, "The Commission does not regard your actions here as measuring up to the standard of responsibility it expects of its licensees. This matter will be considered further in connection with the next application for renewal of license of Station WTTG-TV."

In the case of the WTTG-TV show, a number of third parties had been present and could testify to the fact that sequence had been violated. In Henkin's case, he had kept a tape recorder going during the interview—this is Pentagon policy, which this incident quite specifically justifies—and could therefore produce a complete transcript of what had been said by both parties. A man who has no such records, and few people do, is completely defenseless against misleading editing. As television critic Lawrence Laurent of the Washington *Post* put it, "He appears to be making statements—by himself—showing his face to the viewer in close-ups that magnify statements. No matter how much of his thought has been deleted he is seen and heard making statements. 'You said it. I saw you.'"

3

On April 7, 1971, Congressman Harley O. Staggers, chairman of the House Committee on Interstate and Foreign Commerce, and of its Special Subcommittee on Investigation, issued subpoenas to CBS demanding the record of the production of *The Selling of the Pentagon*, and to NBC for the record of the production of a conservation-oriented documentary called *Say Goodbye*, which had shown men in a helicopter shooting and apparently killing a mother polar bear. "Polar bears," said the narrator, "have two advantages over other threatened animals: they have nothing we need and live where we can't. For them, life is good. At home in a hostile climate, the polar bear for centuries has taken for granted its freedom in the Arctic. But no more . . ." The cubs of the apparently dead polar bear were shown wandering disconsolately on the ice, and the narrator intoned, "Grieve for them . . . and for us." But the men in the helicopter had merely shot an anesthetic into the animal, so it could be tagged for research purposes, and in the original film, not shown on the air, the bear had later been observed lurching to its feet and walking off.

Most commentators have felt that Staggers' subpoena to NBC was a camouflage to conceal a single-minded pursuit of *The Selling of the Pentagon*, and they may be right. But Staggers' rural West Virginia constituency is big on hunters, and the man in his office who drives visitors to the airport because the Congressman has delayed them by being late for an appointment spoke of the fraud in *Say Goodbye* rather than of Pentagon matters. NBC, in any event, had no trouble with the Staggers subpoena, because the program in question had been independently produced by David Wolper rather than by the network, and had simply been slotted into a time period purchased for the purpose by Quaker Oats. All inquiries were passed on to Wolper, who supplied without raising any constitutional questions the material demonstrating the deceptive nature of this relatively brief section of his film.

At CBS, however, the subpoena was big trouble. There were a few men at the Pentagon who had given information on a confidential basis, and the producers had a moral obligation not to reveal their names, which were, of course, part of the documentation in the files. Also in the files were the reams of internal memoranda the show had generated. Some of the comments from senior executives would be ammunition for the enemy. Other comments would be upsetting to Cronkite, because it had been quite impossible for the memo-writers to avoid having some fun with "Walter's" involvement. And the details of how the show had been changed before airing, while they would certainly protect the network against some of the loonier charges launched by the radical right, are the kind of information no outsider has the right to know. Staggers' subpoena was a little vague in its delimiting of what CBS was required to produce; at its worst, it was a fishing expedition that would yield many brightly colored fish, few of them of any conceivable relevance to the lawmaking powers of the Congress. On April 20, by appearance of counsel, CBS refused to honor the subpoena.

Some of the internal problems at CBS seem to have come to Staggers' attention, because five weeks later he amended his sub-poena to apply only to the film and tape from which the material actually broadcast on the program had been selected. Not only were the CBS News files now excluded; all film shot for the show but not used at all could be kept in New York and away from the prying eyes of Congressmen. "We're out to find the truth," said Staggers, a pink cherub with white hair, one of those country-bumpkin Congressmen who are always foxing city slickers. "All we want is the films from which they took something, from which they eliminated. We think the people ought to know. The government gives them a license, protects them against anybody else who wants to broadcast on those frequencies. That gives us the right to find out whether they're telling the truth or not; we're elected by the people to find out." And his committee counsel, Daniel Manelli, small, black-haired, much more urban, added, "The truth is not always complicated."

The Selling of the Pentagon was by no means the first CBS program for which Congress had issued subpoenas, and the others had been obeyed. A House Appropriations Committee had investi-gated charges against *Hunger in America*, and Staggers' own committee had subpoenaed the complete records both of a show on marijuana produced by WBBM-TV, a CBS-owned station in Chicago, and of a projected but never produced show on an abortive invasion of Haiti. These two had been rather hare-brained operations, involving in the case of the marijuana program a pot party at Northwestern staged for the cameras and in the case of the Haiti invasion what amounted to a very marginal CBS subsidy to Haitian revolutionaries (a few hundred dollars on misapplied expense accounts). In addition, all the networks had made available without subpoena their outtakes (i.e., the portions of film not used on the air) on the riots that accompanied the 1968 Democratic Convention in Chicago. There were undoubtedly many other cases going back in time in which the networks and stations had informally made available to Congress, to the Department of Justice and to grand juries film from their files that had never been aired.

But the Congress had not always been heeded. Long before *The Selling of the Pentagon*, Fred Friendly told an interesting story of a refusal of outtakes by CBS at the time when Senator John McClellan was feuding with Secretary of Defense Robert McNamara over the awarding of the contracts for the TFX experimental fighter-bomber. *CBS Reports* had just done a one-hour interview show with McNamara, condensed from about three hours of film, and McClellan wanted to look at the film or the transcript. "My position on this and all other interviews," Friendly wrote, "was that the material we published was public, but that the 'outtakes' were the equivalent of a reporter's unused notes and therefore privileged. . . . Senator McClellan and his staff made vigorous protests. . . . Stanton not only said no, but went down to see the senator and explained our policy. At one point we were so concerned about a subpoena that I had the unused film removed to my home."*

---

* One should note in passing that privilege is extended to a reporter's notes only in a handful of American jurisdictions—no other country knows anything like it, and it is by no means necessarily a good idea. In Canada, the Special Senate Committee on Mass Media (which on most issues pronounced opinions that were a cross between Nicholas Johnson's and William O. Douglas') looked at this problem and came out with a strong recommendation against:

"Communications between lawyers and their clients have been privileged

In response to Staggers' second subpoena, Stanton came to the hearing room himself, and declared that CBS would not provide outtakes and he would not "under compulsory process" answer any questions about *The Selling of the Pentagon.* Counsel Manelli, ranking Democratic Congressman J. J. Pickle of Texas and ranking Republican Congressman William L. Springer of Illinois all gave Stanton a hard time. He distinguished the current case from the pot party and Haitian matters with the explanation that those had involved criminal or possibly criminal activity. Moreover, there had been no question of investigating editing activites in the Haiti case because nothing had been aired at all. In any event, one cannot waive First Amendment rights; the fact that CBS had yielded to subpoenas before did not mean the network was obliged to do so again. Circumstances had changed: the executive branch had been leaning hard on broadcasting and on the press. In this case, the government itself was investigating the probity of a broadcast critical of the government; it was hard to think of anything

---

since the sixteenth century, and there are other areas (doctor-patient, priest-communicant, husband-wife) where qualified privilege has been extended. But we can't accept the argument that these relationships are analogous to that between newsman and informant. In the common law, it is generally accepted that four fundamental conditions must be present to justify privilege: first, the communication must originate in a confidence that it will not be disclosed; second, this element of confidentiality must be essential to the full and satisfactory maintenance of the relationship; third, the relation must be one which, in the opinion of the community, ought to be sedulously fostered; and, fourth, the injury which would be caused to the relationship by the disclosure of the communication must be greater than the benefit thereby gained for the correct disposal of litigation.

"None of these criteria seem to apply to the newsman-informant relationship. In normal privilege, the identity of both parties is known, and it is the communication itself that is protected, not the identity of the informant who made it. Normal privilege is extended for the protection of the informant; but 'newsman's privilege' seems designed primarily for the protection of the reporter. Finally, the newsman can assert privilege in connection with *any* information, whether it be confidential or not; traditionally, privilege may be asserted only with respect to confidential communications.

"Besides, journalism is a profession where no clearly established standards exist; it is hard to see how the public interest could be served by extending this protection when you don't know *whom* you'd be protecting. Our opinion— which we believe is shared by most journalists—is that we should leave things the way they are."

more likely to have a "chilling effect" on the necessary freedom of the press to criticize government activity.

By then, a large community had rallied round CBS. The network had circularized broadcasters, newspapers and schools of journalism and secured large numbers of supporting statements, not to mention letters and telephone calls to Congressmen. Breaking their own rules about when programs had to be broadcast to be eligible for certain prizes, several universities and the Emmy committee had given awards to *The Selling of the Pentagon*. CBS counsel and some eminent law professors had publicly declared their belief that Staggers' subpoena was unconstitutional and unenforceable.

The most serious weakness in the CBS position was the fact of what had been done in the Henkin interview. Both Stanton and Salant in their public statements (if not in private conversation with Congressmen and others) had defended all the editing on the show, but the manipulation of the Henkin interview really was not defensible. When it became clear that Staggers was indeed prepared to force the issue with a contempt citation against Stanton and CBS, the network moved to repair its self-inflicted damage by promulgating new rules to govern editing. To prohibit prospectively what had been done in the Henkin incident would confess something CBS was not prepared to confess, so the new rules did not specifically forbid the splicing in of answers to questions other than the one asked on the air. Instead, they required that should such things occur, "the broadcast will so indicate, either in lead-in narration, bridging narration lines during the interview, or appropriate audio lines." This is in effect, of course, a prohibition: even the most imaginative reader will find it hard to conjure up the image of a television narrator saying, "The interview you are about to hear presents as answers to our reporter's questions answers Mr.——— in fact made to other questions."

Unfortunately, the Henkin dilemma went a little deeper than that. Congressmen were disturbed by the possibility that what had happened to Henkin could happen to them, and they would never be able to prove it. CBS in the past had consistently refused to provide anyone with the full transcripts of interviews from which excerpts had been aired, and other interviewees would not have Henkin's resources. The new CBS rules therefore provided that "Transcripts of the entire interview will be made available to the interviewee after broadcast, upon request of the interviewee."

Later, during the Congressional debate on the issuance of the Stanton subpoena, letters from several CBS newsmen and news executives would be introduced into the record. "Examination of film out-takes," William J. Small, manager of the CBS News Washington Bureau, wrote, "even by fellow professionals, can frequently be a poor judge of the actual editing without much more information." Daniel Schorr wrote that "to try to compare an edited product with the raw material in retrospect without realization of the pressures and needs of the moment is to invite oversimplified and erroneous judgments. To have to live with the constant prospect of such judgments being made would be to live with a form of subtle but real coercion." Burton Benjamin added that "It would stultify decision-making—and in television that process must be swift and certain. It would vitiate the final product, for a producer, knowing that every decision was subject to official review, would tend to take the easy route. . . . It would, in my view, eliminate the appetite for investigative reporting. Much of the investigation has traditionally involved the government. The proposed system would involve a review by the very people you are investigating."

But by the time these letters were sent, CBS had already given away most of the game. There is, of course, a ponderable difference between automatically giving the government outtakes of interviews with people hostile to government policy and giving only those involving the government's own people. But the opposition expressed by Small, Schorr and Benjamin to permitting the government to look over the editor's shoulder had already been overridden within CBS. Assuming that the new rules are for real and not merely part of a tactic to ride out a storm, anyone who feels himself aggrieved by the use made of his filmed or taped interview will have guaranteed access to the transcript, and a free field for second-guessing the editors.

CBS was now ready to challenge Staggers in the forum where he was most likely to be beaten—the House itself (despite brave words to the public, CBS counsel offered no serious hope that the courts would refuse to enforce Staggers' subpoena if the House voted contempt). Staggers' position was weak on two fronts. The first, and most important to the public, was that *The Selling of the Pentagon* had been, after all, a public service which persuaded the department to abandon some bad habits. And among those who would vote for a contempt citation against Stanton were a number who wished to punish CBS not for contempt of Congress but for airing this particular show. Even those who were angry at what had been done to Henkin understood well enough that many of their colleagues strongest in denunciation of CBS would not have been much upset at similar treatment to, say, Bobby Seale.

More important in the House itself, however, was the narrowness of the ultimate subpoena Stanton had refused to honor, and the trivial significance of the information it could produce. Thanks to Henkin's tape recorder and DoD investigations, the subcommittee already had nearly everything that CBS had been ordered to produce. To exert the contempt powers of Congress to compel the production of documents that were already known seemed unwise to thirteen members of the parent Commerce Committee (out of thirty-eight who voted on the issuance of the citation).

Despite press report to the contrary, however, these thirteen dissenters on the parent committee did not line up behind CBS. "Some of the general criticism leveled against broadcast news reporting these days is well-founded," their report said. "Our dissent is not an endorsement of the past conduct of broadcast journalism. In fact, we feel that the physical and technical limitations of the medium and the questionable practices of the past may force Congress at some future date to formulate a more effective national policy in this area to safeguard the public's interest. However, this is not at issue here except that we might lose some of our authority to act properly in the future by acting improperly here." With friends like these, CBS would never be in dire need of enemies.

In the end, the cogency of the minority report—plus pressure from the broadcast and newspaper fraternity all over the country—

convinced the House not to act on the Commerce Committee's request for a contempt citation. On July 13 a motion to recommit the matter was carried by a standing vote of 151 to 147; when Staggers demanded a roll call, the vote became 226 to 181, and the episode was over.

4

Except that such episodes are never over: their echoes resound. In the world of noncommercial television, for example, the new CBS rules for editing speeches and interviews, and for supplying interviewees with full transcripts, have provoked a management crisis. Whatever First Amendment rights CBS may have, the tax-supported Corporation for Public Broadcasting has none: its charter from the Congress requires its service to be "fair, objective and balanced." Clearly, any grants from CPB to the local noncommercial stations that produce documentaries would have to be conditioned on the use of procedures at least as self-effacing as those now announced by CBS. But many of the producers of documentaries at the noncommercial stations see themselves as crusaders, and the imposition of the new CBS rules seems to them the hand of Big Brother clutching at their throats.

Big Brother is in fact loose, and looking around. In the works at the House is a Truth in News Broadcasting Bill (submitted, incidentally, by a Congressman who voted to recommit the contempt citation), which would write into law a slightly tougher version of the new CBS rules and would hold *licensees* responsible for its violation, so that each network might have to perform a considerable song and dance for its affiliates before they would clear time for any documentary. Under the Fairness Doctrine, the FCC can inquire into any and all broadcasting practices for the purpose of compelling networks and stations to carry, at their own expense, replies to their broadcasts, and the FCC's power to do so was affirmed by a unanimous Supreme Court in the *Red Lion* case in 1969. In his last opinion as a circuit judge, before becoming Chief Justice, Warren

E. Burger wrote that "Broadcasters are temporary permittees—fiduciaries—of a great public resource and they must meet the highest standards which are embraced in the public interest concept." And the "public interest concept" is something that gets defined by Congress, administrative agencies and courts, not by broadcasters.

The chance that something wrongheaded will be done has been greatly increased by an almost desperate erosion of trust occasioned by the handling of this trivial dispute about the editing of *The Selling of the Pentagon*. One section of the public believes the program was a collection of lies about the defense establishment; another section believes there was nothing at all wrong with the program, but the military bullies and their allies in the Nixon Administration tried to push CBS around. Except for the Washington *Post* and maybe *Time*, none of the major news media even attempted to examine what was troubling some of the better men on the Hill. (The *New York Times Magazine* carried a flip and ignorant denunciation of Staggers in particular and government in general from the "committed" correspondent Robert Sherrill.) Congressmen looking at a record that demonstrated an instance of clear wrongdoing were told by broadcasters and journalists from all over that the Henkin interview had been (in testimony from Stanton) "fairly edited" and (in a letter from Salant) "in accordance with customary journalistic practice." The proposition that these guys are all fundamentally untrustworthy, never far from the surface of a Congressman's mind, gained increasing authority as the dispute wore on. At least a third of the Congressmen who spoke in the House *against* citing Stanton for contempt also spoke about the news media in terms of irritated distaste.

"The hidden issue here," says Hartford Gunn, president of the noncommercial network service, "is editorial responsibility." In the end, there is no substitute for a professional, attainably objective job of reporting, editing and documentary construction. Especially when the damage is done on a regularly scheduled public-affairs hour, "equal time" to reply is, as Gilbert Seldes once wrote, "an empty formula." Seldes challenged "the idea that a stranger appearing on a program to answer an attack delivered by the master of that program inherits the program's audience, prestige, and hold on the affections of the audience. Quite the reverse is true: he is psychologically an interruption that may be resented. He is attacking someone who has enjoyed the favor of the audience, and he is depriving them of what they are accustomed to have."

"I am as tired of being Caesar's only wife as you are," NBC News president Reuven Frank told his staff in a memo written in the aftermath of *The Selling of the Pentagon*, reminding everyone of NBC rules against "deceptive practice." But broadcast news and public affairs really must be Caesar's wife, even at the loss of some vivacity. A reply to a documentary, given inevitably by someone personally concerned or committed to one side of an argument, can never answer the statements of a news broadcaster who has built an image of impartiality and public service.

Hope that public opinion can police deceptive news practice is an obvious casualty of the *Pentagon* affair. Far from expressing concern about the distortions in the Henkin interview, the university and intellectual communities presumably most attentive to these matters gave the show prizes. A man who agrees passionately with the point of view he believes he finds in a television program is no more likely to examine the technical background of its production than a lover is to inquire into the use of cosmetics by his beloved. David Brinkley once observed that nobody ever accuses a news service of bias on *his* side, and while this comment is not quite true —Seldes made his comments on the inadequacies of "equal time" in the context of the McCarthy-Murrow dispute—it will cover the vast majority of comments about a public-affairs show.

Because interest in substance is so much greater than interest in technique, nobody has ever successfully defined "editorial responsibility" or figured out a way to guarantee its presence. While the *Pentagon* pot was brewing in America, the BBC in England got into worse trouble than any American network has ever known. The source was a weekly public-affairs show called *Panorama*, which chose to celebrate the first anniversary of the 1970 election by a program presenting former Labour ministers now reduced to the status of a "shadow cabinet" on the opposition benches. The Labourites accepted an invitation to appear under the impression that they were to participate in a serious discussion of the differences between being in power and being in opposition. Instead, the *Panorama* producers and host David Dimbleby stressed the salaries and perks of office—the official houses and cars and conveniences lost when an election is lost. There were films of families moving out, lights being turned off, etc. A rock group played mournful ballads in the background. The title of the program, not announced until all the interviews were in the can, was *Yesterday's Men*.

To add injury to insult, former Prime Minister Harold Wilson, already bedeviled by a split in his party over Britain's entry into the Common Market, was presented as the one man in his government who had made a good thing of his year in opposition, by writing instant memoirs of his tenure at 10 Downing Street. Interviewing Wilson before the cameras, Dimbleby insisted on knowing how much *The Times* had paid for serial rights to the memoirs, and refused to take none-of-your-business for an answer. On Wilson's violent remonstrance to the Board of Governors of the BBC, this set of questions-and-answers was eliminated from the film before broadcast. The actual recutting was done while the press waited two hours in a screening room to see a preview of the program, which guaranteed that the whole story would come out immediately. After waffling for a few days the Board of Governors took a reasonably firm stand in support of the ethical probity of what the BBC producers had done, though several of them are known to have neared apoplexy in private conversation when asked for their personal opinions of the show. And some months later the BBC appointed a "Broadcasting Council" of old men, comparable to Britain's long-established Press Council, to handle complaints against news features.

CBS showed great courage in fighting off the Staggers subpoena. If the contempt citation had gone through the House, it is more than possible that the FCC would have felt itself constrained to deny regular three-year license renewals to the CBS-owned stations, because people and organizations convicted of contempt of Congress for actions related to their broadcasting activities are dubious licensees. But a little more courage, properly placed, would have avoided the whole sordid dispute. What CBS News should have done when the Defense Department brief surfaced was to send letters of apology to Henkin and probably to Colonel MacNeil. The statement announcing the dispatch of such letters could have stressed the difficulties of editing under time pressure and the obvious unimportance to the show of the matters on which CBS editing had been unfair. What got the best Congressmen angry and what was discreditable to CBS News was not the editing of the Henkin interview—even Caesar's wife can be allowed erasers on the ends of her pencils—but the subsequent defense of something indefensible. A year later, at mounting cost to its reputation (and not only in Congress), CBS News was still defending the Henkin editing.

5

Publishing, Justice Holmes wrote in 1893, while still on the Massachusetts Supreme Court, can be compared to "firing a gun into a street." It was Holmes's view (he wrote in dissent) that a publication had strict liability for any damage done by its contents, and anyone damaged by any falsehood printed about him could collect automatically, regardless of whether the error was maliciously intended, honest or accidental (in the case in question a mix-up of initials had held a solid citizen up to ridicule because a drunk with a similar name had misbehaved at a police station). To Holmes, widely regarded as the greatest of American libertarians, the right of the individual to his reputation was more important than freedom of the press. But the Supreme Court in the 1960s destroyed the right to reputation of anyone in public life, holding repeatedly that without proof of actual malice or "reckless disregard" for the truth (a proof which is almost impossible under any circumstances) nobody who held public office could collect for even the most false and damaging libel. The late 1960s saw the efflorescence of an underground press with the viciously slanderous habits of the pamphleteers of the Weimar Republic; the early 1970s saw people wondering why the young had no heroes. . . .

There are real conflicts of rights in these situations, and a case can be made for the proposition that "the public's right to be informed" takes precedence over all else. Personally, as a writer who has been sued for "libel by omission" (i.e., failure to credit somebody with an advertising slogan she wrote), I have no trouble recognizing the good in court decisions that protect me against the possibly nasty consequences of mistakes I may make. But the virtual disappearance of the libel law leaves nothing but conscience as a constraint upon a journalist who dislikes a person, an attitude or a policy. And there are always going to be lots of consciences too fragile to withstand the pressure, especially at a time when failure to say what you would dearly love to say, just because you have no evidence to back it up, gets criticized as "self-censorship."

Fairness Doctrine punishes the broadcaster without making the victim whole, and because it is an administratively operated remedy it is too easily abused—either deliberately by a hostile administration or mindlessly by a bureaucracy believing that any power it has ought to be used as often as possible. As administered, moreover, it rewards those *legitimately* attacked as well as those who have been libeled—the FCC cannot be an "arbiter of truth" and grants rights of reply regardless of the evidence behind an attack. What is needed instead is not a law but a custom, a tradition of public apology by the author of a libel, ideally in the time or place where the offending statement appeared. The punishment then falls as it should upon the reputation of the reporter, and the victim receives a full measure not only of restitution but of revenge.

The thing is not impossible. In London a few years back, Thames Television's 6 P.M. *Today* show carried a report on a riot at a secondary school, by a correspondent and camera crew who picked up a couple of likely-looking boys on a street not far from the school, and shot a filmed interview in which the boys gave a highly circumstantial description of the trouble and its cause. But neither of the boys had in fact been a student in the school or a participant in the trouble: they were just larking around. The next night, apparently at the insistence of the Independent Television Authority which licenses the British commercial stations, the reporter got on camera and apologized to the school and to his viewers for his sloppy work. Take by contrast a recent error in the United States, in which NBC correspondent Carl Stern, discussing on the *Nightly News* show a possible Supreme Court appointment for Senator Robert Byrd and the Senator's record at the American University Law School, mentioned "much publicized charges some years ago—never proved—that Congressional employees wrote his term papers." In fact, no such charges had been made (in public, anyway). When the Senator protested, he got a personal letter from an NBC official—*not* from Stern—regretting the broadcast statement. Senator Byrd read the letter into the *Congressional Record*, but the audience of the *NBC Nightly News* never knew it had been sent. Surely the proper course would have been for Stern himself to get on the air and tell his audience that he had misled them, that there had been no "much publicized charges" against Senator Byrd.

American journalism, both print and broadcast, has always been most reluctant to utter retractions in any form but private letters—but it used to happen much more often, before the Supreme Court eviscerated the libel laws. Obviously, no outside organization except perhaps a court after a trial should have the right to compel publication or broadcast of a retraction; equally obviously, no executive or legislative agency should ever have anything to say about such matters. But a press council, of the kind long established in Britain and formed in Minnesota in 1971, could look into charges of serious inaccuracy or misconduct, and could *publicly* suggest the need for an apology if its members felt such a step should be taken. The hunch in this corner is that such a press council should be composed mostly of working journalists (unlike the councils in Britain and Minnesota, which are top-heavy with bosses and public members), because their standards would be higher. The humiliation inherent in making an apology would be considerably worse for the offender if it came about as the result of a public rebuke by his peers, and to avoid the danger of such a rebuke, many reporters would grit their teeth and make their apologies quickly. Such cases would be infrequent, and obviously most complaints would come to rest in the huge gray area where no action is ever taken. But there ought to be a gray area; and ultimately a reporter unable to control his biases will step beyond it into conduct that should be rebuked.

The existing situation—where press and broadcasting expose everybody else's mistakes but never admit their own—is untenable in the absence of effective libel laws, and will become catastrophic if the courts or the legislatures grant a "newsman's privilege" by which reporters alone in the world would become entitled to conceal the evidence (if any) that supports their attacks on others. As I have noted elsewhere, control of professional performance is the great challenge to the increasingly specialized societies of the last quarter of the twentieth century. Education, law and medicine are all being pushed to develop standards of accountability; it is impossible to see why journalism, print or broadcast, should be immune.

The documentaries are what television *does*—everything else is more or less forced upon the medium by events or availabilities. The documentaries lose money, which does not recommend them to management. Fred Friendly in his book recalled his shock at his first CBS stockholders' meeting, when Chairman William S. Paley noted that unanticipated public-affairs programs (incidents in the space effort, something to commemorate the death of Winston Churchill, etc.) had reduced profits by six cents a share in the quarter just past. And documentaries make trouble in Washington, which everyone in broadcasting has reason to fear, regardless of the political coloration of the Administration in power.

In print, Friendly defended the editing job on *The Selling of the Pentagon*, but he rather gloomily told some friends that the troubles following that broadcast illustrated how carefully a network (or, he says, a book) must triple-check everything done on controversial subjects. Nobody can be that careful, and the kind of controls all the networks are now beginning to impose on their producers (quite apart from the growing reluctance of the network managements to air controversial documentaries at all) could kill the most interesting activity in the medium.

A private press council—an appellate body charged with responsibility for common-sense judgment on procedures—could protect the future of the documentary. "Second-guessing," to put it another way, is both inevitable and legitimate. The question is, who is to do the second-guessing? The task before the news divisions of the networks—and the editors of the newspapers—is to develop sensible postaudit procedures by which they can handle these problems, in public, themselves.

CHAPTER 11

Local Television and
the Meaning of Diversity

The foundation of the American system of broadcasting was laid in the Radio Act of 1927 when Congress placed the basic responsibility for all matter broadcast to the public at the grassroots level in the hands of station licensees. That obligation was carried forward in the Communications Act of 1934 and remains unaltered and undivided. The licensee is, in effect, a "trustee," who must qualify under standards laid down by Congress and the Commission, and must give an account of his stewardship at stated intervals. His duty to the public requires faithful adherence to the principle of his trust—to serve the public interest in the community he has chosen to represent as broadcaster. His is a proud and, at times, difficult calling.
——Interim Report by the Office
of Network Study, FCC, 1960

One of the things true about television world-wide is that the less production you do, the more money you make.
——HOWARD THOMAS, managing director,
Thames Television, London

Please do not touch the salads or the pizza. They are for a commercial.
——Sign on the icebox in the lunchroom
at WTOG-TV, Channel 44, Tampa

1

It is eleven o'clock on election night in a busy studio at WISH-TV in Indianapolis, the lights bright on the anchor man's elevated desk, the ceiling dark in the big area above the steel grid that holds the lights. Cameras on three-wheeled dollies track thick snakes of heavy cable along the gray floor. Four secretaries and news director Lee Giles are busy pulling on circular black paper strips with white numbers, to bring the most recent figures for each local race onto a big free-standing announcement board. The figures come to the girls from another girl at a telephone, speaking with the WISH reporter at Network Election Service, the joint venture of the broadcasters and the newspaper wire services set up (despite raised eyebrows at the Justice Department) after the debacle of the 1964 Goldwater-Rockefeller California primary, when the AP count published in the newspapers showed Rockefeller still leading nearly twelve hours after teams working for the broadcasters had tallied all the vote and correctly awarded the delegates to Goldwater.

The figures over the telephone are gross figures, most recent tallies. In addition, one of the four teletypes pounding away on a bench across the studio is a line from NES, relaying to the stations and newspapers the county-by-county totals (and numbers of districts missings) in the extremely close Senate race between Hartke and Roudebush. Bob Gore, the WISH-TV political analyst, a young man with shining cheeks (since moved on to WMAL in Washington), pulls off the sheet from the teletype and sits down with a visitor along the anchor man's horseshoe desk. Nothing is live on the platform: WISH is broadcasting the CBS feed, and the activity in the studio is preparation for the next burst of seven minutes in the half-hour, when CBS will return the air to its local affiliates for their own coverage.

"Porter County," Gore says, with that glorious enthusiasm of a man who has really studied and mastered a lot of arcane detail now suddenly important to others (Here! Here!), "Porter County is Republican. It should be Democratic, if you look at who lives there, but it's Republican. Now, La Porte County is Democratic. Tippecanoe County is *very* Republican, but we have an odd situation there, because the local paper endorsed Sprague, the Democratic candidate for Congress. . . ." He pores over the sheet, looking at what has reported, what hasn't—there are forty-odd counties in Indiana. Hartke is leading in the tally by about two thousand votes, and Roudebush has seemed to be catching him. Gore says, "There are more districts out than you'd expect from up around Gary. That county down in the southern section with all the missing districts— that's farm country but real South, it's Democratic." Then he straightens up and squares his shoulders and says, "I'm going to call Hartke the winner."

Gore goes out through the big doors of the studio, and reappears behind glass in one of the small control rooms on the side, where he can be seen talking eagerly into a telephone. A few minutes later Cronkite glances at a piece of paper on *his* desk, looks back to the camera and says, "WISH, our affiliate in Indianapolis, has called Hartke the winner in Indiana." He hesitates. "Bill Gore," he says, "our political analyst there, is a first-rate man. But Vote Profile Analysis says that race is still too close to call, and we'll have to leave it at that."

Bob Gore returns to the studio, half-glowing, half-shaken. "My mother's in New Jersey and she always watches Cronkite," he says. "And here's *Walter Cronkite* saying I'm a first-rate man . . . and he gets my name wrong."

2

Broadcasting to a narrow audience looks like an obvious contradiction in terms. The essence of the medium would seem to be the simultaneous availability of its message over the whole great field toward which the seed is thrown. And American television has indeed developed, brushing aside sporadic incidents of government resistance, in the direction of a national service built on the simultaneous or (because of time zones) briefly delayed transmission of three network feeds. Local programming could not compete against national programming, which not only offered greater rewards to the more popular talent but also commanded national resources of publicity and celebrity.

Network affiliation has been the fundament of profits in television. Of the 123 stations that reported profits of a million dollars or more in 1970, only 8 were independents; of 77 independent commercial stations active in the country in that year, 54 were losing operations. A network-affiliated television station in the VHF band in any of the nation's 40 top markets is what the men in the network station-relations departments call a "money machine."

Different network affiliations have different values in different cities: the apparent homogeneity of the Nielsen rating is deceptive. ABC's nighttime ratings at WEWS Cleveland, for example, are almost 40 percent better than the ratings the network receives nationwide; CBS nighttime ratings run 20 percent lower at WTOP Washington but 30 percent higher at KDKA Pittsburgh than their average for the country as a whole. NBC is very strong at WBZ Boston and KPRC Houston, pathetic at WCKT Miami. Nobody knows why. It can't be the ownership of the stations—Westinghouse, which has interchangeable executive parts, people moving around all the time, is a big winner for CBS at KDKA Pittsburgh and a big loser for the same network at KPIX San Francisco, a winner for NBC at WBZ Boston and a fairly weak No. 2 for the same network at KYW Philadelphia. All the network affiliates do less well than the national network ratings in New York, because the independents and the noncommercial station are strong; and they also do less well in San Francisco, because people in San Francisco watch less television than people elsewhere.

At night, except for movies and athletic events, the network feed occupies 58 minutes and 40 seconds of every hour, leaving the station 1 minute and 20 seconds to sell on its own account. The contract calls for the station to carry everything the network sends during this hour, but cheating is not entirely unknown. Jack Reilly of The Mike Douglas Show remembers his early days in Omaha, when networks were giving stations only thirty-two seconds to sell between network shows; and Reilly's employer was taking seventy-two seconds: "I got a call one day from the Blackstone Hotel, and the voice said, 'This is Mr. Goodson. I see you're cutting the credits off my show. If I see you do that again, I'll tell the network to cut you off.' I told my boss, and he called the Blackstone, asked the clerk, 'Is there a Mr. Goodson registered there?'—and by God there was."

and he also runs the darkroom, developing films for the news shows.

"The only thing I haven't done around here is sell," Wuertz says. "I'm farm director, and assistant news director. But it's interesting. Tomorrow at nine-thirty I'm interviewing a man with one of the world's best collections of barbed wire—he has nine hundred varieties. It's something that's just picked up around here recently, collecting barbed wire."

Station manager Brown, casual and tweedy, has been at KAYS-TV since 1961; before that, for seven years, he had been a barber ("I've also mixed cement and driven a truck"). Schmidt, who was then an on-the-street salesman as well as president of the station, was one of Brown's customers at the barbershop, and one day he said, "How would you like to carry a briefcase for me?" Brown makes most of the non-network programming decisions (largely on the basis of what he can sell: "We carry Notre Dame football Sunday morning; I have a friend here in town who has a good business, who's a Notre Dame graduate"), but he puts most of his time into selling to local advertisers: "We're after the telephone-directory dollar, the matchbook-cover dollar. We want half of every advertiser's budget, not all of it. And we outreach everybody around here—you have to buy fifty-seven weekly papers or seven radio stations to cover our area." He also appears occasionally on camera reading some of the local commercials he has sold.

News direction is in the hands of Bob Chaffin, a chunky young man who worked as a disc jockey in a Washington, D.C., radio station and for broadcasters in Topeka, Salinas and Hutchinson before coming to KAYS in 1969. His basic tool is the telephone: KAYS buys a WATS line (Wide Area Telephone Service) from AT&T for all of Kansas. "I try to call every county seat, every sheriff's office, once a week, and I use the weekly papers as a backup and a source for leads. We put a slide on the news show— 'KAYS Pays Cash for News Tips.' Then, I have a man in Great Bend who has a grudge against the Great Bend newspaper. He has police monitors and state trooper radio in his living room, and when something happens he calls all the radio stations and us." Chaffin has two part-time men who can go out and take pictures or interview people. Slides are sometimes preferred, because the newsman on camera, by pushing a button in his desk, can control the operation of the slide carrousel that selects the pictures shown on a screen beside his head. And color slides can be developed in-house, while color film must be sent to Denver.

Schmidt himself wanted to be a newspaperman, or maybe a sports announcer. Coming out of Fort Hays College, he got a job with KAYS, then just a radio station, reading the 7 o'clock morning news. He kept it up for eighteen years. "It's a great self-discipline," he says, "getting up and doing seven o'clock news. And it's a great way for people to get to know you. They say, 'There's a go-getter.'" Schmidt rose to be manager of the station, continuing all the while to do the 7 o'clock news. Ultimately, he found allies in town and bought the property.

In 1970, after commissions, KAYS took in over $351,000, of which $87,000 came from the network (which pays Schmidt $75 per hour for prime-time transmissions), $125,000 came from national advertisers (three-fifths of it sold, often as part of a package joining all of Kansas, by the rep firm of Avery-Nodel, two-fifths of it by Schmidt himself), and almost $140,000 from local advertisers, who include local makers and wholesalers of industrial farming and oil-field equipment. The total for the year is roughly what the CBS station in New York grosses every three days, but KAYS is expected to deliver a service not greatly different from that in the metropolis.

It is interesting to note the priorities that enable Schmidt to survive—indeed, to profit—on that kind of revenue. First comes the microwave link to Wichita. At the beginning, KAYS got its network shows (then ABC) by putting up a receiving tower about sixty miles away to take ordinary broadcast signals off the air from Wichita. The service was technically no better than fair. For a wire to Wichita, the telephone company wanted $5,500 a month. Usually, under complicated contracts, a network pays to connect up its affiliates, but obviously KAYS wasn't worth $66,000 a year in line charges to any network. Schmidt finally built his own microwave link, for $85,000; with subsequent improvements, the total investment is now, he estimates, about a quarter of a million.

The microwave is good for more than network shows. Schmidt does not have to own any big library of movies; he can plug into KTVH and use its movies (paying, by contract, one-fifth of whatever KTVH paid). Before the microwave link, KAYS did not carry *Mike Douglas,* because the charges for mailing and handling the tape were greater than the available local revenues could justify; now *Mike Douglas* comes in, with everything else, on the microwave. The construction of another microwave to Topeka, paid for by the cable system, enables KAYS to take from KTWU not only the occasional basketball game but also *Sesame Street,* which Children's Television Workshop sells to Schmidt for $10 a show, for use on his satellite in western Kansas.

Schmidt did not go to color until there was used equipment to be bought; then he picked up for $7,200 each a pair of used video recorders that had cost their original owner $83,000 each four years before. The studio was designed, Schmidt says, "to make every variable a constant." The news announcers prepare and control their own slides and films, and can if necessary change the lighting from dials and switches at their desk; when graphs are to be shown on an easel, one of them goes over and moves the cards; when the weather map is displayed, the newsman not on camera works the camera. A technical staff of four operates and maintains the equipment—a task that requires twenty hours a week on the aging tape recorders alone.

Local shows are very important to KAYS; Schmidt estimates that half of his local revenues are generated "between 9:59 and 10:30"—i.e., in and around the local night news—with another fifth in the early evening news. "Don't get me wrong," Schmidt says. "We would have no audience without the network. But noon is our biggest audience in daytime, except for *As the World Turns,* and people jump from the other networks to our early-evening news block." Schmidt is also helped by his ownership of a radio station with the same call letters; though the selling operations are separate (and, indeed, competitive), radio people are always available for emergencies.

The one-story brick building, bumped up half a story for the necessary high ceiling in the studio, stands at the edge of town, and across the fence from the parking lot three saddle horses graze gravely, and come to the fence when Schmidt emerges from the office. They are his horses; he loves to ride. Himself a descendant of the German farmers who settled this area in the late nineteenth century, bringing their wheat and their Catholicism (and forming later a Democratic knot in a Republican state), he knows his community and everyone knows him. His local advertisers even know what they're buying: "You should see my list for Tuesday night. The local guys don't pay too much attention to the ratings books, but they know when people are watching."

4

Though local television is a very American idea, and European broadcasting has always been based on centralized national service, the most thoroughly local television station this tourist found was in the Channel Islands, minicommonwealths of the British Crown located in the bay of Mont-Saint-Michel off the French coast. (As a measure of their localism, some of the commercial time is sold for help-wanted ads.) The origins of this station are odd, the way most things connected with the Channel Islands are odd. Laws of the British Parliament do not become laws in Jersey, the largest of the islands, until the Jersey legislature approves them (this means, most significantly, that residents of Jersey don't pay British purchase tax or income tax). The Independent Television Authority had been set up by Act of Parliament in 1954, and empowered to build television transmission towers about the country for Britain's first commercial broadcasting service. Except for one slope on the northernmost island of Alderney, homes in the Channel Islands are too far away to receive television broadcasts from England. When ITA came around to build a relay tower which would permit the small Channel Islands market of thirty-odd thousand households to be added to either the southern or westward region for commercial service, London learned that Jersey had not approved the act establishing ITA, and would not do so unless the islands were given a television region of their own.

It is to be feared that greed was a significant motive here—the man heading the group that planned to start a separate Channel Islands TV region was a local Senator, and these were the days when Lord Thomson of Fleet was talking of a commerical television license as a license to print money. Having pushed ITA into accepting the idea of a separate Channel operation, the Senator went for technical advice and assistance to one of the successful regional operators in commerical television—ABC, an English theatre chain (not related to television's ABC on the other side of the herring pond)—and ABC engineering talent designed studios for Channel. "This was a very good deal," says R. K. Killip, Channel's general manager; "we went in on the back of ABC expertise and buying capacity. I was then working for ABC, in Liverpool, and I was part of the package. I did nothing to persuade anybody here that this operation would be prosperous; all I ever said was that it would be viable." (Killip himself, a matter-of-fact, rather muscular businessman with a long face and a shock of white hair, was attracted to Jersey by the viability: on spring or summer evenings, he and his wife and one or more children get into the boat and sail over to a French fishing village for dinner.) In the event, the Senator died not long after Channel went on the air in 1962, and the surviving board was less insistent on profits, which was a good thing, because there haven't been any.

KAYS in Kansas and Channel Television generate roughly the same amount of revenue—$351,000 for the Kansas operation, about $385,000 for Killip & Company in 1970. KAYS gets most of its programs from CBS, Channel from the ITA contractors in England. Yet both are significantly local in their orientation, relying on local advertisers for two-fifths or more of their revenue, and on the local scene for substantial chunks of program.

Actually, Channel has more money to spend than Schmidt does. Schmidt had to sink something like half a million dollars into transmitting and microwave equipment and towers, and must pay all his transmitting expenses—salaries to engineers, maintenance costs, an electric power bill of about $16,000 a year—while ITA built and operates the Channel receiver on Alderney and transmitting tower on Jersey, for a fee Killip describes as "a peppercorn" ($250 a year). Channel has to pay for the ITA programs it carries, but only at a rate of $10 an hour; Channel's revenues available for program operations must be close to $100,000 a year greater than Schmidt's—in an economy where wages are considerably lower. And until late in the 1970s the Channel service will be black-and-white only.

Channel's studio is a room about thirty by thirty, two stories high, with a homemade grid of fixed lights. "Works fine," says Brian Turner, Channel's large, aggressively efficient operations manager, "as long as you don't want to do variety shows. We just light the areas and move the people over to them." One engineer can watch both the studio to his left and the telecine chains in a room to his right. The sound man sits in a slot surrounded by tape machines and a disc player—"as we couldn't breed a sound man with four arms," Turner says genially, "we made everything very convenient for the only kind we could get." Probably the most sophisticated piece of equipment in the house is an RCA sound-tape unit which stacks four cartridges and plays them automatically in synch with film on the telecine chain, permitting Channel to do professional-sounding commercials at absolute minimum cost.

Channel comes on the air in its own right (and its technical staff comes to work) at 4 P.M., with a ten-minute filmed children's program acquired from another ITA television station or from America. Local programming starts at 4:10, with *Puffin's Birthday Greetings,* a feature of a kind older Americans will remember from the radio days of their youth. The puffin is a rather silly sort of flying penguin-like bird native to the Channel Islands, and the station's mascot. Two hand-puppet puffins, one for use on camera and the other, somewhat smarter, for live appearances, have been made for the station. An announcer sitting at a desk in a little room, facing an unmanned, fixed camera, holds the hand puppet in his left hand, and reads birthday wishes to all children in the Channel Islands whose parents notify the station of the impending event. Within a ten-minute time constraint (or, sometimes, to fill ten minutes), the announcer reads from the letter something that describes the child; and then the puffin, which has a movable eyelid, blinks as many times as the number of years signalized by this birthday. It is an extremely popular feature. "Kids come by after school," says Brian Turner, "and say, 'Please, sir, may I see Puffin?' Takes all one's patience not to say, 'Get away.' But that's local television. When we do something wrong, they don't write—they pop around, or call up."

Six o'clock is local news time, ten to fifteen minutes a night of events in the Channel Islands, watched by 85 percent of all viewers at that time. Channel employs the services of eight reporters and a features director who does news stories himself; plus one sound-camera crew and as many cameramen for silent film as desired. Something more than half the time on the local news is taken up by film clips, usually silent with voice-over by the reporter who covered the story. The reporter normally does not appear on camera. "I think people get bored looking at you," says features director John Rothwell.

On Mondays and Fridays, news is followed by a five-minute feature called *What's On Where*, a list of events at Channel schools, churches and such, prepared (mostly from letters sent to the station) and presented by an attractive girl, a former airline stewardess, working on a part-time, free-lance basis. Tuesdays there is *Police File*, a ten-minute "crime information programme" written and presented by a part-time official of Jersey's uniquely complicated criminal-justice system, telling the public what the police know about any recent criminal matters in the island and what more they would like to know, please.

The station's major weekly effort is a 7 o'clock half-hour special on Fridays, which covers in depth and sometimes creates local news events. This program did not begin until fall 1970, and Rothwell had considerable difficulty convincing Channel's board (which must approve all local enterprise involving a significant expense) that it made sense to slot such a program opposite what was then the most popular of all BBC shows (*The Virginians*). By spring 1971, Channel's *Report at Seven* had taken leadership in its time period in the local ratings, and Rothwell for '71–'72 was allowed a budget of

$125 a week for special expenses (i.e., costs other than salaries) associated with the program.

Some of these programs have emphasized British national issues, pegged to significant "mainland" people who happen to be visiting the Channel Islands. One featured a discussion with the managing director of ITA itself, on problems of sex and violence on television; another rested on a long interview with Enoch Powell, leader of the isolationist, anti-immigrant wing of the Conservative Party. In 1971 considerable time was given to the complicated Channel Islands end of the overriding British political issue of the year: joining the Common Market. But it is the purely local material that has built the audience: reports on the background of a greenhouse workers strike in Guernsey, the landing fees at the Jersey airport (the highest in Europe) and fares on flights to the Channel Islands, a plan for the retraining of Jersey teachers, the functioning of the Jersey Arts Council, drugs in the Channel Islands, a new Economic Survey of the islands, the Guernsey philatelic center and the effort to make money by printing stamps, the operation of the Jersey Health Scheme and the question of whether doctors were giving patients credit for the $1.25 per visit paid by the government or simply pocketing the money on top of their usual fees (most of them were pocketing it). A particularly imaginative show presented a packaged tour to Majorca, to show the tourist industry of the Channel Islands (which produces more than half the islands' income) why the competition was doing so well.

Every so often, Channel has done an especially adventurous show based on local interest. The most successful in British national terms was a documentary commemorating the twenty-fifth anniversary of the liberation of the islands (the only part of Britain to suffer Nazi occupation); entitled *The Bitter Years*, this program was bought and broadcast by nine of the other fourteen ITA affiliates. One program Rothwell had to abandon and turn over to ITN, the national news operation: a story he uncovered and tracked for three months, on a fraudulent "Bank of Sark" chartered on Guernsey to a group of American crooks, who took something more than $10 million out of the tills of other banks by the circulation of worthless paper.

Perhaps the favorite of all Channel documentaries locally was a show on a Jules Verne–style race from the Post Office Tower in London to the Empire State Building in New York. One of the participants was a Channel Islander, and Rothwell assigned a young reporter, Marcel le Masson, to go with him. "I had a shoulder pad made to hold the camera," Le Masson recalled rapidly, pausing on his way out of the office to catch an effort to salvage a fancy motorboat sunk in Jersey waters the night before. "When he started running, I started running, down the M-1 at 110 miles an hour, onto the plane—eight hours, fourteen minutes, eleven seconds—and we cut a half-hour program out of it."

Rothwell, who is lean and slight, under forty, from the Midlands and still a little middle-class-angry in a Midlands way, feels especially proud of his "ombudsman" shows. "Somebody called," Rothwell recalls, "and said, 'Please look into the price of cement; there's a fiddle going on: distributors are raking in a profit.' And so they were. There was another call about the price of shoes. We don't have purchase tax here, so shoes should be cheaper here than in England, but there was a suspicion that our chain stores were charging English prices. Turned out they were not, but the British Shoe Corporation, which has stores here, was. Another call charged that the Jersey coal company was buying a grade of coal inferior to that used in Guernsey, and charging more. We broadcast a long, interesting reply on that—the Jersey company said that it did much more sifting of the coal, which cost more and resulted in a better product. I see a good deal of merit to giving a good reply—in the newspapers, we never really found out what the other side was."

It should be noted that what Rothwell and his staff of eight are doing is considerably more courageous than the "action line" sort of program fairly common on American television. These Channel programs assume not that an individual has had bad luck in dealing with a normally reasonable business or government, which can make itself almost the hero of the story by Doing Right on camera, but that the operation challenged is as a whole unreasonable. Sometimes the results are just fun, as when Channel revealed a secret vote of the Jersey legislature to poison the pigeons in the square fronting the government offices. (One viewer wrote, and the letter was read aloud on camera, that he was particularly angry at this decision because as an old man one of his few remaining pleasures was to watch the politicians leave their offices and hope that the pigeons would drop something on them.) But one letter read on the air, from a farmer complaining about police mistreatment of one of his laborers, provoked first a libel suit (which lost) against the letter writer and then a formal investigation of police behavior in Jersey. Often enough a story is embarrassing to storekeepers, whose business is needed. "Lots of guys advertise," Rothwell says rather airily, "but we pay no attention."

Last and least of Channel's local services is a nightly 11:55 program of news and weather in French, which runs five minutes or so and concludes the day's broadcasting. Officially, the language of Jersey is French (English is merely "permitted" in the legislature, though nobody there speaks anything else), and out in the farming sections of the six-by-ten-mile island there are a number of people whose language is a French-based patois. There are also an unknown, probably small but not negligible number of Frenchmen in Normandy and Brittany who watch Channel Television in preference to their own national service. After the French announcers say good night, Channel announces its departure from the air and delivers a performance of "God Save the Queen" to accompany a clip of Her Majesty on horseback, performing one of her numerous duties. The screen goes white, and two minutes later, in an ITA trademark, a soft female voice says, "You won't forget to turn off the set now, will you?" and that's it.

In one aspect of local coverage, Channel is no different from KAYS and all the other local stations large and small: there is almost no presentation of local dramatic and serious musical talent. "Some years ago," sales director Phil Mottram-Brown recalls, "we did two plays in the patois, but it's almost impossible for us to mount the coverage this sort of thing needs. And anything you do has to compete with the very expensive network material." In theory, and this is a pretty good theory, local television should serve national television as the night clubs once served vaudeville and radio, but it doesn't; indeed, television kills local professional entertainment on the air as on the ground. "The regions haven't been able to produce," says Howard Thomas, managing director of Thames Television in London. "People want to work in London." That's a harder saying than Thomas knows.

5

One-fifth the size of the next smallest ITA operation, Channel airs more really local coverage than any other ITA region. In London, the Thames Television 6 o'clock show, rather ambitiously called *Today,* is clearly a feature program, hosted by an amiable Irishman named Eamonn Andrews, best known as the host of the English version of the old tear-jerker *This Is Your Life.* Though the episodes presented on *Today* commonly do have some local reference—the two West Indian children abandoned in a washroom were left in a London washroom, Sandy Wilson is singing some of his own songs in a London night club, the author of the new book about Prince Philip is in London for its publication—the fact is that the programs could be put on a national network without change. And Thames would not dream of denying it. "You can't," said one of the men who works on *Today,* "cover London." What's news in London will be left to the national *News at Ten;* nothing not important enough to warrant national presentation is considered important enough to appear on television at all.

Few American stations could afford to adopt this attitude: local coverage is among the goodies American owners promise to supply when they apply to the FCC for their licenses. But little KAYS puts more time into local affairs than almost any American big-city station. Television in Wichita, which is to Hays more or less as London is to Jersey, is dominated by KAKE-TV, probably as deeply committed to the importance of local enterprise as any station in the country. It was the first ABC affiliate to become No. 1 in audience in its market (in 1956: Ollie Treyz, then president of the network, came out for the ceremony, which featured a cake-baking contest won by a lady who put figurines for the entire Lawrence Welk band atop the cake). Its president and general manager, Martin Umansky, escaped New York in the 1930s to study journalism at the University of Missouri and moved west to become news announcer, DJ, salesman and factotum of a Wichita local radio station, building from there. He believes passionately that "it's the local stations that make the network. Their standing—their image in the community [Umansky grimaces with distaste while using the word 'image']—will determine the ratings of many network shows."

KAKE has developed over the years a number of annual local specials. Its election coverage is extraspecial. "We do it as a party," Umansky says. "I got the idea from having been a kid in New York and remembering the excitement of election night, the bonfires, the dancing. We have a band. We feed people—all the politicians come. We do our own tallies for the state—we have six hundred women from the League of Women Voters, who call in the results. In 1970 we plugged into a Boeing time-sharing computer in Seattle." On a less significant level, Umansky turned the selection of the local Miss America candidate into an annual station promotion (indeed, the girl was crowned "Miss KAKEland" until the 1970 contest, when the winner was "Miss Wichita"; Umansky said, "We've grown up enough"). When Century II, Wichita's combination convention hall and arts center, refused to book the touring company of *Hair,* KAKE's fifteen-man news staff did a two-hour special on the center and the show. Other programs have been pegged to the needs of the local symphony, the Community Theatre (which Umansky long served in a thespian capacity), the Singing Quakers of Friends University, a Model Cities Black Art group (Umansky is chairman of the Kansas State Arts Council). Part of the news show on Sunday night is "Consumer Scene," a run-down of recently uncovered consumer frauds; KAKE protects itself from possible commerical wrath with a preceding announcement that "This feature is compiled by the Sedgwick County Attorney's office, which is solely responsible for its contents."

There are eighty stringers scattered around the state who send news items and film to KAKE, and if the story is important enough, news director Paul Threlfall, who has been with the station since Umansky opened the doors in 1952, will fly out in his own plane and pick up the film. But week by week the Wichita station airs less of purely local reference than goes out from Hays. The news department fills an hour and 25 minutes a day—15 minutes at 7:45, 15 minutes at noon, 25 minutes at 5:30 and half an hour at 10—but 40 percent of that is taken up by national and international stories brought to Wichita on the ABC network's "DEF" (Daily Electronic Feed) and other chunks are rewritten from the AP wire.

The problem is that a news editor's judgment of what is worth broadcasting grows increasingly severe as the size of the market rises. At Channel, a lady may call in to report that a budgerigar (parakeet, to Americans) just flew in her window; the man who takes the call relays word of this event to Rothwell; and he invites her to bring the bird to the studio that night—Channel will show her and the bird and invite applications from anyone who just lost a budgie. On Jersey, and even in rural Kansas where viewers are scattered over literally thousands of square miles, people feel they have neighbors, and what happens to your neighbor is always sort of interesting. A Wichita station faces a situation of town v. country; the suburbs aren't interested in what happened today in town; and the only neighbors the city-dweller has are the people whose homes he can see. KAKE's satellite in southwest Kansas, KUPK, puts the local citizenry on camera as actors in commercials, to "have the viewers talking the next day about the commercials as well as the programs." In Wichita there wouldn't be any talk—it would just look like amateur commercials.

6

Though "best" statements mean no more here than they mean anywhere else, most television professionals would rank WCCO-TV in Minneapolis–St. Paul at or very near the top in the over-all class of its operations. Of the fourteen top-rated shows in the Twin Cities in 1970, seven were the different nights' editions of WCCO's 10 o'clock news, *The Scene Tonight*. When Umansky lost his anchor man to a rival in Wichita, he sent a team up to Minneapolis to copy the four-men-at-the-desk pattern of the WCCO *Scene*, and paid for a license to use the format on KAKE.

News director Joe Bartelme, a small, businesslike man who looks too young to have been with the station since 1957 (following some years of experience at Cedar Rapids), commands a staff of thirty-seven, fifteen of them photographers and TV cameramen. There are six trucks capable of originating live remotes. When the Minnesota legislature is in session, WCCO has a camera team as well as a reporter on the spot at all times. "But often we don't get anything to use," Bartelme says. "Material is a problem. We have forty-five minutes on *The Scene Tonight*, seven nights a week. And the fact is that there isn't that much local news."

Whether or not the news department feels there is forty-five minutes' worth of news at 10 o'clock, WCCO-TV is going to run a forty-five-minute show. It had special dispensation from CBS to delay *The Merv Griffin Show* by fifteen minutes. As of fall 1970, minutes on *The Scene Tonight* were selling for $800 each, while minutes in the station's slots on Merv Griffin were selling for $100 each. Running the news show fifteen minutes longer is worth more than half a million dollars a year in revenues to WCCO-TV, at virtually no increase in costs, because the news department would have to be the same size anyway. Bartelme's solution to his materials problem, not uncommon in the larger stations, has been what the trade calls the "mini-doc," i.e., the brief documentary that runs serially through several news shows. "Right now," Bartelme said, ticking off projects, "we're doing two parts to slot into the news show on unwed fathers, a three-parter on slumlords." One mini-doc has been routinized: "We send a reporter to a neighborhood and have him stay there three, four, five days, then interview people. Twice a month we run a five-minute section of *The Scene* called 'What's on Your Neighbor's Mind.'"

But there turns out to be something seriously and interestingly wrong with all this. What is on your neighbor's mind, when he is being interviewed by "the television," tends to be what the media have already put there: people don't think the need for a traffic light by the school or the smell of a leaky sewer near the brook will interest television viewers. One tends to get not local events or (to use the FCC's favorite word) "needs," but national issues or problems that can be given a local twist. (Indeed, WCCO-TV sent a reporter-cameraman team to South Vietnam in 1969 and 1970, and to the Middle East in 1971—entirely commendable enterprise, of course, but not much to do with local television.) The quest for the big audience yields automatically a bias for the big issue, and the bias is unconsciously reinforced by the reporter's and editor's own sense of greater importance when the story he seeks seems to have national significance. It is also, incidentally, safer: a New York station has nothing to lose by a nationwide study of water pollution with interesting footage on the death of Lake Erie, but Mayor Lindsay might be mad as hops about a show specifically pointing up the city's uncontrolled dumping of raw sewage into the Hudson and his administration's inability to pick sites for treatment plants and actually build them.

In June 1971, KYW-TV, the Westinghouse station in Philadelphia, pre-empted the entire NBC Saturday night schedule to present a three-hour special on the planning of and fighting about the proposed Bicentennial Exposition for that city in 1976. It is not a criticism of an able station staff and management to point out that the idea for this program, an obvious one for a television station in Philadelphia, was presented by Jack Reilly, whose connection with KYW is that he produces *The Mike Douglas Show* (for Westinghouse) in its basement—and the only reason Reilly had come up with it was that his seatmate on a train from New York one afternoon was a member of the Bicentennial Commission who said gloomily that he was hoping to get some attention for the project on one of Philadelphia's unaffiliated UHF stations.

Group W takes seriously the local responsibilties of its stations, and requires each of them to produce a half-hour local public-affairs program every Sunday afternoon, plus another half-hour in prime time once a month (plus considerable "cultural, educational and/or informational" programming in other Saturday and Sunday daytime slots). KYW has a weekly "Report From . . ." program on which prominent state and local figures, up to and including governors and mayors, are interviewed to find out what's on *their* minds. General manager John M. Rohrbach, who came to KYW from the company's Baltimore station, likes the fact that his offices and studios are smack in the heart of town, "not like being on Television Hill in Baltimore, or out on Soldiers Field in Boston—or on City Line Avenue, where the other two network affiliates are, here in Philadelphia. I'm much more conscious of *people* here." Included in the station's 43 "managers" (out of 198 employees) are a top-level area vice president and a community affairs director whose prime responsibility is to keep up with "the community." But the nation's best public-school music program got shut down in a budget crisis without any special attention from KYW-TV or its rivals, to take a not insignificant example: it was more exciting, somehow, to catch the protest demonstrations about the budget cuts in general, which are part of a national story.

At WMAQ-TV, NBC's owned station in Chicago, local and national interests are mixed from the beginning. Looking at it one way, the station has the use of the entire 120-man NBC Midwest news staff; looking at it another way, the station has no news staff of its own. Program director Harry Trigg came to the station in 1949 from the Goodman Theatre (then a nonprofessional teaching operation, now a Ford-funded professional regional theatre), and he has two and a half hours a week for "the arts," but it doesn't work. "It's horrendous," he says, "in terms of digging up the people, finding the ideas, the scripts, things for them to do." An annual feature, called *The New Performers*, takes twenty-six Chicago high-school kids, "six up front, twenty in chorus and supporting," and puts them to work to supply an hour's entertainment, "our show, our material." Trigg is happiest with some semidocumentaries by Scott Craig, one of three producers on the WMAQ staff. Among them were a history of slavery in Illinois ("black actors and script") and an hour based on the columns of Mike Royko ("episodes—from what happened to the children of Bonnie and Clyde to racial injustice to the Alderman who thought he and his colleagues were the best judges of what should be in school libraries"). Unfortunately, Trigg adds, "none of these things can be routined: local stations' big efforts are all specials." One thing that is routined at WMAQ is *The Kup Show*, two hours late at night on Saturday with columnist Irving Kupcinet earnestly seeking enlightenment or jokes from a mixed bag of literary and entertainment figures seated in soft chairs.

Every Sunday night at 10:30, WMAQ broadcasts a half-hour public-affairs show—but "we don't call it that," says general manager Bob Lemon; "it's like telling people not to watch when you call something 'public affairs.' We call it *Sunday Night Special,* and over seven years we've covered a wide range of this community's problems." Major efforts—like a ninety-minute special on the court system with Howard James of *Christian Science Monitor*—may cost WMAQ as much as $60,000. The station hopes to get back some of the cost by selling such programs to the other NBC-owned stations (as KYW expects to spread the cost of its larger documentaries over the other Westinghouse stations), and sometimes this works.

WMAQ, then known as WNBQ, was the first local television station to discover what a profit center the station's own news-sports-weather show could be. William Ray, now chief of the FCC division of compliance, was its news director in those days, the late 1940s and early '50s, and he remembers it as done with mirrors—no cameras on the streets, no news sources but the telephone and the teletype and the most recent editions of the papers—plus the savvy of an experienced journalist at the mike and the first production values applied to televised weather prediction: a young Japanese drawing isobars with a grease pencil on the far side of a translucent plate printed with a map. When Ray left in 1959, he says, his weatherman on camera was earning $125,000 a year. . . . The reason Ray left was discouragement at his inability to persuade NBC in New York to give him a budget for outside cameras and reporters, though the streets were beginning to crawl with camera teams from the local CBS affiliate. "They said we were still first in the time slot," Ray recalls, "so why spend the money? They found out why later, when they stopped being first in the time slot. I quit when I learned the network's hatchet man, who went from station to station cutting costs, was coming to Chicago. There would be bursts of interest from New York—one day, I remember, there was a wire—gotta put on three public-service half-hours immediately, or our renewal is going to look bad. I have no doubt that's the way a lot of people still operate their stations today." At about the time Ray joined the FCC, in 1961, WMAQ pulled up its socks and began looking around the city.

Like all local stations, WMAQ has found that audience reaction is most visible after editorials (when WCCO-TV dropped editorials from *The Scene Tonight* in 1970, because its editorialist had decided to run for Congress, mail dropped by two-thirds). In 1969 WMAQ tried to capitalize on the interest in editorials by running a set of five "on basic problems in our society" and then combining the five into a Saturday night show. Mobile units were put on the street in the next few days to get public response to the editorials, and there were so many comments recorded that the station ran two ninety-minute shows and one thirty-minute show, on Saturday afternoons, presenting thirty-second clips of Chicagoans responding to what the station had said. "I know it's good," Lemon said in fall 1970, "it's the essence of television—and it's like potato chips, once you hear one of these responses, you can't stop. But it didn't get an audience, and while we're doing it again for editorials on *The Troubled City,* I don't think we'll get an audience this time either."

7

When television was young, expenses were much smaller and profit potentials were known to few, a number of unaffiliated stations in the larger cities attempted ambitious programs and extensive local coverage. Symphony orchestras were telecast on commercial channels in Chicago, Cleveland and Minneapolis. In New York, Channel 13, then WNTA, produced a *Play of the Week,* including full-length dramas by Shaw, Brecht, O'Neill, Chekhov, adaptations of Turgenev, greatly increasing total audience for each and recouping the costs (but not, alas, much more) by broadcasting the same play five straight nights. KTLA in Los Angeles pioneered the large-scale use of remote units—news on-the-spot live, the TV trucks cruising the streets, listening to police and fire calls. Some of this tradition survives. For others, the most remarkable moment on television was probably the moon walk, but for those of us who were in Los Angeles on February 6, 1971, the high point of the medium will always be the earthquake coverage from the KTLA helicopter, the zoom lens revealing the cracks in the dam, the death of a pickup truck under a collapsed freeway bridge, the horror of the rubble that had been the wing of a hospital—all while the ground was still rumbling and bouncing under our feet. To give Los Angeles its due, these pictures were carried by all the city's television stations, KTLA asking only for a super (white lettering superimposed on the picture) reading "From the KTLA Channel 5 Helicopter." It is hard to imagine such cooperation, even in the face of natural disaster, in either New York or Chicago.

But in most cities the remote capabilities of the independents are now otherwise occupied. These stations have one advantage over network affiliates, and only one: their evening hours are theirs to plan, and they carry the baseball games in summer, the basketball games in winter. These, plus the kiddie cartoons in the later afternoon, are the only sure audience they have. In general, even if the cameras were not implanted at the ball park, the resources of these stations for local coverage are much slighter than those of the network affiliates—not only are their average audiences much smaller (rarely as much as a fifth of the audience to the weakest network affiliate), but they have to pay for their programs instead of being paid. Often enough, they are expected to spend rather foolishly what resources they do have, competing with the networks in areas where competition is flatly impossible. Thus, Forum Communications, seeking the license of WPIX, Channel 11, in New York, accused the incumbent ownership of illicitly identifying yesterday's film from Eastern Europe as today's film. In the course of this ugly and inconsequential dispute over news honesty (which served to obscure the fact that Forum was more likely to provide more interesting and more local programming than WPIX was offering) nobody seems to have asked why WPIX should have been attempting to cover Eastern Europe at all.

And there is another model for operating a local station, represented in its purest form by WTOG-TV in Tampa–St. Petersburg. The station went on the air in January 1969. It is a UHF operation in a market where no UHF has made money since the first VHF arrived (but Tampa started on the FCC maps as an all-UHF market; and here, unlike the situation almost everywhere else, the oldest sets are already equipped with UHF converters). Though there are three VHF network affiliates in Tampa, one of them, the ABC outlet, is what WTOG general manager Howard Trickey calls "a crippled V." The ABC tower is low, awkwardly placed and low-powered, but the FCC has on three occasions refused to permit its removal to the general tower farm, because a stronger Tampa signal on Channel 11 would interfere with—would "short-space"—the Channel 10 signal from Miami.

WTOG picks up some network shows that the local affiliates don't carry, notably the CBS movie Friday night and the NBC movie Saturday night, when the Tampa affiliates are carrying their own movies and taking all the revenue. In spring 1971 the independent was also allowed to carry the Sunday night *CBS News*, which the local CBS affiliate had refused to clear. But mostly WTOG must rely on syndicated shows—programs the networks have already broadcast, now made available for second or third or even later runs. And it doesn't get first crack at these. "We couldn't afford to pay the syndicator the price the V's pay," Trickey comments. "The film salesmen come to us fourth. But, then, we figure that when he comes to us, he really wants to make a deal."

WTOG has no news staff at all, and does no news shows. "Why does everybody feel you've got to have news?" says program director Loren Mathre, a large blond young man who came down to WTOG (as did Trickey) from KSTP Minneapolis–St. Paul, which has the same ownership. "It's because of what you promised when you got the license. We are in a market with two stations doing a substantial, good job on news. What's the reason for us to copy? We didn't promise. We're an unaffiliated UHF. We just promised to try to stay alive."

Trickey feels that WTOG has stayed alive—and, indeed, become profitable—because it has thought through its problems and spent enough money: "We came on with new equipment—top drawer—the best that money could buy. Then we could attack the market aggressively—we didn't have the usual hat-in-hand UHF posture; we walked in the door as advertising pros, and we got respect." The tower is fifteen hundred feet high over flat country, and broadcasts at 1.8 million watts, which is giant power. Each of the two RCA color cameras cost $83,000, and the values, for local advertisers whose commercials are made at the station, are real enough.

A few of WTOG's plans have not worked out. Trickey had expected (on the basis of previous experience in Green Bay, Wisconsin) that an independent could steal some audience from the affiliates with coverage of local sports. "But," he says, "many high schools and small colleges here believe the rights to their games are as valuable as the National Football League. By the time you get done with the cost of the rights and the remotes, you're just trading dollars." Programming has concentrated on the juvenile community because, in Mathre's words, "The easiest thing to do is get the kids. We start with the old *Lassie* show, now known as *Jeff's Collie*, then we have cartoons and *Lost in Space* and *Batman* and *Mr. Ed* and *Patty Duke* and *Munsters* and *Star Trek*—that's a pretty strong line-up, with good flow."

Now and again, WTOG has attempted to do something with local talent, but, Mathre says, "The best I can say is that it's uneconomical. We can produce a local country-and-western show for $250, or buy one for less than $50, and the one we buy is better." Provided someone else will take care of the actual program, however, WTOG is more than willing to perform local service. The event Trickey is most pleased with was a meeting of the local school board held in the WTOG studios before WTOG cameras, pre-empting the entire evening schedule. The station supplied telephone girls to receive calls and relay questions to the school board members; the issue before the board was busing, and the audience was large. During the 1970 elections, WTOG made fourteen half-hours available to the League of Women Voters, for interviews with candidates. For some of these half-hours, Mathre manned a camera himself. "You can learn how to run our cameras in four minutes," he says; and Trickey adds, "With the new zoom lens, you don't even need to be smooth on dollying." The other cameramen are often college kids employed part-time at the station. "We move them around," Trickey says. "They're more interested in learning a lot than in joining a union."

8

"The men who own local stations," says Walter Cronkite, "have the mentality of movie exhibitors." But the fact is that even stations which care about local service—the KAKEs and WCCOs and WMAQs and KYWs—don't do the job well. Indeed, it is more than possible that—un-American as it may be to say so—this is a job that cannot be done well in large cities. No big-city newspaper covers its territory anywhere near so capably as an average small-town daily. Moreover, we have now had half a century of judging newspapers by the attention they pay to national and international stories, remote political and economic events. People in the news business no longer pride themselves as they once did on their knowledge of and influence in their own community. Local is identified with trivial, and condemned: "It was inexcusable of Channel 2," Fred Friendly wrote, "to turn its 11 P.M. news into a local edition, when millions of New Yorkers who couldn't get home in time for Cronkite at 7 P.M. were watching their first news of the day." The 1969–70 Du Pont-Columbia *Survey of Broadcast Journalism* expressed pleasure that "a growing number of stations showed a willingness to go overseas for news and public affairs footage during the year." But why should a station be praised for allocating scarce resources to overseas stories while its own backyard grows rank with unharvested news? The tendency of both newspaper and broadcast news to concentrate on remote events must be a major cause of the "alienation," the "feeling of powerlessness," now noted by the same professors who for years were violently critical of the papers for their stress on little local stories.

Wallace Westfeldt of the *NBC Nightly News* feels that the local news shows are getting steadily better, and Du Pont-Columbia has found a number of programs to praise—a series of daily film reports by KPRC-TV in Houston, detailing the failure of the city to supply essential services, including water and sewer lines, to depressed Negro neighborhoods; an exposé by WSB-TV in Atlanta of the local numbers racket; a tough show by WJZ-TV in Baltimore on blockbusting real-estate agents; a show by Chicago's WMAQ-TV on a bank paying less than standard interest rates on savings deposits; a program by WJXT-TV in Jacksonville on crookedness in the police department of Jacksonville Beach (this station was also responsible for the document that blew the whistle on the racial bias of Judge Harrold Carswell, a Nixon nominee to the Supreme Court); a ninety-minute special on WHYN-TV in Springfield, Massachusetts, on the quick deterioration of a new public housing project; a documentary by WIIC-TV in Pittsburgh on a training school for juvenile delinquents, threatened with a fund cutoff by the state legislature. (The Du Pont-Columbia *Survey* also commended KVOS-TV in Bellingham, Washington, for a series called *Our Northwest Environment*, taking special note of the fact that Bellingham is "a small city." The comment was New York provincialism at its purest, for KVOS-TV is essentially a Vancouver station, with a studio and, more important, a selling office in the Canadian metropolis.) *Broadcasting* reports a major investigation of police corruption by WHAS-TV in Louisville; *TV Guide* a similar study by WDIO-TV in Duluth.

Yet all this "investigative reporting," expensive and sometimes courageous as it may be, is a poor substitute for attention to the unobvious physical, economic and social changes that are determining what will happen to the area the station serves. The list of significant local stories now ignored can be as long as one wants to make it: unionization of local teachers and other municipal employees; deterioration of public transportation; growth of suburban industry; diminishing land values in the central city; changing roles of local fraternal organizations; conditions of local railroads; costs and effectiveness of local health care; changes in local air and water quality (*not* reports of demonstrations "to save the Earth"); coordination among area police forces; the decline of the concert business; maintenance and use of local parks; energy supply and the local electric company; suburban election campaigns and the issues on which they are fought; etc. really ad infinitum. Often there is not much here to "investigate" as the word is generally used, and there are no "issues." But there is a great deal to report, to analyze and to understand, before people less well informed than reporters ought to be can begin identifying and screaming about an "issue."

Unfortunately, the FCC, while promoting localism to beat the band, has been discouraging the use of imagination in local coverage. Stations are not to know their city and its hinterlands as a newspaper would. Instead—though decrying the networks' reliance on ratings, their insistence that they "give the people what they want" —the FCC has required of local stations that they maintain contact with all "community groups" to ascertain "community needs." But this is simply a feedback loop, because the "community groups" to which the FCC will listen are mostly church-related, foundation-supported, federally-encouraged or otherwise linked with similar groups in national alliances, invariably seeing their needs in relation to national needs. The easiest way to satisfy some of the most pressing community leaders (and by far the cheapest) is to set aside a few jobs and contracts for the boys and give some time at some hour none too easy to sell anyway, which will permit the leaders to talk to their friends via broadcast facilities. The FCC at license renewal time will ask only for the number of hours devoted to community problems, the number of community groups consulted, the number of community needs met. Even at Westinghouse, which lives not only in accord with the FCC but virtually in an odor of sanctity, the way the rules work is beginning to produce major frustrations. "All that interests them in Washington is quantity," says Frank Tooke of KYW; "whether or not there's any quality to what you do doesn't interest them at all." No doubt the television industry has discriminated grossly against Negroes, and there were many too few black faces before or behind the cameras (WRC-TV in Washington, the NBC-owned station, seems to have the only local evening news show with a black anchor man), and "community" intervention has mitigated (though not cured) the disease. But the device used to achieve the result—making license renewals contingent on "community" consultation—has operated to limit investment in local programming and the growth of local professionalism.

Everywhere in local television a special case of Parkinsonism makes costs rise faster than benefits. In the big cities union rules require a team of four to go out on local assignment—a reporter, a cameraman, a sound man and a driver. (London unions have gone even further: Thames Television can cover only at a price of six men on the story, the argument being that big-city professionalism requires two cameras with two sound men to permit separate camera angles and smooth transitions.) Where television cameras are used, there must also be, by union rules—no longer by technical necessity—a gaggle of engineering people in the truck, men to man the lights, etc. Color adds dramatically to costs but only marginally to income. Popular identification of the program with the anchor man and the suave "eyewitness" reporter diverts any increase in revenues to salaries for on-camera personalities rather than to improvements in coverage. And news stories, like comics, don't *last* as long on television: people get bored, which discourages the investment of time.

9

In theory, public television should fill some of the holes in local television coverage. Most commercial stations live on national spot advertising, and sell their minutes to local retailers on a very hard-nosed business-benefit basis. But the public television station is dependent on community support, either through appropriations from the city council or school system or state legislature (80-odd percent of revenues, on the average) or through local contributions. And some public television stations do try, mostly through live coverage of school board or city council meetings, and through interview programs in their studios. Most of them have only minimal contact with what performing arts may exist in their communities. The 1970 Report of the Corporation for Public Broadcasting mentions a four-part series on *The Folk Music of Arkansas,* produced by KETS in Little Rock with camera teams roaming the Ozarks; an original musical play produced by WMVS in Milwaukee; and a Vermont State TV series on *The Sights and Sounds of Vermont,* including a fiddlers' contest and a lumberjack roundup.

The neglect of local professional performing talent by public television has been tragic and unnecessary. Chicago, Washington, Houston, Minneapolis, Milwakuee, Seattle, Oklahoma City and many others have repertory theatre companies which survive only through aid from the Ford Foundation, and in one stunning grant in the early 1960s Ford gave the nation's professional symphony orchestras a lump sum of more than three times their annual box-office revenues—and all the educational television stations have been living off Ford largess from birth. But because the bureaucracy at Ford divides up the arts and television in separate watertight compartments, the Foundation has never made serious efforts to bring together the television stations that desperately need material and the performing companies that desperately need audience.

Few public television stations have the money to produce news shows, and it seems fair to say that most of them wouldn't know how to go about it if they did have the money. The first important exception was KQED in San Francisco, which stepped into the void of a newspaper strike in January 1968 with an hour-long program called *Newspaper of the Air,* employing reporters from the striking papers. That fall, with Ford money, KQED built on this experience to create *Newsroom,* a format in which reporters who have covered a story are subjected to questioning about it by a city editor who sits inside a big horseshoe table ("in the slot") and other reporters who sit along the outside of the table ("on the rim"). Mel Wax, the city editor, is also in effect the show's producer, determining which stories will get air time. On stories within easy driving distance of San Francisco (which includes the state Capitol in Sacramento) the presentation is by the reporter who was there himself, and the men on the rim are reacting more or less cold, like the viewer. Wax himself introduces the story and the reporter, and may ask a question or two. The story is most likely to be presented on film, sometimes with sound on the film, sometimes with a voice-over by the reporter; but often enough a slide lecture is the medium.

At KQED the room where the show is shot really is the room where the stories are written, though the horseshoe table itself, at one end of the room, is unoccupied until air time. The flavor of the ensemble can be savored by a visitor, because the room is approached from the second floor of the reconditioned warehouse KQED uses as its studios, and the stairway leading into the *Newsroom* cavern has a landing at the top. Reporters wearing real green eyeshades and typing with two fingers sit at chewed-up wooden desks, sometimes looking into space, sometimes chatting with copy girls, rarely communicating with each other. Wax sits at a desk in a rear corner, writing his own copy and looking at that of his reporters. The line-up is firm at 5 o'clock for a 7 o'clock show that runs a full hour. Because everybody is to be on camera, not just the anchor man, voltage rises all over the big room as air time approaches, cameras are jockeyed into position around the horseshoe table, copy must be cut to length.

How good a job *Newsroom* does is a matter of dispute. Because the show is "designed to be an alternative to news programs offered on commercial television," national and foreign news are covered as well as local news, and here the ignorance of the men on the rim—the fact that they have not covered the story but have merely been reading about it, like the foreign-affairs editors of a news magazine—sometimes makes the format unfortunate. Some intelligent San Franciscans in 1971 said the program had become "too ideological," which seemed to mean that the *Newsroom* reporters continued to stress race troubles and student troubles and war protests after viewers had become bored with them; audience was dropping. But the fact that *Newsroom* reporters had beats and covered them meant that questions could be answered at the rim with some authority; local politics and "social process" stories were undoubtedly covered better by the KQED *Newsroom* than by any commercial news show. The Ford Foundation liked the format, and has funded similar programs for Pittsburgh, Dallas and Washington;

Fred Friendly said in 1971 that the one in Dallas was the best of all, "because *Newsroom* works best where the newspapers are worst." Ford also helps with the bills for a wholly local news show, *The Reporters*, on WGBH in Boston, offering a young, eager, predictable, not quite professional group of newsmen.

KQED has six mobile units, and uses them. It devoted an entire day to protests in the Bay region after the Cambodian invasion, mixing half a dozen live remote pictures in its control room, and has covered live such interesting events as the riot in the Berkeley City Council chambers when the question of buying the police a helicopter came up for consideration. There have been symposia on subjects like marijuana and suicide, and sometimes they run all through one broadcast day and into the next. Large efforts have been made to put "the community" on camera, talking about itself, and even to put the production of programs in the hands of ordinary people "outside." General manager Richard Moore says, "We regard KQED as more like a social agency, a communications center linking up the various parts of the community."

The problem is that all this unorganized personalization can reduce to theatrics and become dreadfully dull; and Kierkegaard was right in his perception that the worst thing in the world is to be bored; bored once, twice (or more often) shy. And meanwhile KQED has conveyed little sense of what is actually happening— to the port, the financial community, the retailing establishment—in the most beautiful but also most rapidly deteriorating and dysfunctional city in America. What the Black Student Union thinks of racism in the choice of secretaries by the chemistry department is on any scale much less important—and after the third time much less interesting—than the impact of containerization on the port or of computerization on the banks. KQED may illustrate a danger that neither money nor dedication to local coverage is enough: without professional program judgment, a simple-minded search for relevance and community participation will lead to irrelevance and community neglect.

In early 1972, in response to a funding crisis, KQED cut *Newsroom* back to half an hour.

CHAPTER 12

Public Television and the Meaning of Diversity

The strongest objection to the more trivial popular entertainments is not that they prevent their readers from becoming highbrow, but that they make it harder for people without an intellectual bent to become wise in their own way.
> —RICHARD HOGGART,
> assistant director general, UNESCO

Public broadcasting is the source of countless words and images performing many services . . . helping teachers in classrooms, exercising young minds at home, enriching leisure through hobbies and crafts, improving professional skills and developing new professionals—providing an arena for the clash of ideas, a cultural center in the home, an extension of the municipal hearing room, a ticket to what's beyond the here and now and an open invitation to the individual to grow.
> —Annual statement, Corporation
> for Public Broadcasting, 1970

If what interests you doesn't interest other people, then maybe, to quote Noel Coward, you shouldn't be in show business.
> —AUBREY SINGER, features director, BBC

1

Television came to America shining in new hopes appropriate to a new medium. The second half of the twentieth century was to be the era of the common man; television, as Pat Weaver put it,

313

would make the common man the uncommon man. More people would see a televised *Hamlet* in one night than all the theatre audiences of history put together. The great scientists would explain their discoveries; the great composers would write new operas for the home screen; theatrical and historical resources would join together to recreate the past for a public eager—as the American public always is—for self-improvement. Meanwhile, the cameras ranging the globe would make an end to provincialism, and bring the real world, all of it, to the living room. All this would serve the self-interest of advertisers as well as the public good; the costs of the service would be more than paid by the proprietors of the advertised brands. In the end, of course, nobody would pay them, because the mass production of goods sold by television advertising would reduce costs more than advertising increased them. Or so it says here.

But the advertiser, as noted, went looking for audience; and the audience, while willing to try anything once, opted once again not to improve itself. Comedians, sports, westerns, mysteries, goo of various varieties, finally the quiz shows gave serious observers of the medium an increasing malaise. The FCC had from early on reserved a large number of channels for noncommercial operation. After the quiz-show scandals, it began to seem extremely important that these stations become a viable alternative to commercial broadcasting. Though the FCC in 1952 had put aside 242 channels for educational stations, by the end of 1959 only 44 were on the air, and most of these were broadcasting only a few hours a day.

Typically, the educational station was controlled by a school system or a university, and its programming bias was toward whatever instructional efforts the Ford Foundation or its subsidiary Fund for the Advancement of Education was willing to finance. There were teacher-training programs in Texas, college courses for credit at Penn State, a complete junior college program ("TV College") run by the Chicago City Junior College, school subjects for classroom use in Dade County (Miami), Florida. Mostly it was one teacher speaking to one camera, and dullsville. The apex of this development, if the expression is permissible, came in 1961, when

Ford equipped two airplanes as TV transmitters and set one or the other of them to circling every day four miles above the town of Montpelier, Indiana, broadcasting two channels' worth of third-rate instructional programs to schools in a heavily populated area of more than one hundred thousand square miles.

As late as 1960 educational television was very small potatoes. The Ford Foundation Annual Report for that year showed seven grants for "Activation of noncommercial channels." The largest of the grants was to the University of Connecticut, for $34,550; the smallest, to Tri-County College in Saginaw, Michigan, was for $18,000. Those seven stations were *all* the new noncommercial channels started in 1960. The quality of the equipment with which they opened service can be guessed from the size of the grants—and in a few instances can still be experienced by discouraged viewers.

Here and there, however, educational stations had secured a fairly broad base of community support, and were moving on to more ambitious, not specifically "educational" programming. WGBH-TV in Boston, sponsored by a consortium of fourteen institutions, was moving from a BBC Third Programme kind of radio to an extensive television service; and while its transmitting tower was underpowered and in the wrong place (to the south, when all the commercial sticks were to the west: antennae set to catch the commercial stations produced unsatisfactory pictures on WGBH-TV), the station did have Channel 2 to play with. In San Francisco a committee of volunteer enthusiasts, brought together by the American Association of University Women, had grabbed off Channel 9, KQED, for noncommercial use, started it with two two-hour bursts of evening programming a week, finally built it into a multiple-purpose programming entity, with drama, cabaret, chamber music, education, controversy, how-to, local events, etc. KQED is a membership corporation with a board elected by vote of all those who give more than a minimum annual contribution. In 1971 the budget for the station was over $4.5 million, and something more than a quarter of that was raised from 45,000 members (3 percent of the households in the reception area), who contributed at least $14.50 a year to become members. Among the ideas first brought to the screen by KQED was the televised auction, with celebrity auctioneers displaying to the camera items contributed by well-wishers, to raise money for the educational station; such auctions are now annual events in most cities of the country. William S. Hart of WYES-TV in New Orleans describes his "Bid-By-Phone Auction. . . . This marathon lasts nine hours a day for a solid week. The governor, the mayor, churchmen, celebrities and prominent businessmen act as our auctioneers."

As early as 1954 the Ford Foundation began giving money to a service designed to help these isolated educational stations exchange information and programs. National Educational Television and Radio Center (later just NET) was based in Ann Arbor, and performed almost exclusively technical chores, duplicating and distributing first films, then tapes supplied by stations which took payment, usually, in the form of other stations' programs. By 1960 Ford had put $14 million into this venture, and sometimes wondered why.

The breakthrough in noncommercial television came in 1962, when a self-perpetuating nonprofit corporation raised the money to buy Channel 13 in New York. Educational television lost its state-college, school-system image and became something the New York–based national media started to discuss more seriously. Congress responded with the first appropriations to help build and equip noncommercial stations, and Ford escalated its support. NET, with an annual $6 million Ford grant, became in 1962 a New York–based producing organization, supplying filmed programs, documentaries and cultural entertainment to noncommercial stations all over the country. It was still not a network, however: AT&T interconnection charges were far beyond the resources available. Except in the Boston-Washington corridor, where a group of stronger stations had established an Eastern Educational Network, NET programs were telecast by the "affiliates" as the tapes arrived in the mail. The most popular programs could be on the road as long as six months.

Two unrelated events in early 1966 changed the focus of non-commercial broadcasting: McGeorge Bundy left his job as Special Assistant to the President to become president of the Ford Foundation, and Fred Friendly resigned as president of CBS News to protest his network's refusal to pre-empt daytime entertainment to carry George Kennan's testimony on Vietnam before Senator Fulbright's Foreign Relations Committee. The morning the story of his resignation broke, Friendly received a telephone call from the White House, where Bundy was in process of cleaning out his desk. The new president of the Ford Foundation had been browsing in the history of his organization, and had found records of vast sums poured into noncommercial television (by 1966, almost $130 million) with extraordinarily little to show for it. He felt the need for a television adviser, and Friendly was apparently now at liberty. The two men dined (for five hours) that same week in New York, and presently Friendly was on salary as a consultant to Ford and Bundy's office was decorated with three TV sets, each permanently tuned to one of the networks, just like the office of a television executive.

In Friendly, Bundy found an overpowering salesman still driven by the original conception of television as "seeing at a distance" and committed through many budget wars to the proposition that good television cannot be bought for cheap. Bundy felt two prime missions at Ford: to "do something" about race relations in America and to create an alternative television service. WGBH-TV had been a pet of sorts at Harvard when he was dean of the faculty, and Lyndon Johnson had thrown his worst temper tantrums in response to what he considered slanted news on the tube. Bundy was interested in television.

By the summer of 1966, Bundy and Friendly had half-found, half-built their bombshell. In response to an FCC invitation for suggestions about what should be done about a system for communication by satellite within the United States, Ford submitted a wholly unexpected proposal. The satellite, the Foundation urged, should be dedicated to television use (no channels reserved for telephone or telegraph or data communications). It should provide free interconnection service for noncommercial television stations. Its operations would be considerably less costly than the AT&T line-plus-microwave system the networks were renting; the difference between the high Bell System prices and the low satellite costs should be a "people's dividend" from the space program, to subsidize educational television.

Ford's proposal was greeted with shouts of pleasure, all over the country, but after a while the shouting died down a little. There were technical difficulties with some of the specific suggestions in a proposal that had been drafted in less than a month—twenty-four hours a day, seven days a week, as the August 1 deadline neared. The costs had been figured wrong, and the people's dividend was more likely to be below $30 million than, as Ford had calculated, over $60 million. And, worst of all, the managers of the educational stations were not entirely sure why interconnection would be so important to them. They were local operations, locally supported; did they really want to be part of a network, like the commercial stations? What was the point?

In December 1966, Ford amended its proposal somewhat, and simultaneously announced that it was making a grant of $10 million for "remarkable demonstrations of the power of a national educational television service." This would be a two-hour program, to be supplied every week on Sunday nights, starting in fall 1967, to nearly all the nation's noncommercial stations (which had been dark on Sunday nights, for budget reasons). It would "pull together the intellectual and cultural resources of this country to speak directly, once a week, to the great issues of the day in every field of action."

Ultimately the new program was christened Public Broadcasting Laboratory. Its producer was Av Westin, who had been one of Friendly's most efficient and alert assistants at CBS News; its anchor man and commentator was Edward P. Morgan, who had been a leading news figure for ABC, especially on the radio network. Cultural presentations were supervised by Lewis Freedman, who had been director of programming for Channel 13 in New York. Opening night was November 5, 1967, and it was well attended, thanks to major advertising in most metropolitan newspapers and reams of advance publicity. By the end of the first two-hour program, the audience had left, never to return. On Broadway the show would have closed like a door; on television, with Ford largess, it lasted two years, demonstrating not "the power of a national educational television service" but the truth of Sol Hurok's observation that "If the public doesn't want to come, *nothing* can stop them." By its last months, PBL was drawing considerably less than one percent of the Sunday night television audience.

An inquest on PBL, while long overdue, is by that token out of date. The topic for the first program—race relations—was predictably wrong: people were getting programs about race relations every week on the commercial networks. The "cultural" part of the show was a pathetically childish dull play done in white-face by a semiprofessional Negro company. The nonfiction part was built around a staged black-and-white confrontation at a theatre-in-the-round, which went on and on and on. Later shows were better paced, more professionally produced, even, on rare occasion, more sophisticated. But all suffered from three systemic weaknesses: they were so good for you they tasted like medicine; they were so fascinated by television they looked very much like off-focus copies of what the networks were already doing (the fascination extended to advertising: deprived of the possibility of spotting the program with commercials, the producers interrupted their program with one-minute anticommercials to tell you how bad products were); and they all came opposite Ed Sullivan and *The FBI* and *Bonanza*. "Bad as it was," Pat Weaver said recently, "that show might have drawn an audience if they'd had the sense to start it at seven o'clock, opposite the kiddie shows on the networks." Believing their own propaganda about the American public's dissatisfaction with commercial television, the promoters of PBL sent it out to do direct combat with the most popular shows on the air. It lost.

Even before PBL went on the air, the Ford Foundation sponsored occasional NET nationwide interconnections, the first of them being the coverage of Lyndon Johnson's 1967 State of the Union Address to Congress, followed by a live discussion of what the President had said, the discussants being college professors and public figures rather than newsmen. It turned out that the professors and the public figures had little more to say than the newsmen were saying, and said it less well. Nevertheless, Bundy and Friendly stuck by their guns, insisting on interconnection as the essential need of public television, and the satellite proposal (grounded in the bureaucracy of the FCC, where it still remained in early 1972) as the most efficient device.

This position had already come under strong but not publicly admitted attack from the Carnegie Commission on Educational Television, a high-prestige fifteen-member committee headed by James R. Killian, chairman of the corporation of MIT. Like Bundy, this committee, set up in 1965, had a bias toward the educational world of Greater Cambridge. In addition to Killian, the representatives from the Harvard-MIT galaxy included former Harvard president James Conant, Boston-based Edwin Land of Polaroid, Franklin Patterson of Tufts (en route to the presidency of Hampshire College)—and, as assistant to the chairman, Stephen White, television critic for *Horizon,* a former New York *Herald Tribune* foreign correspondent who had been working at MIT as aide-de-camp to Professor Jerrold Zacharias, managing the multimedia and common-sense aspects of the Physical Sciences Study Committee course for high-school students. White wrote the Carnegie Report.

The most clearly permanent of the suggestions the Carnegie Commission made in January 1967 was a decision on desirable nomenclature. "Educational television" was a terrible name for a service, an invitation to people to stay away. Killian and White, moreover, were not very interested in the use of television for instructional purposes: they wanted to know what could be done with these precious noncommercial frequencies during prime time. To describe the service that would bring prime-time programs to the people without advertising, White coined the brand name "Public Television." That much has stuck.

So have the idea and name of the Corporation for Public Broadcasting, an autonomous body formed by Act of Congress late in 1967, with a board of fifteen appointed by the President and confirmed by the Senate. Hereafter, reality departs from recommendation. Carnegie wanted CPB to be funded by the federal government at about $100 million (at least $60 million) a year, with the money to come from a dedicated trust fund, derived from a tax on the sale of new television receivers. Instead, CPB has an annual direct appropriation from Congress, at the mercy of annual votes; and the largest appropriation to this date is that of fiscal 1972, $30 million in direct grant plus another $5 million to be matched by private sources, with the dollar worth about 75 percent of what it bought in 1966.

Most important, Carnegie wanted CPB to be essentially a service agency for local stations. These stations should be interconnected for the distribution of programs that could be taped, and then used at any time; and occasionally for the national display of an event or an extraordinary production by any one station. But the ensemble would not be operated as a network with a planned nationwide schedule of programs. Arthur L. Singer, who was Carnegie's staff liaison with the Commission, told a meeting of the National Association of Educational Broadcasters in summer 1971 that "The Carnegie Report considered the advisability of a fourth network, and rejected it as a solution.* The Public Television system," he continued, "has assumed the posture of a fourth network, with what are really insignificant variations, and is now operating exactly the way it was assumed, a few years back, a fourth network would operate."

In fact, the law establishing CPB forbids it to operate a network, and the "Public Television system" to which Singer refers is a separate entity, the Public Broadcasting Service (PBS). This is a membership corporation, with the stations themselves as members. It has a board of eleven, six of whom must by charter be managers of local stations. Its income derives entirely from grants by CPB. PBS will operate the interconnection facilities, which in 1971 were still being built (the network operated originally through facilities leased from the Hughes Sports Network), and will pay AT&T its

---

* Very bluntly, too: "Ordinary networking of taped or filmed programs, insuperably linked with the concept of the single signal, appears to the Commission to be incompatible in general with the purposes of Public Television. It presupposes a single audience where Public Television seeks to serve differentiated audiences. It minimizes the role of the local station, where Public Television, as we see it, is to be as decentralized as the nature of television permits. Public Television is justified in reconsidering the best uses of interconnection in terms of its own needs, rather than imitating thoughtlessly the familiar manner in which the commercial networks use it."

annual ton of flesh. The size of the AT&T bite was negotiated in a three-cornered way, with the FCC sitting in as friend to public broadcasting and rate supervisor for the Bell System's long lines division, and the final price, for interconnecting 110 stations, was set at $4.9 million, less than a quarter of what the telephone company charges each commercial network. Because public television is still in a growth stage, AT&T agreed to reach that final figure gradually, and in fiscal '72 its charge to PBS was $2 million.

With money from CPB, PBS commissions "production centers" to make "national television programs" for the network. There is no requirement that these production centers be local stations, or even nonprofit corporations, but in fact the pressures to get the money for the stations are considerable. These pressures made it necessary to collapse NET into New York's Channel 13, which was done, slowly and very painfully, in 1970–71, though NET was permitted to continue the use of its logo (the three letters with a roof on top) on the programs made for PBS. In 1971, Children's Television Workshop, which receives only $2 million of its $13 million budget from CPB, was the only contract producer for PBS not part of a station operation. CTW supplied the network with seven hours of programming a week, and its shows, *Sesame Street* and *The Electric Company*, accounted for more than half the total daily audience of public television. Ten hours of PBS network service came in prime time, Sunday through Thursday nights; five hours from NET, the rest divided among WGBH, Boston; KQED, San Francisco; KCET, Los Angeles; WTTW, Chicago; WETA, Washington; WQED, Pittsburgh; and the South Carolina State system that officially produces William Buckley. The total funds allocated for national programs in fiscal '72 were $23.5 million, of which $12.1 million came from CPB, $9.2 million from Ford, and $2.2 million from corporate "underwriters" like Mobil and Xerox, which have paid to purchase from the BBC programs like *Civilisation* and *Masterpiece Theatre*, and receive in return a telecast card of thanks.

To the public, these networked programs (promoted by a million dollars' worth of advertising a year under a special Ford grant, and talked up by numbers of public-relations representatives) are public television. For some station managers, and for some who were connected with the Carnegie Report, they symbolize a victory for Friendly and Ford over Carnegie and Congress. "The local station is intended to be the key element," says John W. Macy, a very steady man in a gray crew-cut and horn-rimmed glasses, who was chairman of the U.S. Civil Service Commission before he became president of CPB. "That's the Congressional intent, if you read the record." But in fiscal 1972, CPB committed only $4.2 of its $40 million to general support of local stations. There were at the start of that period 203 such stations eligible for grants; the average grant was about $20,000. (Friendly points out that Ford under his guidance gave $26 million to a selection of stronger local stations between 1967 and 1971.)

There is, however, another side. Nobody is more committed to the localism of television than Richard Moore, general manager of KQED in San Francisco. Back in the early 1960s, KQED did weekly string quartet recitals, presented dancers, worked closely with the American Conservatory Theatre, San Francisco's repertory company, and put on some ACT productions. "But we gave it up," Moore says. "NET was that much better. Besides, these things are very expensive—you must go under Actors Equity and Screen Actors Guild rates, tie up all your facilities. The unions here have caught onto the fact that there is such a thing as public television. You can't do a drama for less than $150,000." With few exceptions, the local stations don't want to do drama. They don't want to do music, or dance, or animation, or anything else requiring the employment of professional talent other than journalists. That's for separate production companies, or for the network.

Getting nationally usable programs from the stations has been much harder than CPB expected. "We've given the stations thirteen grants of $50,000 each, for programs," Macy says, "and it's from this experience that I draw my views of what can be expected. It's been very hard to get subjects of broad enough interest and professionalism. We've learned that there isn't that much talent." The Carnegie Report took as an article of faith the proposition that significant performers would emerge all over the country if offered the chance to appear on television: "In the large cities and the universities that possess educational stations, there is creative talent that has never found its way to television. There are performers of high professional skills who do not seek or would not necessarily meet the taste of the commercial mass audience." If this assumption is wrong, then much of the policy Carnegie recommended will also be wrong.

Unfortunately, the assumption can be partly right, and a policy based on it would still be catastrophic. Nicholas Johnson has described public television as "a source of programming ideas, public affairs issues, and technical innovations. It is commercial broadcasting's graduate school, its farm club, its underground press, its research and development laboratory." This is, in fact, nonsense—public television has been none of these. But if that's what it were, who would watch? The program from public television goes out into the identical air that carries commercial programs. On some terms, to some degree, with some audience, it must be competitive. "We still have too many people who think we're amateursville," says John Macy. "We can't afford to do things that aren't professional in appearance."

The educational television system CPB found in being had almost no audience at all. Unfortunately, the system also had awful habits of lying about the size of its audience. As early as 1954, KUHT, the first educational station, began talking about "800,000 viewers" in the Houston area, though the station's audience was in fact almost undiscoverable. The Carnegie Report cited Nielsen data to the effect that during an average week in 1966 6.86 million homes tuned to an educational station, which was 12.5 percent of all households. But the *total* viewing Nielsen found was only 8.24 million hours out of more than 2 *billion* household viewing hours a week—or 0.4 percent of all viewing. Any one *unsuccessful* commercial network show, with a rating of 16, commanded more of the public's time than an entire week's schedule on noncommercial television all over the country.

CPB has paid ARB to rate its programs, market by market, and has then buried the data, preferring instead to cite surveys in which interviewers ask people whether they looked at anything on the noncommercial channel during the last week or month or year.

(Respondents are only rarely asked in these surveys exactly *what* they viewed.) Such surveys produce figures of two-fifths, one-half, two-thirds of the population claiming to look occasionally at a noncommercial station. But in city after city the 1970 and 1971 ARB books (after a much-touted "doubling" of public television's audience) show noncommercial programs with nighttime ratings of 1 and even less, night after night, and cumulative nighttime ratings that rarely go over 2 or 3 percent for the week as a whole. Only in New York and in Boston does public television approach as much as 5 percent of the audience for an evening.

George B. Leonard of *Look* paid tribute to KQED in fall 1970, in an article entitled "Television Is Live and Well in San Francisco," with the subhead, "KQED may just have it—the talent, cash and bounce to transform a sick medium." At about the time that issue came on the stands, KQED ran its own telephone-coincidental audience survey, and found that on the average evening its total draw, for all programs, was 1.7 percent of its area's households. Nationwide, of course, in absolute numbers, there are viewers—even a 1 rating delivers 600,000 homes, nearly twice the circulation of *Harper's* or *Atlantic*. For authors, an interview with Robert Cromie on *Book Beat* has clear selling value, and the print version of *Civilisation* was for a while the No. 1 best-selling nonfiction book in the United States—at $15 a throw. The mail polls after *The Advocates* sometimes produce tens of thousands of replies from people who wish to register support for one cause or the other. (Usually, despite public television's alleged "radicalism," the "conservative" side wins the vote.) But neither nationally nor locally has the impact of public television been significant.

In 1966–67, WBBM-TV, the CBS-owned station in Chicago, ran a weekly program called *Opportunity Line,* offering jobs to the unemployed; the program received two thousand telephone calls a night, and produced nearly six hundred jobs a week for the callers. Such a venture is of course a natural service for public television stations, and two of them hastened to try it: WETA in Washington, which received from twenty to forty calls a program and only a handful of job placements each month; and KDPS in Des Moines, which never placed anybody in a job. (*Job Man Caravan,* on South

Carolina's state-operated television system, has apparently done somewhat better, but no figures are in print.) When Congressmen rose in wrath about *The Selling of the Pentagon*, Hartford Gunn, president of PBS, was asked by Fred Friendly whether he thought his organization and CPB could have survived the broadcast of that show on public television. Gunn's first reaction was: probably not— but then he thought again. "No, I don't think we'd have had much trouble," he said. "Nobody would have seen it."

When James Day came from KQED to become president of NET in 1969, he asked to meet with the producers of NET programs. "There were about fifty-five of them," Day recalls, "and we met in groups of about twenty. We had a dialogue. I asked them, 'What do you watch on television?' And I found they kept their roles as producers and viewers separate—they produced shows they wouldn't watch themselves. They all watched *Sixty Minutes*. I asked them why, and they said, well, they liked the simultaneity, the surprise. I told them, 'But the whole NET schedule next week is a surprise to you guys.' And they said, well, it was Harry Reasoner and Mike Wallace: they're dependable. We haven't been dependable. There's been no interrelationship. We've had a guy in a cubicle working a whole year on one one-hour show, and not caring about anything else that goes on."

"The problem at PBS," says Joel Chaseman, who is in charge of programming for the five Westinghouse stations, "is that they lack consistency. There's not a big enough centralized pot, and the viewer doesn't have that certainty of what he's going to get." Gunn at PBS believes this inconsistency was an inevitable result of NET's insistence on the "anthology" format—a weekly slot for documentaries, mixing cultural and public affairs; a weekly slot for theatre of all kinds; a weekly slot for music and dance. "There's no habit in it," Gunn says fretfully, having been unable to shake his commitments to NET. "When you go from classic drama to avant-grade, or from a rock concert to a symphony orchestra, that's a real wrench to most people. The audience doesn't exist that wants that series week after week, going from rock to opera to symphony —or, rather, maybe it does exist, but it's a really tiny elite."

What the producing stations want from PBS is a time slot and money, with no supervision and no control. "Because I spent twenty years in station management," says Gunn, who ran WGBH in Boston before coming to Washington, "I can sympathize. But I see real problems in audience service and audience-building. It's going to take a massive effort to build an audience base." And the American experience is that network shows—expensive, professional, dependable, even repetitive—are what draw the audience.

These arguments were unfortunately and unnecessarily politicized in fall 1971 through the combination of a bold speech by Clay T. Whitehead, the aggressive thirty-one-year-old director of President Nixon's Office of Telecommunications Policy, and the reactions to it from selected spokesmen for CPB. Whitehead picked up from Singer's speech and the Carnegie Report, and directed his fire at the new National Public Affairs Center under the PBS aegis, which was scheduled to begin producing its own programs in January 1972. "Instead of aiming for 'overprogramming' so local stations can select among the programs produced and presented in an atmosphere of diversity," Whitehead told the annual convention of the National Association of Educational Broadcasters, "the [public broadcasting] system chooses central control for 'efficient' long-range planning and so-called 'coordination' of news and public affairs—coordinated by people with essentially similar outlooks. How different will your network news programs be from the programs that Fred Friendly and Sander Vanocur wanted to do at CBS and NBC? Even the commercial networks don't rely on one sponsor for their news and public affairs, but the Ford Foundation is able to buy over $8 million worth of this kind of programming on your stations."

This would be fair comment by a newspaper critic, though pretty strong stuff—and the critic, noting the request by PBS to the Republican Party for accommodations for 237 people in San Diego during the convention, could go on to raise hell about PBS's plans to spend several million dollars of tax money or tax-exempt money on a coverage of the 1972 conventions that nobody would watch and that could not conceivably offer anything of importance the networks would not have. (These plans were dropped in late 1971 in the wake of Congressional displeasure about Sander Vanocur's $85,000 annual salary.) As a statement by the man who is in charge of drawing up the plans for permanent financing of public broadcasting, however, the Whitehead speech was a threat. CPB took it as such anyway, leaking an internal memo of defiance to the Washington *Post* as a kind of declaration of war.

For Hartford Gunn the situation was tragic, because Gunn believes his worst handicap is simply lack of funds—he could produce programs more likely to catch an audience if the government gave him more money. This may or may not be true—*The Great American Dream Machine* costs well over $100,000 an hour to produce without much expenditure for on-camera talent, but it has won little audience by television standards and deserved even less, being mostly a collection of snotty sophomoric comment on commercial television (always the prime subject of American noncommercial television), advertising, popular taste and domestic customs, made to look sophisticated by jump cuts between very many very short takes, unusual camera angles, distorting lenses, and the like. The PBS claim to greatly expanded public support for a national service would be much stronger if Gunn could show a collection of interesting program ideas now aborted by lack of money.

But the whole subject is extraordinarily difficult. "Nobody knows what the right balance is between the national service and support for the local stations," Gunn says. "Anyway, what is the measure of programming in public broadcasting? Commercial television has profit and loss. We may not like it, but it's a measure. People know where they stand. How do the young professors and the others running this system know where they stand?"

2

Indeed, what *is* the purpose of public television? Nearly all discussion of the subject begs the question by assuming that commercial television is (a) very bad, and (b) unsatisfying to most Americans. Even under these assumptions there remains the problem of deciding which of the infinite variety of subsets in the great set of not-commercial-television should be recommended by the analyst. Still, "anything" being "better" than what we have now, the problem isn't immediate.

Almost any survey of American attitudes toward television finds that the people want "better" programs; almost any search for specific suggestions finds almost nothing that is not already being done. People's imaginations are circumscribed by their experiences. The trap is in the notion that there is one "public" to be served; in the real world there are many publics, and the same person belongs to several of them at different times. Some of these publics are clearly not being served by commercial television. Since *The Honeymooners,* there has been nothing on American television aimed at a working-class audience or portraying a working-class ambiance. (When BBC's *Till Death Us Do Part* was transmuted to CBS's *All in the Family,* the blustering antihero was changed from a docker to a shipping foreman. "An elevator starter!" says Fred Silverman, getting the job description wrong. "That's pretty funny in itself.") Despite the obvious applicability of the medium, and the popularity of public television's how-to show on French cuisine, the mechanically curious hobbyist—the reader of *Popular Science* and *Popular Mechanics*—has not been served on television. Foreign-language instruction is a natural for television, especially on levels above beginner, where television programs and films from foreign parts can be broadcast in the original language, as they sometimes are in Germany and Holland and Scandinavia. The noncollaborative arts, poetry, painting and sculpture, have scarcely been touched by television on their own grounds (as distinguished from interview programs or biographies) and with all respect to Peter Herman Adler and *NET Opera* the history of musical presentations on American television is little short of dismal. (*NBC Opera,* which Adler ran in the 1950s, was more interesting than the NET program has been to date, possibly because it was live television, under great tension, orchestra and singers working together from separate studios, possibly because it was expected to deliver and hold *some* audience. Moreover, on NBC most operas were presented full-

length; on NET most operas have been cut down to ninety minutes.)
WGBH in Boston commissioned a series of nonobjective film treat-
ments to accompany the performance of concert music, with
predictably embarrassing results, *Fantasia accadèmica* . . . wrong
problem, wrong answers.

Serious drama, new or classical, receives little time and less
attention; and when it does get put on, the odds are about even that
it will be mangled by its producers to fit the schedule or by somebody's preconceptions about the audience—or, as in *Hollywood
Television Theatre* or *Hallmark Hall of Fame,* a desire to get big-
name movie stars up on the marquee. Intelligent conversation to be
eavesdropped is made impossible by the celebrity system, the
gimmickry of adversary situations, the display of masturbatory self-
gratification on "radical" programs like WNET's *Free Time.*

If public television were meeting the special needs of varied
publics, the fact that no individual program drew much of an
audience would be unimportant. But except for *Black Journal,* which
loses white audience faster than black audience, what PBS dis-
tributes is intended to have mass appeal; it is rather like what the
networks feed, but "better." It still breaks into half-hour, hour, and
ninety-minute pieces, as though these time units were ordained
from on high, and everything runs the same length every week,
and the reference point is always to the commercial channels.

Fred Friendly has said that public television is necessary to do
the job commercial television ought to be doing but shirks. Most of
the more popular and admired programs broadcast by PBS—*The
Andersonville Trial, Chet Atkins with the Boston Pops, Forsyte
Saga*—would have been entirely plausible commercial products,
except that they would have cost three to five times as much to
produce if their participants and producers had been paid at commercial rather than educational rates. The real ratings hits have
been telecasts of silent movies, at the wrong speed. "I found the
popularity of *Forsyte Saga* rather sad," said John Boor of Seattle's
KCTS. "What it means is that people want from educational television what they get from commercial television." This is not a
criticism of the programs or of public television; it means that some

way is needed to get such programs onto commercial stations, where they would be entirely suitable and much more heavily viewed. They don't seem to do public television any good, and it doesn't help them.

Sometimes the insistence that public television will be "better" becomes more than a little patronizing. Benjamin Britten's opera *Owen Wingrave* must be set by the cameras in the context of his festival, and the singers must tell us all how great it is; Jim Day must interview every Englishman he can find who is appropriately impressed with the profundities of Galsworthy's treacle; Alistair Cooke must stand in awe of *Masterpiece Theatre*. Ambitions soar. Reuven Frank of NBC News regretted recently that "we have built up through silly promotion the idea that a documentary will tell you everything about a subject—but it can't be: one hour of television is only five thousand words of copy." Yet Jim Karayn, vice president and general manager of the new PBS Public Affairs Broadcasting Center, tells the *New York Times* that he expects to go beyond "the headline capabilities of television journalism, to go further in really zeroing in on what is happening in this country—and why it's happening." Hartford Gunn says, "If somebody will give us the money, I'm willing to do far-out things. I don't see why we can't do a situation comedy, and do it better than the networks." The Greeks had a word for this sort of thing: they called it "*hubris*." The gods punished it.

Of course, the gods are always punishing the weak. Arthur Singer suggests that "a man who will enjoy himself dizzy watching the local high school football game can be heard complaining a few days later that the Super Bowl on national TV put on a second-rate performance. The local symphony orchestra sounds great in the town auditorium, but let a few false notes be sounded in Carnegie Hall and the audience begins to walk out." Without admitting it in so many words, the Carnegie Report had proposed that public television yield the values of professionalism to others and cultivate smaller gardens. Macy and Gunn and Day, while they have their own disagreements, object that you can't get an audience that way: all television channels are Carnegie Hall. If you have no audience, you have no impact anyway. Even the people originally most enthusiastic about performing on your channel will stop coming around after a while, especially if you can't pay them—and except in the three largest markets neither commercial nor noncommercial television can hope to pay people a living wage to produce any considerable number of programs for local use only. In the fourth largest market, Philadelphia, the noncommercial station, WHYY-TV, Channel 12 (VHF), dismissed its entire local production staff in summer 1971; the station held no PBS contracts, and there wasn't enough money available locally to pay production costs.

In the television framework alone, the Gunn-Macy-Day argument is irresistible, as an argument. (In a larger framework, considering the obligations of the electronic media to generations of young artists deprived of the opportunities once offered by the touring companies and local concert series and night clubs the media have destroyed, there is much to be said for using public television to give the young exposure, experience, and support regardless of the audience they draw.) The difficulty with the argument is that the PBS approach as currently practiced does not provide the audiences either. The decision to go to an 8-10 schedule in 1971–72 was an obvious disaster, because the commercial networks, reduced to three hours a night starting at 8, were sure to be blasting off at the same hour with the biggest rockets they could find. (In fairness to PBS, Gunn wanted desperately to program 9-11 in 1971–72, but the stations of the Eastern Educational Network were committed to launching their own 10 o'clock news show, and informed PBS they would delay half of each night's feed to the next evening at 8 if the public network insisted on carrying out its plans. That would destroy the national advertising campaign designed to draw audiences, and Gunn gave in.) Surely, the time for public television to program its most generally attractive shows should be early-fringe opposite the commercial news or late-fringe opposite the talk shows (or both: public television, seeking cumulative rather than instant audience, can repeat a great deal); and the time when the networks are firing their heaviest artillery should be the time when public television reaches deliberately for audiences specialized by location or by interest. At the very least, each new PBS season should open in the spring, when the networks are sagging back to reruns, rather than in the fall, when everything on display at the networks is more or less fresh produce.

Most of the time, for most of the people, public television is going to be less important than commercial television. The newspaper and magazine critics who keep proclaiming that public television is the most important thing on the air are doing its cause a serious disservice. But they will keep doing it; and the people of public television, who have so few other rewards, will keep lapping it up. Very sad.

3

On one thing everybody is agreed: public television must be a place for experiment; as the Carnegie Report put it, "Public Television possesses a great advantage over commercial television: it can enjoy the luxury of being venturesome." CPB has funded a National Center for Experiments in Television, in San Francisco, under the aegis of KQED; in 1971 the Center picked up an additional $300,000 from the Rockefeller Foundation. It is well housed, in the thousand-square-foot studio Metromedia built for its UHF station KNEW and then gave to KQED when this operation proved hopelessly uneconomic. The director of the Center is Brice Howard, a stout guru in a billowing red polo shirt and steel-rimmed glasses and great bushy beard, who was a producer for NBC and the *Hallmark Hall of Fame* before turning his mind to experiments. He is a humorless man who laughs often—very serious, very profound:

"The camera is a transducer. There is a point where the light becomes electricity. When you begin to realize that the field in front of the optical system is important only in terms of the flow of electrons, and that whatever the field may be, the system transfers it to the same size of screen . . . We want to go back to Plato's *Republic* and ask those sorts of questions, relating to that surface. What *is* reality? . . . We're out of the bag of making the image representational. The motion picture world is very adept at that, but that's not what the electron is about."

The studio is hung with bright bits of cloth and shiny gray sheets of Mylar; TV cameras are on stands and hanging from the high ceiling; monitor screens stand on high and low tables. ("We're different from conventional studios: here the cameras are fixed and everything else is on wheels.") Different-colored lights bathe different parts of the big room. Against one wall a battered couch with a coffee table serves as a place to receive visitors; across the way is a small electronic workshop with a wooden worktable and pigeonholed wall for parts. Two of the television cameras are facing into each other, generating a neurotically repeating spiderweb of test patterns. There are tape recorders for musical accompaniments to whatever anyone wants to do, and at least one is always going. A staff of ten ("any one of whom can make every piece of equipment work") putter about, everyone keeping his own counsel.

On this day in early 1971 the staff is breaking in two "interns"; a new pair is sent every month, for four to six weeks of exploring experimentation, by two of the nation's public television stations. Very young and more than a little bewildered, the two interns (one male, one female) are turning dials behind Sony cameras and looking through viewfinders while two almost equally young hosts talk earnestly in their ears. The Center's press statement says, "Predictably, reactions to the intern program have been both personal and introspective. But David Dowe [KERA, Dallas] was able to sum it up quite simply: 'This has probably been the most valuable experience of my professional life.'"

Since its founding in April 1969, Howard says, "The Center's output has been a record of one hundred hours of tapes of our work, of what we are attempting to discover." Ninety minutes of this got on the air in 1970 as *!Heimskringla, or The Stoned Angels,* which brought in Tom O'Horgan and the La Mama troupe as well as Howard's "video-space experiences" to play games with the Viking discovery of America; it was not very well received. "But we are going to have to start producing," Howard says, troubled, "or we won't survive."

Quite a lot of what Howard says is over the line that separates meaning well from meaning nothing at all. Music becomes electricity at the microphone just as much as light does at the image orthicon, and the sole importance of either phenomenon is that electrical signals are transportable in ways that sound and light are not. (The photon hitting the eye has neither more nor less meaning then the electron leaving the camera tube.) But Howard is far from the only man in television to be taken in by what Jonathan Miller recently called McLuhan's "system of lies," and anybody who has to live on foundation grants acquires a manner useful for the purpose.

The Center has not been entirely without influence. Among those hanging around was Robert Zagone, executive producer of *San Francisco Mix*, the most experimental of the PBS programs for 1970–71, a series of moving-picture montages to illustrate what KQED's Moore called "irreducible human gestures"—Searching, Loving, etc. It began as an hour show, with bits individually made by a staff of thirty to forty people, but that didn't work very well, partly because nobody could handle the politico-emotional problems of choosing the one-minute and two-minute bits, partly because an hour of such material was, Zagone says, "a lot for the viewer," partly because the idea itself was weak. "We found a lot of things," Moore says, "for which that nonlinear form is not handleable." For the second half of its season, *San Francisco Mix* went to half an hour and became a little less experimental ("It was never experimental really," Zagone says; "it just wasn't a panel show"). The production operation was broken into six autonomous groups, each of which received six weeks to do its own half-hour show. The new low-light cameras were used to catch the streets at night for what became voiceless stories, less aggressively "television," more interestingly an employment of pictures. But Zagone was sick of it by the end of the year: the weekly slot exhausts the avant-garde just as much as it does the stand-up comic.

Howard himself was especially excited by the work of twenty-one-year-old Steve Beck, a young engineer with hair to his nipples who had transferred from the University of Illinois to California at

Berkeley to work at the Center. Beck was inventing a picture generator which would eliminate the camera from the televised image as *musique concrète* had eliminated the musical instrument from the phonograph record. He was working at the shop in the corner, soldering resistors to sockets in a mass of spaghetti wiring and turning dials; and the monitor camera facing the reception couch was tuned to what he was doing. Sinusoidal blobs in various colors chased across the screen, alternating with lightning bolts, slowing down, rising, falling, speeding up. "Ooh," Howard would say, "that's a good one—he's never done that before. And you know, there's no camera—no camera. This tool is going to make it possible for the human organism to express itself in ways never possible before. The instrument is clearly a harmonic of the artist."

4

In spring 1971 young Beck's image generator was still rudimentary, capable of few effects, very much being built on the workbench. And while it is obviously important to encourage young engineering talent, Howard in waiting for Beck to come up with a finished product was wasting what he seemed to consider precious time, because he could have bought a ready-made image generator from the Research Service of l'ORTF, the French national broadcasting organization. The ORTF device is a green metal box a little larger than a standard office typewriter; for full use, the customer would also want a briefcase of modules that can be plugged in to increase the maneuverability of the images. The inventor is Francis Coupigny, a round, cheerful, curly-haired engineer who heads a staff of fifteen at the Service, working under a general mandate "to renew and better artistic conditions in the total audiovisual domain."

Among Coupigny's other recent projects have been a *truqueur universel* (universal trickster; though English is supposed to be a rich language and French a restricted one, the best operative translation seems to be "special-effects machine"); a device that automatically makes animated cartoons from a relatively small number of drawings (this one is by no means perfected, though its results have been aired); a special Super-8 film camera with associated telecine chain to allow the use of very small hand-held cameras and ambient light in the making of television programs; and a simple mixer-blender which allows amateurs to put together into one package film, stills, titles, animation, and anything else they may have, without help from technicians. Howard is not to be blamed for not knowing about Coupigny's work, which has had no publicity at all, even in France. "I prefer," Coupigny says (in French), "to have a quiet life and do my own work."

Coupigny has been at the Research Service since the mid-1950s, when Pierre Schaeffer, director of the Service, convinced him that although he had been trained as a mechanical engineer he should work as an electronics engineer. It says a great deal for Schaeffer's position that he was able to hire into a specialized job somebody without the necessary pieces of paper—and, even more remarkable, give Coupigny the same freedom in choosing his own staff. L'ORTF is surely the worst bureaucracy in broadcasting—only 30 percent of its $380 million annual budget goes into programs. But among the expenses the French broadcasting system has undertaken is about $3 million a year for the two hundred people who work for Schaeffer. In 1970–71 they were given only forty hours of air time for their programs over the entire season, and half of that was after nearly everyone in France had gone to bed (contrary to tourist belief, the French are early-to-bed; television audiences drop drastically after 10 o'clock). Nevertheless, the programs are there, nearly all of them on film, and even when they don't get air time at home they win prizes at festivals and sometimes sell to the BBC or one of the German state systems; and Schaeffer's people busily carry them around in cans to university seminars; and they have an influence.

Schaeffer is an angry young man of sixty who looks younger, a compact, arrogant fanatic with burning eyes that belie a consistently matter-of-fact, even bored manner. Though the recipient of the most exclusive and demanding math-physics education France can offer (he is a Polytechnicien, as all the press releases from the Service point out), he is essentially an autodidact in the fields where he has made his life's work. The type is really rather American, including the entrepreneurship which took Gaullist gratitude for services rendered in the Resistance and parlayed it first to the direction of the newly liberated radio in Paris and then to the ORTF research operation. While working in radio, Schaeffer became fascinated with the possibilities of the then-new tape recorder, and invented and named *musique concrète*, the use of nonmusical noises for the purpose of musical composition. This plunged him into the musical avant-garde, where he has remained.

Schaeffer's interest in television came rather late, in the 1960s (the Service was founded in 1960, and at the start most of its work was in radio); he came to television, in fact, through readings in intellectually fashionable English-language cultural sociology, especially Harold Lasswell. He is not particularly a visual person, and the programs he has been responsible for producing tend to be visually sloppy (up to and including the sight of moving mike booms and of cameramen with mickey-mouse ears of film cartridges on their cameras charging across the viewer's field of vision to set up the next shot). Some of his programs have been criticized in France for failure to exploit the medium qua medium. One of the most successful Research Service shows, for example, was *Conteurs*, in which traveling cameramen caught some of the old people in each French province and had them tell the camera and the mike, and several other elderly locals sitting at the same table, agreeing, heckling, contradicting, some of the juicier legends of their part of France, in some of the juicier regional dialects. "It was a big hit," Schaeffer said (in French). "But those imbeciles," he added, nodding in the direction of l'ORTF, which resides in a round tower a few blocks away, "said it was too much like radio."

Another successful Research Service show was *Vocations*, an intellectual adaptation of Allen Funt's *Candid Camera* idea. (The French were crazy about the original *Candid Camera*, too.) Each of the twenty programs in the *Vocations* series—which had to be completed before the first show was aired—presented a leading figure in one of the larger worlds, law, art, music, theatre, psy-chiatry, business, labor, etc. The show was described to the inter-viewee as a discussion of his work and his attitudes toward it, and it was suggested that there be a twenty-minute warm-up, going over some of the questions that would be asked, before the cameras rolled. In fact, the cameras were on during the supposed warm-up. Then the victim was told the show was on, and the same questions were asked again. *Then* he was told what had been done to him, and asked to react to the fact that he had exposed all the differ-ences between his private replies to questions in apparently casual conversation and his public replies to a camera; and that was filmed, too. In some cases, the answers had been considerably different— in most, the manner had changed substantially—from the first to the second part of the program. These programs, which can reason-ably be described as pandering to dirty curiosity, drew the largest audiences anything from the Research Service has achieved; and made Schaeffer no friends at all.

In fact, of course, whatever the ethical problems, *Vocations* deals seriously with the Platonic questions Howard likes to talk about back in San Francisco; they are typically French questions anyway. Schaeffer is fascinated with the sea changes undergone by reality as the points of reference change, and one of the standard exercises in his Wednesday night "teleclub," which draws a hundred or so students and television people to the Research Services offices, is the exposure of different films made from the same footage of the same events by differently motivated editing. At one of these seances observed in summer 1971 the multiplicity of viewpoints was further illustrated by having two separate discussions—one in a closed room for some of Schaeffer's senior assistants and visiting dignitaries from journalism, and one in the little auditorium itself. The discussion in the closed room was televised into the auditorium, with three separate screens showing three different angles on the discussants; and in the interstices people in the auditorium volun-teered *their* views, which were also televised back at them by a camera set among the receivers on the stage. Schaeffer himself sat with the audience, addressed as "Pierre" by one and all, and shown in close-up by the camera whenever he made a contribution. The films under discussion were different cuts of a UNESCO conference about television (what else?); and Schaeffer was disappointed that the students in his group all preferred the most "objective" of the editings, the one that told you most clearly what was going on and the sequence in which statements had been made, leaving the viewer to form his own judgments. The bright lights required in the auditorium by the camera employed (Schaeffer does not waste Coupigny's talents on his teleclub) made the evening just short of unbearable.

Programs like *Conteurs* and *Vocations* go in the prime-time slot in the first network. The late-night slot on the second network has featured *Un Certain Regard* . . . , a series on what Louis Mollion, Schaeffer's director of programming, calls "the great good living people" (the list includes Jean Piaget, Konrad Lorenz, Peter Brook, Jean Monnet, Margaret Mead, and Picasso) and "the great things—ecology, computers, changes in medicine, the brain, education, etc." None of this, obviously, is especially experimental. But there are many other Research Service efforts which do not make the first network in prime time, and only rarely appear even on the second network late at night, and which are very unusual television. For example:

- Variety in the streets. One of these is in the can, presenting performances by a poet, a professional storyteller, a prize-winning young cellist from the Conservatoire and a pop group, before the surprised shoppers and stall-keepers of an open-air market. The camera alternates between the entertainers and the audience, collecting spoken and visual comments. "The public," says Mollion, "is an element of the spectacle." Others projected for this series, if the money arrives, will take similar mixed bags of entertainment to a suburban center to pick up workers entering the Métro and children on their way to school; a swimming pool; a Prisunic (five-and-dime); and others.
- *Poetic Varieties.* "Everyone thinks of poetry as a statue," Mollion says. "We shall make it move." The central device will be

to treat modern poems as chansons to be sung to music written for the occasion, mixing them up with more popular older poems that have been set in the past.

- "Reality-fiction, to reinvent the theatre." Actors recreate their old performances, talking about them as they do so: "We make the *telespectateur* confront a real documentary which is at the same time a fiction."
- Contemporary music in rehearsal. Schaeffer feels that the presentation of a concert-hall concert is hopeless, especially if the concert is orchestral: "The conductor is simply grotesque, the violinist looks like an idiot." Instead, the Research Service records on film the rehearsals for a performance, and pieces together a one-hour show, preferably with the composer present and commenting. Several of these shows have been on the air, l'ORTF regarding them as Schaeffer's *Fach*. Among the composers already presented in this way (to audiences Schaeffer estimates at about ten thousand viewers per program in all of France) are Varèse, Messiaen, Stockhausen, and John Cage. Schaeffer, incidentally, will not televise opera: "It is the television of grandpapa, all over Europe."
- *Public Oblige*. In effect, a broadcast audience research session, with viewers discussing, preferably with the writer in attendance, a dramatic program they have just seen. Mollion would like to mix some such program of ongoing viewer reactions into the televised coverage of the next French Presidential election, but doubts, with excellent reason, that he will be given the air time.
- Novels *not* adapted for television. This simply presents a narrator reading a text—"We do not change a *word*," says Mollion—over five or six broadcasts. The first book being done this way is Marcel Aymé's *En Attendant*.
- Architecture series. This one derives from an unexpectedly successful show, on the French pavilion at the Venice Biennale—"*completely* revolutionary," says Mollion. Aside from this one exploration of the guts and the surface beauties of a building, the series is still a project.

Ideas such as these grow out of noodling sessions at the Research Service, and outside producers and directors are often employed to make the actual program. Among the more important recipients of Schaeffer's commissions is Peter Foldes, a soft-spoken young Hungarian refugee who worked five years at the Service and is now free-lancing an unusual talent for cinematography. Foldes has done a good deal of animation, usually in a variant of the German Expressionist style of half a century ago, and has made several programs with a computer, supplying the machine with a small number of drawings, then manipulating what the computer does with them by means of a standard computer input panel. As of late 1971 there was still no machine that could do this sort of work in Europe, so Schaeffer sent Foldes to the Computer Image Corporation of Denver, Colorado, where he has done several mostly violent short subjects. Neither of Foldes' completed computer-based shows has been accepted for broadcast by l'ORTF, though one of them (about a boy who grows up to be a great giant and eat people and crush everything in his path) was broadcast by BBC. Encountered in mid-1971, Foldes was at work applying his special techniques to commercials for Norelco, Philips of Holland in general, and a collection of beers.

The show with which most Frenchmen would identify the Research Service is a series of animated cartoons, originally broadcast on the second network just before the evening news, and then rebroadcast to mounting controversy on the first network, presenting the life and times of *Les Shadoks,* birdlike invaders from another galaxy who are very stupid and very, very logical, in the manner of the put-upon French schoolboy. Also in the cartoons are the Gibis, a fat breed of wise but often outlogicked natives, plus a bad dog. "Nobody knows why he is bad," says Jacques Rouxel, the inventor of the strip, "but there can be no question that he *is* bad."

Rouxel, who is in his middle thirties, is a product of the best French business school, and was working as an account executive for a French advertising agency when the idea for *The Shadoks* came to him. Its source, he explains, was the comic strips in the American newspapers, which he had discovered as an adolescent studying at the French Lycée in New York. Television, he thought, could use something similar, a daily, very brief cartoon. He came to the Research Service with the idea, he says, because he had heard about the Animographe, Coupigny's machine to eliminate much of the drudgery of animation. Using the machine, an animator can make continuous-motion pictures with only six to eight different frames per second, as against twenty-four different frames in conventional technique. Each of the 102 episodes in Rouxel's first *Shadoks* series lasted two minutes, which required a fairly high noise level, and an almost Walter Winchell rapidity in the patter reporting the doings of the creatures or the sayings of Professor Shadoko ("This is a sieve; note the three elements of the sieve: the exterior, the interior, and the holes"). Rouxel believes that the machine-gun-burst quality of the televised *bande designée* makes it impossible for a number of episodes to be strung together successfully, but Munich and London are doing just that, Munich in the original French and London in an English translation.

"*Shadoks* polarized the French," Rouxel says with considerable pleasure. "The intellectuals and the children loved it, and everyone else hated it." But television is a medium where love turns out to be stronger than hate (indifference being stronger than either), and after the first year's run on the two networks l'ORTF commissioned another 102 episodes, the new series to run three minutes each. Now that the strip was respectable, the Animographe, which Rouxel had found disappointingly jerky in results, was abandoned, and a force of ten animators was hired (outside the nexus of the Research Service) to turn Rouxel's story boards into episodes.

Not the least of the attractions of Schaeffer's Service is its housing, a splendid eighteenth-century château with two-acre garden in the heart of the swanky 16th Arrondissement. The technical people are housed in the carriage house, the littérateurs and the studios in the main building, which retains its noble staircase but not much else from the days of glory. The halls are littered with filing cabinets, floodlights on stands, rolls of cable, bookshelves of film and tape. Apart from the first-floor auditorium and studios, which keep the old proportions of the mansion, all the rooms have been subdivided by partitions to make more offices and control rooms; Schaeffer himself uses only the corner quarter of what was once a master bedroom. Everybody changes offices and titles every few months anyway, because Schaeffer is always making new and more logical organization charts. "Sworn enemy of the sclerosis of the *fonctionnaire*," as *L'Express* puts it, "Pierre Schaeffer has institutionalized a kind of antimanagement."

Visitors waiting to see Schaeffer sit in the littered hall, opposite the claustrophobic elevator (deep, narrow, low-ceiling, all in dark red—"just right for carrying coffins," says one of Schaeffer's people), on kitchen chairs with aluminum tube legs and plastic seat and back. Streams of men and women, usually arguing, pass back and forth, and sometimes odd sounds of nonmusical instruments issue from opened doors. (Mollion, who has to oversee the production of real programs, secretes himself at the other end of the château, in a room that was once a small studio, which gives him a soundproof window in the wall between himself and his secretaries.) The work of the musical research groups continues, and concerts as well as lectures are given at the auditorium in the château, at UNESCO, or at one of the big halls in the headquarters tower of l'ORTF, only about ten minutes' walk away.

Schaeffer's Research Service is a little more fragile than this sounds. Though the ideas for programs are the most exciting in the world, the programs themselves often fail to reach (let alone hit) the target; like a lot of other things in France, the Research Service talks better than it is. There are important issues related to what television *is* that just don't interest Schaeffer at all—though he will support Coupigny's experiments because he believes in experiments, his tastes would permit him to accept a television service that consisted entirely of broadcast film. Though he has televised some live theatre (most of it American—the Living Theatre, the Open Theatre, the Bread and Puppet Theatre), Schaeffer has not worked on the problem of adapting stage performances for broadcast purposes—surely the most important technico-artistic problem of the medium. He gets a little defensive about this. "The French public," he says, "prefers films and the theatre of the boulevards. *I* am not happy about the presentation of theatre on the little box, but the public is." He can be calmly manic and calmly depressed in the same paragraph, the one continuous thread being a steady dislike—not unreasonable—of the organization that supports him: "L'ORTF is the worst milieu in the world—monopolistic, commercial, snobbish, and bureaucratic at the same time."

Mostly, Schaeffer's problem is that he doesn't get air time. But what he dreads most of all is the real possibility of a third network that would be turned over to him and his friends. "Now the French sit down before television and don't change the station," Schaeffer says, "and when we do something intellectual, everybody in Paris talks about it the next day. If there were a third network for the elite, it would be like the radio service of France-Culture. It has one percent of the audience. Ninety-nine percent of the people never hear it."

CHAPTER 13

Cable Television:
Hick, Hook, Hoke, Hooey Us

Cable communication . . . can, to begin with, deliver the full range of mass entertainment and information services now being delivered by commercial and non-commercial television. . . . It can provide commercial services that open-circuit television and radio cannot: neighborhood entertainment and information associated with national or neighborhood marketing and merchandising services; marketing and merchandising services not associated with entertainment and information; data transmission; message services of various kinds including fire alarm, burglar alarm, surveillance, meter reading, and the like. It can perhaps serve enormously in providing or supporting public services: above all formal and informal instruction but also health services, welfare services, employment services, consumer education services, library services, community development services, and no doubt others that can be identified; within this general area might also be listed the services that the system might be able to provide to the political process by its enlistment in political campaigning and the services (or disservices) it might provide by making possible instant polling of a populace. Finally, the system is itself a research tool for the social scientist.

—ARTHUR L. SINGER, vice president,
Alfred P. Sloan Foundation

I am the person in this organization who actually has to go out and do these things that other people dream of and speculate on. I do not see the large-scale implementation of many of the technical developments of which cable television is possible. . . . I could have a computer terminal in my home right now. . . . But I don't feel the need for a computer terminal in my home. I frankly don't, and I think the marketability of many of these services has been drastically overrated.

—ISRAEL SWITZER, chief technical officer,
McLean-Hunter TV Cable Ltd., Canada

My friends on the production side keep saying, "Twenty more channels
—look at all the programming you're going to need! Let's build studios!"
But that's *nonsense.*
                    —GORDON KEEBLE, Keeble Cable, Toronto

1

Of any given phenomenon, it may safely be said that this, too,
will pass. You will die and so will I, and so will everyone mentioned
in these pages. New tastes, new technologies, new talents—new
times, my masters! new times—will create new institutions in
television as elsewhere. Someone will make a lot of money out
of it, too, and someone else will go broke. For what will happen
is always more or less unpredictable, and it is only in the engineering
sciences (if there) that men can invent on schedule.

What exists today in broadcasting is the result of past patterns
of choice, and each choice occurred within its own pattern. The
dominant pattern, as William Stephenson has pointed out, has
been one of convergence—that is, the mass media are mass because
large numbers of people share a *Weltanschauung* that sends them
off in one or another of a limited number of directions. A crucial
aspect of any choice, however, is what the economists call its
"opportunity costs"—i.e., the value to the individual of the other
choices that have to be forgone. The fewer the channels, the less
the opportunity costs of watching any one of them, the greater
the convergence of choice. Right?

Right.

Next: more channels, higher opportunity costs of watching any
one of them, greater diversity. Right?

Maybe. Not so fast, young fella.

Okay, wise guy.

Certainly something of the sort happened in radio, with the
growth from about three thousand stations in 1954, when television
was already in the ascendant, to seven thousand stations in 1971.
Networks disappeared as a significant force in the gathering of
audiences, and stations became specialized by "format," each appeal-

ing to a relatively restricted group. With a few exceptions (most notable among them, WCCO in Minneapolis, which offers diverse programming and draws almost four times the audience of the second most popular station in that area) the radio broadcaster accepted the idea of a "fragmented" or "fractionalized" audience, and each tried to grab its own fraction. The 1971 *Broadcasting Yearbook* lists 37 stations which describe their programming as "100% black," 12 that broadcast at least an hour a day in Polish, 54 that broadcast at least 24 hours a week in Spanish. It is rare today for a radio station to start service without an announcement of a specific format to be followed—country/western, rock, pop, talk, news, classical—and the FCC will inquire officially into any change of format proposed if the ownership of the station changes hands.

Format radio meets what Lee Loevinger, while an FCC Commissioner, called the "considerable indication that most of those who watch television or listen to radio habitually want a consistently homogeneous kind of programming." But television has not provided that homogeneity. Except for a handful of Spanish-language stations in New York, Los Angeles, Miami and the Southwest, independents have not found it worth their while to challenge the local network affiliates with programs from which any substantial section of the total audience would be automatically excluded. Old movies draw bigger audiences than the most calculated appeals to smaller groups. In any event, the FCC has not yet permitted format television: to secure a license, the local television entrepreneur must pledge to carry a varied program diet. Looking at the differences between the breadth of radio and the narrowness of television, most observers find their source in the small number of television channels.

In *On the Cable*, the Sloan Commission Report on Cable Communications, issued in December 1971, Stephen White—the same Stephen White who had a hand in the Carnegie Commission Report on public television—contrasted the "television of scarcity" required by the limited number of available channels with the "television of abundance" that would come when television programs could be delivered to people's homes through the coaxial cables of what others have called a "wired nation." Probably as many as forty channels can be carried simultaneously on a cable; by paying for the service (usually $5 a month, though prices have been rising), a family can receive better broadcast TV pictures on all channels plus any other data transmissions—from cablecast programming not available on the air to medical diagnosis by computer—that may be coming from the "head end," the cable equivalent of a broadcast transmitter.

Some of this cheerful prediction is the stock-in-trade of the cable promoter, businessman and idealist both. (One of the broadcasts in Ford's PBL disaster had presented a vision of how happy everybody would be in East Harlem when instead of getting a lousy broadcast entertainment service for free each household would pay $5 a month for a cable television service that would also permit people to watch the doings at the local school board and community planning board.) But White had been around the television world, and knew that television programs (unlike radio programs) cost lots of money to make. The abundant television service he foresaw would rest not merely on the provision of extra channels, but also on special *extra* payments people would make for special programs —opera, theatre, foreign films—that cannot be offered by conventional television because the audiences they draw are not large enough to tempt advertisers to pay the costs of providing them. Not cable alone but pay-TV via cable would release the communications genie from his bottle to fulfill his mission of doing good for mankind.

For many of these ideas, this is the second time around. Pay television, making possible the profitable service of minority interests and minority tastes, had been the wave of the future in the late 1950s and early 1960s. *Life* gave a six-page spread to the opening of a pay-TV system in Bartlesville, Oklahoma, in 1957; pay-TV, said an article in *Atlantic Monthly*, "will represent a wonderful coming of age for the talented writer, director and producer." Investment adviser Manny Gerard remembers that some time in the early 1960s a special report from the Stanford Research Institute said pay-TV would be a $3-billion-a-year business by 1970.

In those days, advertisers were paying about 2¢ per household per hour for all the commercial messages broadcast during that hour. A charge of as little as 50¢ an hour would yield the broadcaster as much revenue from a pay program as he could get from an advertiser-supported broadcast with an audience twenty-five times as large. Pay television was possible off-the-air; the signal could be scrambled at the transmitter, requiring a special descrambler before it would make pictures on the television set. A few pay-television experiments were authorized in the UHF band by the FCC—the most ambitious was an RKO-General effort in Hartford, Connecticut —but the Commission was very edgy about it. Obviously, the biggest hits at the home box office would be the shows that were also hits for free: every economic rationale would lead the pay-TV operator to steal from broadcasters, which would mean depriving the public of much-loved attractions. Congress would not like that.

All such hassles could be avoided—or so it seemed in the late 1950s and early 1960s—by eliminating the broadcast operation and simply delivering programs for pay through wires run to the subscribers' homes. The first ambitious venture of this kind was Canadian, in the Toronto suburb of Etobicoke, run by the Famous Players subsidiary of Paramount Pictures. At its peak in 1962, before a monthly service charge was introduced, the Etobicoke system enrolled 5,800 subscribers, who bought programs by putting coins in a meter. The most popular attractions were the Toronto Maple Leafs away games, not then available on either government or private television, and nine Canadian professional football games a season; most of the rest of the programming was movies. But there were also a Broadway musical live from its theatre, an off-Broadway *Hedda Gabler,* and a performance of Menotti's *The Consul.* Jack Gould of the *New York Times,* always encouraged by the presence of culture on the tube, wrote of *The Consul,* "It is not too much to suggest that seeing the program, with Patricia Neway's superb tour-de-force in the heart-rending evocation of the human spirit under trial, must rank as one of the most civilized experiences in viewing that can be imagined." Movies cost $1 each; the sports attractions and Broadway productions cost $1.50. Paramount and its subsidiaries lost something more than $6 million in the not quite four years of the Etobicoke experiment, and abandoned it.

By far the biggest push in this direction came in 1964 in California, with an enterprise called STV (Subscription Television), a joint venture by an astonishing assemblage of promoters—the Los Angeles Dodgers and San Francisco Giants; a fat old-time movie operator named Mattie Fox; the computer-cum-engineering firm of Lear, Siegler; Sol Hurok; Reuben Donnelley, which sells the Yellow Pages for the telephone company; the investment banking house of William R. Staats—and about $16 million of money from the stock-buying general public. They hired as president none other than Pat Weaver, long gone from NBC, tired of the advertising business and eager to be back in programming.

Like the Paramount system at Etobicoke, Weaver's STV would offer three channels of television piped in through an otherwise unused dial position, and would charge people for what they used (Weaver's first act as president was to revoke the old plans to assess a $1 weekly service fee). But there would be no cash box on California television sets; STV had a better idea. The STV cable had two-way capability. In addition to sending the picture, the STV "head end" would send out queries every ten minutes to the cable selector box on each customer's set and receive replies as to whether or not the set was taking an STV program at this time. These replies from the set would come back to the STV computer in the form of a three-digit binary code (i.e., 001 or 100 or 011, etc.), which gave the possibility of eight different replies, only four of which (the three channels plus "OFF") were required for the pay-TV operation. Four others were thus free for some future extension of the service; some day soon, Weaver thought, the lady of the house would take a few minutes a day looking at the STV merchandising channel to see what the department stores had on sale, and would order her desires through push buttons on the cable box on her set.

Ultimately, Weaver thought, the STV system would hook into videotape recorders in the subscriber's set, allowing him to place an order each night for almost anything he might like to see the next evening. "You want a special stock market report," Weaver said earnestly, "or you want to see Maria Callas' debut as Carmen at La Scala, or to take a course in nuclear physics—all you'll have to do is make a phone call." The phone call would set in motion a chain of automated devices, by which the tape recorder in the home would be synchronized to a tape machine at the STV head end; at some time in the middle of the night, both tape decks would start spinning like dervishes, recording the two hours of the next night's program material on the householder's machine in perhaps six minutes of electrical connection. There would be Kabuki Theatre from Japan by satellite, Sadler's Wells, any Broadway show. . . .

The salesmen were out in the neighborhoods selected by Donnelley for the introductory period—substantial neighborhoods of private homes, a little above the California average income but not so much so that they couldn't serve as reasonable demonstrations of the market potential. Before the first wire was on the poles, Weaver had commissioned his first production, a tour of a Mexican art exhibit with a commentary by Vincent Price. He talked about what a pleasure it was to have three channels: the three networks were all competing with each other for the same audience, while he could serve three different audiences. Going into service in July 1964, Weaver expected to have 60,000 California homes wired into his system by the end of that year, at least half a million in San Francisco and 700,000 in Los Angeles connected to his programs and his computers by 1970.

This project ran into bad luck from the beginning. Pacific Tel & Tel, very skeptical, demanded that STV pay the full costs of the wiring in advance; the telephone company would then graciously permit this investment to be recaptured by means of monthly credits on the STV bill. The Lear, Siegler equipment was extremely expensive; the boxes to be installed at the customer's television set cost $90 each, of which only $5 would come back immediately through the installation charge. The most important examples of that summer's programming would come almost free—both the Dodgers and the Giants, as co-venturers, had agreed to take their payments in stock during the first year—but the hardware ate up much of Weaver's budget for programming other than baseball. Worst of all, the movie theatre owners formed an aggressive Citizen's Committee for Free TV, which got enough signatures to place an initiative on the ballot that November, to create a state law that would prohibit pay television. The costs of fighting major referenda in California run well into seven figures, and STV was pushed onto the defensive.

In the end, the bird wouldn't fly. Both ball clubs did poorly, which limited the interest in paying to watch them (even though both had previously kept their home games off free television, preparing for the glory days when they would tap a box office in the home). Weaver did not have the time or the money to get significant programming for the opening months. The political opposition was savvy, and though Donnelley had polls purporting to prove that most Californians were in favor of pay-TV, the initiative carried on election day. A year or so later, STV won a decision by the California Supreme Court, declaring unconstitutional the law written by the initiative, but by then Weaver was gone, and so was pay television.

2

In hindsight, it seems odd that STV did not offer broadcast signals as well as programs-for-pay as part of the wired service. Even in Los Angeles, which is mostly pretty flat, cable connections to the antenna terminals of a television set can give a better color signal than any aerial (especially for UHF channels). To some extent, the proprietors of STV felt no desire to use their wire to improve broadcast signals; they liked the idea that cable pictures would be better than what the customer could get off the air. But beyond that there lay a terra incognita of the law. There were some cable systems already in operation, in rural areas and the mountain states and the valleys of Pennsylvania, but their right to distribute broadcast programs to their subscribers was by no means secure.

Even the Supreme Court, which does not always look beyond principle to the facts of a case, might well prove reluctant to establish situations like the one in Hays, Kansas, described two chapters back, where a broadcaster would have to pay to rebroadcast to the public a basketball game being televised by a distant station—but a cable system owner could simply pick up the signal from out of town and distribute it to his paying customers without any cost to himself.

Then the Supreme Court in 1968, very surprisingly, in an impossibly artificial opinion, did decide that cable systems had no copyright obligations to broadcasters or to the owners of programs taken off the air. Cable systems were taken as extensions of the listener's set rather than of the broadcaster's transmitter. "Broadcasters perform," wrote Justice Potter Stewart in *Fortnightly Corp.* v. *United Artists.* "Viewers do not perform. When CATV is considered in this framework, we conclude that it falls on the viewer's side of the line." Cable operators could not alter the transmissions in any way (they could not, for example, black out the broadcaster's commercials and substitute their own), but so long as they merely relayed programs into homes they could not be required to pay under the Copyright Act of 1909.

These copyright problems had not arisen in the early days of cable television, then called "Community Antenna Television" (hence CATV), because broadcasters were delighted to have the effective range of their signals extended by some clever radio repairman who put an aerial on top of the hill and fed the resulting reception through cable into the otherwise blocked-off valley. But during the 1960s cable systems began to import broadcasts from distant stations, via newly perfected microwave relay systems, and local broadcasters panicked. It was hard enough to make a living in, say, Wilkes-Barre, Pennsylvania. If the Wilkes-Barre cable systems began importing New York and Philadelphia stations, major league ball games and big-time movies as well as network programs, the Wilkes-Barre broadcasters might as well shut up shop.

Before 1959 the FCC took the position that cable systems were outside its jurisdiction. But microwave relays were inside, and in

1962 the Commission denied permission for a microwave relay that would be used exclusively to bring distant signals to cable subscribers in a town where the local broadcaster had a struggle. Commissioner Kenneth Cox, a Kennedy Democrat from the State of Washington and a fighting gamecock of a regulator who was the most generally admired man on the Commission even by those who hated his views (he was not reappointed by President Nixon when his term ran out in 1970), established himself as the prime spokesman for the beleaguered broadcasters and against the advancing cable operators. Under his urging, the FCC in 1966 asserted complete jurisdiction over all cable television systems, and severely limited their activities.

The basic rules of 1966 compelled cable systems to carry all local stations, required systems to black out any distant channel carrying a program to be shown that day on a local station (guaranteeing the local station an exclusive in its area on all network shows), and forbade any cable system in any of the hundred largest markets to import any distant signals at all without special hearings and specific consent of the Commission. Cable systems in small towns and rural areas could import distant signals without consent, provided they did not "leapfrog" a nearby station to bring in programs from farther away. (This rule could be waived if a case were made for doing so: a cable system in Carlsbad, New Mexico, for example, was allowed to import the Los Angeles independent stations, leapfrogging Phoenix.) In December 1968 the rules were changed, and cable systems in the top hundred markets were authorized to import distant signals, provided the originating station gave consent. This turned out not to be very useful, because for much of its programming the originating station had bought from the copyright holder only the right to broadcast on its own frequency in its own locality; legally, it couldn't give anybody permission to use the signal elsewhere. Nor was this legal restriction unreasonable, by the way—the syndicator of a film series or a movie company would lose some of his chance to make a sale in the cable system's market if his same program was already coming in from far away on the cable. Indeed, it is not yet certain that Justice Stewart's reasoning in

Fortnightly applies to the importation of distant signals: the question is still in the courts.

Under FCC restrictions, cable spread rapidly through towns and smaller cities where only two of the networks were available via broadcast facilities, but the cable operators bogged down in the cities. The number of subscribers continued to grow—from a million in 1964 to nearly six million at the end of 1971—but the rate of growth was not spectacular by television standards, and a number of failures to sell cable to homes in big-city suburbs indicated that somewhere cable would hit a ceiling.

Meanwhile, pressure was building on the FCC rules—much of it from inside the Commission itself. When the rules were promulgated, someone had to be put in charge of enforcing them, holding the hearings, issuing the waivers, and chairman Rosel Hyde turned to the liveliest of the Commission's hearing examiners, a skeptical Harvard Law graduate named Sol Schildhause, to head a Task Force on CATV. To Cox's amazement—because Schildhause was generally on his side of the political fence, a liberal Democrat with a tendency to believe that government makes better decisions than private parties can—Schildhause became a crusader for cable television, processed waiver requests rapidly and reported them favorably to the Commission and made speeches lauding the future of cable. What interested Schildhause, however, was less the distant signal than the possibility of originating different kinds of programs, serving minority interests, on the many channels a cable system could offer.

From the beginning, many cable operators marginally supplemented their delivery of broadcast programs by "Cablecasting," on an otherwise unused channel, some virtually costless program like a running news teletype or stock market ticker or weather gauges. But in some areas where, as Schildhause says, "it's hard to sell bread-and-butter distribution," CATV operators deprived of the chance to import distant signals began offering more ambitious cablecast programs. The Commission at first refused to permit the sale of advertising to help pay for these programs, then reversed itself and permitted advertising "in natural breaks"—that is, at the beginning and end of a program, or between the acts of a play, or during the intermissions of a sporting event. In fall 1969 the Commission required all CATV systems with more than 3,500 subscribers to originate "a significant amount of programming" after April 1, 1971.

There are two ways to originate programming: go for the sort of thing that has done well on broadcast television (movies and sports), or look for what's different about cable. The more difficult it is to sell the service, the more like broadcast programming the cable services will be. In Manhattan, where most householders can get some reception through rooftop aerials, the two franchised operators, Teleprompter and Sterling, have contracted with Madison Square Garden for the home games of the Knicks and Rangers; and Sterling has invested large amounts in important feature films. On the other hand, places like Altoona, Pennsylvania, where only two network signals are easily received by any antenna a home-owner is likely to put up for himself, no local origination whatever was required to achieve 65 percent penetration (23,000 homes, a revenue of $1.2 million a year in $4 monthly pieces). Thomas Moore, the former ABC-TV network president who now runs the new program-producing Tomorrow's Entertainment division of General Electric, says that "as off-air signals improve, with advancing technology, cable proprietors will have to find better things to offer their subscribers." This may overestimate the savvy and under-estimate the greed of the average cable operator; in any event, Moore does not believe cable will provide enough revenue to support the programs he wants to produce, and he expects to sell cable systems "as an auxiliary to what we do in closed circuit"—i.e., television to theatres, as in the Ali-Frazier fight.

"Our typical format is feature films and outdated syndicated films," says Bill Brazeal, program director for Tele-Communications, Inc. of Denver, which in spring 1972 owned nearly a hundred cable systems in twenty-five states. "The films are expensive—just the freight and handling, the use of the chains and the VTRs—and I don't know what it's worth. Sometimes we get comments from people who enjoyed seeing features without interruptions. Then we also use some local talent. In Enid, Oklahoma, there's a young, attractive librarian who comes down to the studio for a children's hour, bringing a few five- and six-year-olds. The economic value to us is that when little Johnny's on the wire, mama and papa and grandma and grandpa will subscribe.

"In Lake Tahoe the man in charge of our origination has a flair for cowboy stuff. He starts service at 5 P.M. with country and western music, puts himself on the air in a cowboy hat, accepts requests over the telephone, and the damned phone rings itself off the wall—you can't believe it. We covered City Council hearings there—the city had just incorporated, and it was writing a charter, covering questions like the signs on the roads, dumping in the lake, lots of controversy, and we got a big audience. The local newspaper and radio conducted a survey to see how many were watching, and they said it was 50 percent, in prime time on a Monday night. But the City Council eventually asked us to stop carrying—they found they couldn't bang on the table or use the language they wanted to use."

Brazeal is a large, amiable man of about forty; the certificate on the wall of his office testifies to his graduation from a Dale Carnegie course. He came to cable television through an appliance store in Alliance, Nebraska. When a cable operator came to town to talk to the City Council about a franchise, he organized the committee to support it ("for the selfish reason that I like to watch football games"). He was so successful in his promotional activities that the cable company asked if he would like to work for them.

"We have a moral obligation to some communities to provide local services," Brazeal says. "And economically it's good—it gets you more subscribers—I don't care whether it's the flower-and-garden club or the superintendent of schools. It has a cohesive effect on the entire community. We have small-town bowling shows, in a small town bowling is a big deal, and the man with the high series comes down to the studio; he thinks he's a real celebrity, and he is. People like to see familiar faces. Yes, you'll get small percentages, but enough small percentages will give you a majority."

In addition to its cable systems, Tele-Communications owns a microwave service that operates through the Southwest from Los

Angeles through Texas, and in the mountain states across the plains as far as Minneapolis. (It was Tele-Communications that got the FCC waiver to leapfrog Phoenix and pick up from Los Angeles for Carlsbad, New Mexico.) When the last link is built over the mountains near Durango, the company will be able to carry television signals over something not much less than half the country, for no expense out of pocket because its microwave links are never fully employed. The company has bought the antique three-thousand-film stock of National Telefilm Associates (old movies have a special value because they can be run and run and run without paying residuals), and has signed up the Utah Stars of the American Basketball Association for cablecasting of all home games. "We're bringing major league basketball to forty communities through the mountains," says Bob Magness, president of Tele-Communications. "We didn't make the deal till two days before the first game in 1970–71, but we sold half of it to Lucky Lager Beer and Coca-Cola."

Most cable origination relies on what the literature calls "volunteer" (i.e., amateur) talent before the camera, and nonunion engineering behind it. "Out through the systems," says John R. Barrington of Teleprompter, the largest owner of CATV installations nationwide, "we use moonlighting cameramen, high-school kids, retired schoolteachers, as long as we can get away with it." Vic Waters, who runs the program end of Vancouver Cablevision in Canada, says that he covers local soccer games on a basis of " 'You supply the announcer'—that's how you get the community involved."

In theory, universities should be useful; in fact, they haven't been. What usually happens, Brazeal says, "is that cable is tremendously intriguing at the university when it starts. But it's work, takes time and effort to put together a meaningful program; and there's not much audience, and they lose enthusiasm."

By far the most ambitious programming for cable has been done in New York, especially by Teleprompter, which has been franchised for the northern half of Manhattan. The studios are in a ground-floor apartment in one of the incredibly dirty high-rise apartment houses built over the ten-lane highway that connects the George Washington Bridge and the Cross Bronx Expressway. From this master bedroom hung with gray drapes, Teleprompter originates a nightly news'show about nothing but northern Manhattan, reports from elected representatives on the city, state and federal level, political analysis directly from an assortment of the academics and radicals and nuts who populate some of the district. Every night at ten there is an entertainment feature in Spanish, elaborately produced cabaret or comedy from Puerto Rican television stations. There was an expensive semidocumentary entitled *King Heroin* that got rave notices in the *Times*. Remotes include news coverage (of school demonstrations and the like), visits to block groups and block parties, city hearings of various kinds, Columbia University basketball games and Irish football. Teleprompter and Sterling on the southern half of the island have also promised what the FCC calls "public-access channels"—i.e., an opportunity to get on the cable for anyone who wishes to come speak his piece or present his show on his own videotape. In fall 1971 Teleprompter and Sterling access channels were sporadically used by a school for the deaf, some ethnic culture societies and various counterculture groups subsidized by foundations, but the picture was only rarely comprehensible (the half-inch videotape used by the counterculture groups was especially hopeless), and nobody in town knew there were such programs anyway. One group that demanded time on an access channel for a manual of positions of sexual intercourse, illustrated by teen-agers, was regretfully turned down.

The Teleprompter cablecasting channel does ten half-hour children's shows a week, featuring a gallant blonde named Leslie Shreve, who used to work for an advertising agency but now does magic tricks ("Ernie," she says, nodding toward Ernest Sauer, one of two directors on the Teleprompter payroll, "makes me do magic tricks"), and tells stories, and sings, and wears costumes charmingly, and has been doing it all herself, seventy shows a year (the shows are constantly recycled for presentation), since June 1969. "And I do specials," she says. "I did one the other day for Dental Health Week—my grandfather's an orthodontist. I have to use things that aren't copyright, and rewrite everything, and write the music. It's low pay and nonunion, and it's supposed to be part-time, but I find I work here an eight-hour day, and then go home and tape the music and read kid stories—and I've never been so happy in my life."

Three black shows were "in the works" in mid-1971—one called *Hour of the Dream* and another called *Black Phoenix*, backed by Teleprompter money, though not by much of it—program director Bob Bleyer says his maximum budget per show "for talent and props" is $350. What sells the customers, of course, are the one hundred events (including eighty-four Knick and Ranger games) from Madison Square Garden, for which Teleprompter and Sterling pay $5 per subscriber—in late 1971, at an annual rate of about $200,000 for each system—maybe half of it recoverable through advertising sales, the rest gladly written off on the promotion budget.

In November 1971 Teleprompter took a long step toward realizing the dreams of those who see cable as a much more diverse service than broadcast, by televising live from the New York State Theatre in Lincoln Center a performance by the New York City Opera and its superstar Beverly Sills of Rimsky-Korsakov's opera *Le Coq d'Or*. What made the occasion so important was that, thanks to astonishingly sensitive television cameras originally developed for use in the space program, acceptable pictures were gained for the home audience without any change whatever in the performance presented at the theatre.

Television purists could complain that the balances between orchestra and voices were imperfect (they were), and that theatrical lighting with its varied "temperatures" made less realistic television pictures than the all-white lighting of studios. But the result was surely a more interesting evening's television than most of what gets on the tube (this observer thought it rather more amusing than the live production of this work, which is not one of the City Opera's triumphs), and the price to Teleprompter was less than $25,000, including a fee for the financially pressed New York City Center. It must be said that Teleprompter's then dominant boss, Irving Kahn, had special reason to seek good publicity at this time, having just been convicted of bribing public officials to keep the cable franchise in Johnstown, Pennsylvania; and that Sterling refused to put up a penny in cash for the rights to share the cablecast, finally negotiating a deal by which Teleprompter got access to a package of Grove Press films Sterling had bought, plus a specified number of uses of a Sterling remote-origination truck. But firsts are firsts, and the man who pulls off a first can always make a case for the proposition that he will do better next time.

3

"We are light years ahead of the United States in penetration and to some extent technically," says Gordon Keeble, once chairman of the board of CTV, the Canadian network operated cooperatively by a dozen privately owned stations, and now the proprietor of Toronto's Keeble Cable. As of mid-1971, 20 percent of all Canadian homes—27 percent of all homes in metropolitan areas—had been wired for television, as against 9 percent in the United States, including something less than 2 percent in metropolitan areas. The reason for the disparity is that American network television programs are available in and around American cities to anyone who puts an aerial on the roof, while most Canadians can get them only through extraordinary initiative or cable systems. The intellectual leadership of Canada has been struggling to achieve a separate "Canadian identity," and feels a reasoned discomfort about sharing a continent with so overwhelming a cultural influence as the United States. But most Canadians most of the time want American television programs. "I was brought up in the Maritimes," says Michael A. Harrison of Southam Press, "and when we wanted an outing to a big city we didn't go to Montreal, we went to Boston." Murray Chercover, who now runs CTV, says, "The Alberta wheat farmer has a blood brother, all right, but he isn't the Ontario industrialist; he's the South Dakota wheat farmer. The natural dividing lines on this continent don't run east-west, they run north-south."

Even where American programs could not conveniently be brought in cable received a stimulus from the very limited number of Canadian broadcasting stations. Except along the shores of

Lake Ontario, where CHCH-TV of Hamilton functions as Canada's only unaffiliated station, Canadian viewers have at most a choice of two signals, one from the government-owned Canadian Broadcasting Corporation and the other from the private network. Many Canadians can receive only CBC broadcasts, often rather serious-minded and far from universally popular. ("The two national sports," says a cant line in Canada, "are ice hockey and slamming the CBC.") Under these circumstances any cable origination would add substantially to the attractiveness of television, and a number of cable systems began providing alternative programs. The most luxuriant growth was in Montreal, where a second French broadcast channel was long in coming, full American service was unavailable without microwave, and an indigenous culture of great popularity (and considerable value, by the way) provided a chance for varied programming.

The world's largest cable system is Vancouver Cablevision, which pulls in a full American service from Seattle and the Canadian station in Victoria as well as the home city, using merely a moderately fancy antenna mounted atop a small apartment house on the slope of the city that faces the United States. Some 160,000 households are hooked into this cable—61 percent of all private homes and 95 percent of all apartment houses in the areas of Vancouver where the system operates, over some thirteen hundred miles of wire. Twenty-five connection crews are on the street every day (in addition to a staff of thirty maintenance men), and among the services they will perform, gratis, is the removal of the householder's own rooftop antenna, which makes the house look better and assures that the customer will not discontinue the service. In 1970, when the Canadian Radio and Television Commission threatened to restrict the carriage of American signals by Canadian cable companies, Vancouver rose as one man to tell Parliament that such behavior by government servants would be intolerable.

The same company owns a cable system in a low-income section of Montreal, where it has originated programming since the early 1950s, at first just in French (to make up for the fact that most of the broadcast channels on the cable were in English), now in

French, Italian, Greek, Hebrew, Yiddish and Arabic. "The origination pays for itself," says vice president Bud Garrett, "by selling cable." In Vancouver, however, Cablevision waited until 1970, after CRTC had "encouraged" all cable systems to go into origination, before launching its own program service. "CRTC wants us 'to go into the community and let the people talk,'" Garrett says, "and that's just a drain, because here that kind of programming doesn't bring us customers."

Nevertheless, Cablevision has equipped a substantial color studio, every bit as big as that of a small-city television station (except that there are no outsize doors to permit the entrance and exit of automobiles, refrigerators, and furniture: cable systems in Canada are not allowed to make or carry commercials). To run a service from 7 to 11 or 11:30 five nights a week (on a budget of roughly $2,000 a week) Cablevision went looking for someone who had *not* been associated with television, and came up with one of Vancouver's popular radio personalities, Vic Waters, a gray-haired man with a rolling, portentous voice and the strong opinions considered proper to a radio personality.

Among the programs originated in Waters' basement studio is *A Show of Hands*, news of the world of the deaf, including interviews, by people from the Western Institute of the Deaf. (This show started, Garrett says, "because a fellow I knew in the Army had a daughter who was deaf.") There is a weekly program from the art museum, from the two universities in town, from the executive branch of the city government and the legislative branch of the city government, from the Internal Revenue (entitled *$32 Million a Day*); there is a show on Indian affairs produced by an Indian ("Took me a year to find the man," Waters says; "he works for the Department of Manpower and Immigration, and he runs the show"). The luncheon speaker at each week's meeting of the Vancouver Board of Trade is presented that night. The national Film Board of Canada presents an hour each week. Several clubs have their own regularly scheduled program—the one Waters likes most is the World Ship Society of Western Canada, "romantic seaport and coastal stories, sometimes with films or slides, sometimes with memorabilia." Two half-hours a week present recitals by young Vancouver musicians, winners of prizes at an annual Kiwanis festival and top students from the conservatory wing of the University of British Columbia.

Early in 1971, Pierre Juneau, chairman of the CRTC, asked Garrett whether Vancouver Cablevision was measuring the audience to its origination channel, and got the answer, "No—and we don't want to; it would just inhibit what we do." But the other local media have paid the origination channel the one tribute that matters—its schedule appears in the local papers and in the Vancouver edition of *TV Guide*. Waters, while grateful, considers his listings a mixed blessing: "One of my big selling points at the beginning was that our programs would last as long as the material warranted—if it took forty-one minutes to say it, we'd run forty-one minutes; if it took twelve, we'd run twelve. But the paper and *TV Guide* wouldn't carry our schedules unless we programmed in standard hours and half-hours."

Rogers Cable in Toronto has the largest originating staff in Canada—eleven people headed by Phil Lind, a spectacular, modern young man regretfully turning thirty, who came to cable television from graduate work in political science at the University of Rochester. Lind has not tried to measure his Toronto audience, but he has done two surveys of viewing on the cable channel in his company's affiliate in the suburban district of Brampton-Bramalea. One survey measured the audience to a report on that day's staging of the town's annual flower festival, to which Rogers contributed a TV van that roamed the streets, taking pictures of the crowd as well as the parade; for this feature, Lind credits his channel with 72 percent of the viewing that night. The second survey was for the coverage of the finals of the Canadian national high-school lacrosse championships, in which one of the teams was from the local school; for this one, Lind reports 60 percent of all viewers.

Lind feels that Rogers' most interesting feature has been its closed-circuit FM radio (which costs subscribers an extra $1 a month over and above the standard $4.50 fee). He credits the Greek radio program with some two thousand sales to the Greek immigrant community, and allots almost as many more for the Italian radio. There are some cable television hours in those languages, too, and in Portuguese, and Lind has appealed to CRTC to allow advertising on origination channels for the foreign-language shows. "If we can't get permission to carry advertising," Lind says—and he can't— "we'll have to develop some form of pay-TV."

One of Lind's more ambitious efforts was an attempt to cover an amateur theatrical performance at the Queen Elizabeth Theatre. "The director was a professional from New York," he recalls, "and he wanted a couple of hundred bucks, which was okay. We'll pay something for talent—we can even get along with the Musicians' Union—but the step-up fee for stage hands was eighteen hundred dollars, and that didn't include lighting." Among the more interesting ideas Lind has attempted to sell is a channel that would offer a one-day-delayed repeat of the previous night's best broadcast programming, so people who missed something could catch it. So far, neither CBC nor CTV has been willing to approve; and, indeed, it may not be possible for them to approve, because their union contracts or purchase terms provide for the payment of residuals on any rebroadcast, and a repeat on Rogers would probably qualify as a rebroadcast.

"Philosophically," Lind says, "I think most cable programming should be local. Communities are entitled to their own closed circuit—programming from Scarborough, say, for Scarborough; a St. Jamestown news show from six to six-thirty every night. But there's terrific pressure to upgrade and upgrade and upgrade."

4

"This could be like the Agora in Athens or the Forum in Rome," Vic Waters says wistfully; and in the cable world that statement is almost modest. "We have built a means of serving ethnic and geographic minorities," said Irving Kahn of Teleprompter. "Just give us the wire, and the satellite for interconnection. . . . The more of them we serve, the more wire we'll sell. We're going to pay a lot of dough for culture; we'll use the theatre groups of necessity, and that's the best reason. The only hope for this country sits in a little thing called cable; it's the greatest thing since Seven-Up."

The only hope for this country . . . It isn't easy to convey the big picture of cable television as it appears to the owners of cable systems, the enthusiasts at the foundations, the optimists at the universities. But let us try:

In the future, every home will be built around its home entertainment center. This may look like a television set or (more likely) it will be a television wall, which would be illuminated in its entirety by a picture. Sixty, eighty, a hundred channels will feed into the home entertainment center, providing the family with an immense choice of programs. Indeed, there is no reason why the family should be restricted to what the cable system is putting on the wire that evening. There will be video cassettes for rent or purchase, offering a complete choice of everything mankind has done (of which pictures exist or can be made)—or the cable system will be hooked into a computerized videotape library, which allows the viewer to select from a catalogue of tens of thousands, hundreds of thousands of programs whatever meets his fancy or that of the kids, tonight. For each taste, each habit, each hobby, each race or creed or color, each political viewpoint, each curiosity, the home entertainment center will offer a tailored show. Nobody will ever go out at night again.

Nobody will ever buy a newspaper either, or a book or magazine, or patronize a library. Part of the home entertainment center will be a facsimile printer, which will deliver through the cable what now comes from the boy on the paper route. The computer center can supply on the screen or through the facsimile printer (according to taste) anything in print, either in its entirety or in part. The contents of all the world's museums will also be available in the computerized file, permitting everyone to become an expert on every kind of art. That college lecture courses will be among the choices—leading to advanced degrees earned at home by everybody —is something that goes without saying.

The home entertainment center will not be a mere one-way reception device. An alphanumeric computer terminal, much like a typewriter, will come with each installation, permitting the householder to query a time-sharing computer about any subject that interests him, to work out his own problems with the help of the computer's memory core, or to respond as seems best to the questions asked by the lecturer in his home-entertainment college courses. This terminal will also be hooked through the computer to any number of legislative and administrative institutions, obviating the need for representative democracy. Instead, every political issue that arises in the society will be presented through the home screen in the form of a series of choices to be made, and the people by tapping out their views on the computer terminals (*Ja* or *Nein*) will directly control government decisions ("like a town meeting").

Shopping as we know it will disappear. Some dozen or so of the channels will be devoted to the presentation of merchandise by "stores" (really big automated warehouses); when the viewer sees something he or she would like, the computer terminal is there to convey this information to the appropriate parties, meanwhile debiting this household's account in the cashless society for the amount of the purchase. Papa (white-collar Papa) will be able to stay home a lot more, too, because the home entertainment center, coupled into the telephone system, will permit him to "visit" with all the people he now must see face-to-face to get his work done. When the system gets perfected, with holographic pictures realer than real life shown on the television wall, nobody will ever go out again in the daytime either.

It is child's play, of course, to hook such a system into the electric and gas meter, saving the utility company the costs of visiting— or into the pantry, to save market researchers the costs of surveys to find out who's buying what. And the camera(s) used for Papa's business can also provide a perfect burglar and fire-alarm service and baby-sitting, in case anybody does go out. While the family is home, indeed, the cameras can give the authorities as much surveillance of what's going on at this house as they—guided, of course, by the people's instantaneous plebiscites on policy—may consider to be necessary.

. . . Now, now, there; don't fret. Dry those pretty eyes. It can't happen. Honest. No way.

But there is going to be a whole lot of talk.

5

In the first week of August 1971 the FCC made the first tentative motions toward releasing the cable giant from the chains the agency had forged and sending it out to sic 'em. "Cable," said the Commission's letter to Senator John Pastore, outlining the rules it expected to impose in 1972, "can make a significant contribution toward improving the nation's communications system—providing additional diversity of programming, serving as a communications outlet for many who previously have had little or no chance of ownership or of access to the television broadcast system, and creating the potential for a host of new communications services."

The most important of the Commission's regulations was a requirement that cable systems in the top hundred markets offer *at least* twenty channels to their subscribers. On these many channels, each cable company in the cities would be *required* to carry all signals normally viewed within a thirty-five-mile radius of the center of its service area; would be *expected* to offer three "full network stations" plus three independents in the top fifty markets, two in the next fifty and one in the others; and in the top hundred markets would be *permitted* to bring in any two distant stations. (In the markets below the top hundred, previously open to as many distant signals as the cable company wished, provided nobody was "leapfrogged," the proposed new rules forbid all importation of distant signals. This sort of 180-degree turn is not unfamiliar to students of the FCC.) When the distant independent is carrying a recent film or a newly syndicated series for which a local broadcaster has bought exclusive rights, cable systems in the larger markets will have to black out the distant station.

In addition to its carriage of broadcast signals, each cable company in the top hundred markets would be required to set aside three free nonbroadcast channels for noncommercial use—one for education, one for state and local government, and one for "public access," first-come, first-served. To meet the requirements of the public-access channel, the cable system will apparently be required to maintain a studio, with a cameraman who can turn a camera on this piece of the public that wants access and an engineer who can direct the resulting signal to the right channel on the cable. Apparently the seeker for access will be required to pay out-of-pocket production costs (unless he wants only "a brief live studio presentation not exceeding five minutes in duration"); also apparently— the Commission's final notice runs sixty-four pages and is very detailed but on many matters not very informative—the cable operator will be permitted to operate his "origination cablecasting" channel on an amateur, cost-free level.

In addition to these free channels, the cable operator will be expected to maintain channels for lease to anyone who wants to use them for any legal purpose (no pornography and no gambling, both forbidden by the criminal code). Looking down the road, the Commission expects that there will be a strong demand for channels to lease, and that "the cable industry's economic interest may well be found in reducing subscriber fees and relying proportionately more for revenue on the income from channel leasing." To ensure the availability of channels in the early phase, the Commission requires each cable system to provide at least one nonbroadcast channel for each channel relaying signals off the air. In the latter phase, the cable systems will be required to abide by an "N+1 availability" rule, demanding the addition of a new channel "whenever all operational channels are in consistent use during 80% of the weekdays (Monday-Friday), for 80% of the time during any three-hour period for six weeks running. . . . Such an N+1 availability should encourage use of the channels. . . ."

Finally, the cable operators will have to provide in their systems "the capacity for two-way communication. . . . Such two-way communication, even if rudimentary in nature, can be useful in a host of ways—for surveys, marketing services, burglar alarm devices, educational feed-back, to name a few. Of course, viewers should also have a capability enabling them to choose whether or not the feed-back is activated."

6

One need not be a connoisseur of government regulation to know that all this is fundamentally cockeyed in a familiar way. There is the neatly artificial breakdown of the top fifty, the next fifty, etc., the nuts-and-bolts of 80 percent of the time on 80 percent of the weekdays for three hours running; the "studio presentation not exceeding five minutes' duration"; an ominous note about the need to redesign everybody's television receiver to get the most out of cable—and very little discussion of actual experience to date. Where there is such discussion, it can be willfully wrong: "It is by no means clear that the viewing public will be able to distinguish between a broadcast program and an access program; rather, the subscriber will simply flick across the dial from broadcast channels to public access or leased channel programming, much as he now selects television fare." But this is not at all the way the viewer now selects television fare, as the stockholders of TV Guide will be pleased to demonstrate, and what surveys have been done of cable-origination channels in Canada indicate strongly that the public has no trouble at all distinguishing between the $200,000-an-hour program that typifies nighttime television and the $65-an-hour program on cable's own channel.

The oldest cable system in Canada is the one in London, Ontario, started in 1952 by E. R. Jarmain, who had a dry-cleaning business and an electronics hobby. This area then had no television service at all (even now there is only one local station). Jarmain put a parabolic receiver on a hilltop and ran wires to the homes of his friends. In 1960, Famous Players, laying the groundwork for forward motion if the Etobicoke pay-TV experiment was successful, arrived in Ontario with financing for Jarmain, and presently a big selection of American and Canadian signals was pouring into London and the surrounding Middlesex County. By 1970 more than 80 percent of all homes in London were wired.

Since the early 1960s Jarmain has offered its own cablecast programs, community-interest local shows and glimpses into academia, with elements of "public access." When the Television Bureau (TvB) of Canada launched a special audience study into the effects of audience fragmentation on advertising, the cable-origination channel was listed in the viewing diary. "Highly significant in the results of this study," the report concluded, "is the fact that we were unable to find any viewing of measurable proportions to the locally programmed cable channel." In fact, it was worse than that. Ross McCreath, chairman of Canadian TvB, says that in two weeks the 374 diaries kept in the London area failed to show a single entry in any time slot for the cable-originated channel. "This report has been challenged," said the Senate Committee on Mass Media a year later, "but it has not been refuted." And a separate survey of viewing in Middlesex County, done for the committee, got the percentages up to 100 without mentioning the cable channel.

People are not prejudiced against cable origination: Teleprompter and Sterling in New York undoubtedly get ratings of 20 or better among their subscribers for the Knicks and Ranger home games, and a telephone-coincidental survey by Teleprompter during its elaborately produced 1970 election coverage (remote pickups from the political clubs of northern Manhattan, etc.) indicated that 9 percent of those viewing television via cable were watching that rather than network coverage—a result the TPT management considered disappointing, but an outsider must regard as impressive. The heavily advertised *Coq d'Or* drew 16 percent, beating two of the networks. Larry Haeg of WCCO-TV in Minneapolis says that "if local CATV does high-school football and basketball, it will get a major share of audience."

Moreover, there are times—not many, but important—when a community does wish to keep in close touch with its local political or social processes: Brazeal's story about the Lake Tahoe council meetings could be told about other places, too. Writing about CATV in Daly City, Virginia, Nathaniel Feldman of the RAND Corporation noted that "In times of stress, the channel functioned to provide intercommunication among more people than any single building in the community would have been able to hold." The value of having a neighborhood television channel, for those moments when people are overwhelmingly concerned about their neighborhood, cannot be overestimated. Moreover, neighborhood need not be defined geographically for this purpose. In March 1972, Teleprompter picked up a Black National Political Convention from WTTW-TV, the Chicago public television station, and put it on in its telecast entirety for Manhattan cable subscribers.

There are other obvious values to cable. "Cable can create a milieu where you can make mistakes and it won't hurt," says Gordon Keeble. "Wonderful for a young playwright, or a stand-up comic—any talent can try out. I haven't had buyers yet, but look at the opportunity." The National Endowment for the Arts and the foundations that take an interest in these matters could be persuaded, one would hope, to support cable origination of artistic enterprise—the audiences at worst would be comparable to those available off-Broadway. And all over the country, if we are to maintain the talent pool from which superlative talent emerges, there must be a way to handle what Steve White has called "the third-rate contralto problem."

It is certainly possible, though not likely, that cable can bring sustenance and audience to the higher levels of the performing arts, while bringing major league opera and theatre and ballet and musical performance to many who have never had such chances before. In spring 1971, en route to the next season's single *Coq d'Or*, John Goberman of the New York City Center, an ardently bright young man with a splendid brush mustache, laid out for the two Manhattan cable companies a forty-four-night package of opera and ballet at a price of a million dollars for the two companies together, the shows to have production values other than just a camera in the house, the price to be guaranteed for seven years.

But a million dollars was a fifth of the total revenues of the two Manhattan systems for 1971; and while Goberman was right in his statement that "$25,000 is *nothing* for three hours of anything," the cable companies were then paying only about $4,000 between them for each Knick and Ranger game. Opera and ballet would probably sell cable ("make it a cultural resource in the house," Goberman urged, "like the *World Book*, though nobody ever opens it"); but the Knick and Ranger games sold more. And under the new FCC rules, barring a good copyright act from Congress or a reversal of the *Fortnightly* decision, the cable companies will be able to offer movies on distant stations at no cost whatever to themselves.

At $25,000 an opera or ballet for cable rights, the New York City Center would have come out just ahead of the extra payments that would have to be made to the performers and crew. Because nobody was being asked to do anything he wouldn't have had to do anyway to get the performance on stage, the City Center had been required to pay only double the usual night's fee or wages to its unionized personnel. Sales to other cable systems, interconnected by conventional means or, after a few years, by satellite, were going to provide the extra income both for City Center and for its people, who would participate. But the unions had gone along with these arrangements only on condition that each opera and ballet would be telecast *live*; no tapes could be made, and no reuse would be permitted. And the contract with the Musicians' Union (the only one that is public knowledge) called for an additional payment of about $1.15 per man for every additional 10,000 homes added to the cable systems served.

The arithmetic of this contract is highly discouraging. Assuming opera could draw 5 percent of the homes on a cable system (which would be very good over the long run) and that payments to the orchestra run about 15 percent of the cost of producing an opera, the City Opera orchestra contract would call for about $500 in additional payments to the company for every 500 homes watching opera. At $1 per home for stage and talent costs alone, opera would not be viable even on a pay-television basis. These prices probably would come down if cable programming ever became a regular matter. But before cable origination could become important, the unions would have to give up not only this potential bonanza, and not only the restriction to live cablecasting insisted on at the City

Center, but also the conventional television talent contract provisions requiring payment of residual fees for every subsequent use of a program.

For cable must be a repeat medium. What the advocates of minority programming seem not to understand is that most members of each minority are also members of the general audience—and that in television, unlike print, the choice of any one program absolutely excludes the use of the others. A man can read both *Esquire* and *Life;* but he could not see both *Coq d'Or* and the ABC *Movie of the Week* that same night. Opera has an audience of limited size, and if the opera is available only once it will lose some of that audience, quite possibly much of it (because light viewers tend to view the most popular shows), to a more general attraction. "We *must* have a repertory system," says Gordon Keeble. "To justify the costs, advertisers need so much audience. We must be able to say, 'Good—we'll run it over and over again until you get that audience.' To do that, we're going to have to relate talent rates to audience level."

Most commentators on the future of cable have paid little attention to these questions, because the reformers in government and the foundations and the universities are again in thrall to the false notion that the American public doesn't like what is now offered on the tube, and will jump for alternatives: "anything" will be "better." In the past this has been a harmless thralldom, like being in love with a movie star—even an admirable thralldom, because men of good will should wish to believe that public taste rises higher than the average of what is put on the tube. Moreover, the money spent while under this spell, mostly on public television, has added to the stock of decent programs available to more serious viewers. Now, unfortunately, the magic spell could work some harm.

Cable as the FCC was setting it up in August 1971, with more movies in each market (for that's what the distant independents will offer), seemed likely to *diminish* the funds spent for programs. Leland Johnson of the RAND Corporation writes, "Because local audience is generally more valuable [to advertisers] than is the more distant audience, the financial costs of audience lost to the local station are likely to outweigh the gains to the distant station—implying a net reduction in financial resources available for programming. Under these circumstances . . . the 'benefit' of cable growth might well lie largely in providing the public with more channels of worse stuff."

This impending disaster was less the result of errors by the FCC than of irresponsibility by the Supreme Court in the *Fortnightly* case ("We take the Copyright Act of 1909 as we find it," Justice Stewart said blithely, despite general recognition of what Judge Henry Friendly called "the truly superlative ambiguity of its few apparently simple provisions")—and of failure by the Congress to set the copyright house in order. (In fairness to the Court, Justice Stewart also pointed out that reform of the Copyright Act of 1909 has been kicking about in various House and Senate committees since *1955*.) As the situation is now structured in law, *none* of the money the cable subscribers pay for their cable goes to pay for programs; the cable operators keep it all. The cable carries anything off the air for nothing, and in the proposed FCC rules apparently would not even have to make gestures toward providing programs itself. As the Sloan Commission puts it, "The FCC is silent on the issue which we think central: the question of compensation for programming."

If the copyright decision had gone the other way, the cable companies would have made their deals with program suppliers, either via the stations or independently. In carrying distant signals, they could then have blacked out the commercials and inserted their own, giving them revenue directly based on the value of the imported program as perceived in their community. The chances for significant programming originated for cable would be greatly improved if the cable companies had to pay for whatever they carried. At all times, they would have to make judgments as to whether the public served by the cable really wanted any distant signal enough to justify the expenditure, and this decision would be a function of what the local stations broadcast. They would, in short, be in real competition with local broadcasters, rather than simply being parasites upon them. There would be a market, and the market would make more or less automatically many of the decisions the FCC now sweats so grievously to find. Markets do not necessarily make the right decisions, but the fact that they are continuous gives them a capacity to discover and correct errors—and a government can exert leverage on a market much more effectively than it can tell people what to do all the time. As Leland Johnson of the RAND Corporation wrote, "One needs rules that by their nature lead the regulated firms in the right direction without recourse to qualitative judgments by the Commission."

In February 1972, the FCC took a half-step back from the perils of its first proposal by limiting the program content cable operators could import from distant stations into the top fifty markets. Broadcasters in these markets were guaranteed exclusive use of all films and network series they purchased, placing the cable system in the same competitive position as another broadcaster. But FCC rules can change from day to day with shifting majorities on the Commission, and at best the FCC provisions of early 1972 prevent cable from diminishing the income of program-makers. What still must be found is some way to channel some of the receipts of the cable companies into the hands of the program producers and the talent.

All this is in the lap of Congress. The Court in *Fortnightly* was not speaking ex cathedra from the Constitution but merely interpreting a law, and Congress can very easily change the copyright law to secure a different result. One hopes that in considering this issue Congress will remember that the prime need is money for programming—and that the worst scandal of all would be to permit the establishment of a cable television service that charges the public billions of dollars in fees (at 60 percent penetration, $5 a month, the industry's revenues would be just under $2 billion a year) with not a penny of it going to the programs that are the reason people buy a set.

"Many influential thinkers suggest—if forty-two channels of capacity can be provided to serve the public, let's have forty-two channels," Murray Chercover of Canada's CTV told a Liberal Party conference in 1969. "I must point out that we do an imperfect job of filling two channels nationally now." There is clearly a sense in which time on television channels is already abundant—a city with five stations (and the great majority of Americans live in places with access to five or more stations) already offers thirty thousand-plus hours a year of television programming. Unless new sources of funds for programs are found to go with the expansion in the number of channels, the promise of cable is, in a word, a fraud. And the cable proprietors will fight like fiends against any copyright legislation that erodes what they have taken to calling their "right" to import distant signals without paying anything. Even the very modest diminution in the FCC proposals forced in November 1971 by Clay Whitehead (public television's bogeyman, Nixon's director of telecommunications policy) has been denounced as intolerably restrictive by many cable operators, though their trade association (and an equally dissatisfied National Association of Broadcasters) formally accepted Whitehead's proposals.

The Sloan Commission's suggested pay-TV system would, of course, provide programming money. As envisaged by Sloan, the system would involve payment by the month for a special channel rather than payment by program. Entrepreneurs would lease these channels from the cable companies, and sell directly to the consumer. White's example in the Report is opera, and what he proposes is that one-quarter of the thirty million homes Sloan expects to be wired by the end of the 1970s would be willing to pay a dollar or two a month for the chance to receive operas on the little screen. The revenue to the entrepreneur would be $90 million to $180 million a year; assuming that his leased channels and interconnection costs ran half his total revenues, he would still have $45 million or $90 million (the lesser of these figures being well above the total box-office receipts of professional opera performances in America) available for program and profit. The FCC letter to Senator Pastore, with its brief reference to "leased channels" undoubtedly contemplated a similar service, but the FCC lives too close to the political nexus to risk detailing the possibilities.

FCC rules already permit pay-TV on cablecasting channels, subject to the same "antisiphoning" provisions the Commission has established in broadcast pay-TV—no feature films more than two years old, no sporting events that had been carried in this market on free television at any time in the last two years (the new proposals raise this to five years). As of the end of 1971, no cable company had accepted the FCC's invitation to pay-TV. In the absence of routine interconnection (to be provided later by the satellite, maybe), no cable company has a market large enough to tempt program sources. With relatively few homes wired in cities and suburbs, the expanding companies wanted to use any special programming to sell cable at $5 a month forever rather than to sell an individual show for a dollar or two. But beyond that, the men in touch with the market knew that cable subscribers believe they are *already* paying for whatever special programs they receive—certainly in New York the subscribers to Teleprompter and Sterling believe they are buying the Knick and Ranger games. Any effort to slap on an additional charge for programs would be furiously resented.

What makes the Sloan presentation less than convincing is that an entrepreneur leasing a channel from a cable company would be a long time coming to opera as a program source to sell his pay-TV channel to the public. Certainly he would seek to make a deal for the most popular existing network television shows long before he would open negotiations with the Metropolitan Opera. Once a pay system got momentum, mere FCC antisiphoning rules would be unlikely to keep the movies or even the sports events on the broadcast air—by what legal authority does the FCC (by what constitutional authority could the Congress) prevent people from selling their services to the highest bidder? The political actions that would follow are suggested not only by the 1964 initiative in California, but by the violent reaction of Congressmen to the closed-circuit theatre-TV of the Frazier-Ali fight in 1971. It is only by prohibiting pay-TV entirely that Congress would be able to retain for that very high fraction of the public *not* on the cable (at least 40 percent in 1980, even by Sloan's estimates) access to the programming that probably means more to poor people than to anyone else. The New Left has been gung-ho for cable, but the Black Caucus Congressmen, who live closer to the problems of our own little Third World, have already introduced legislation to prohibit pay-TV in interstate commerce.

About one-fifth of the local cable franchises awarded in 1970–71 forbade the use of the system for pay-TV. Perhaps, as a reasonable compromise on the pay-TV issue, the new legislation needed in any event should require cable-systems over a certain minimum number of subscribers to set aside a percentage of their receipts from subscribers to pay for cablecast programming, either originated locally or imported over whatever cable networks may develop.

Getting copyright legislation that will be good for programs and their creators rather than for cable companies or broadcasters is going to be one hell of a job. And it is not made any easier by the fact that the smart operators of the cable industry can keep their pot of gold by persuading the politicians, the intellectuals and the putative communications experts at the foundations to keep *their* eyes on the lovely rainbow.

CHAPTER 14

If There Is No Answer,
What Is the Message?

Every pressure on man from the media will sharpen his incentives, will show him an expanding universe of delights, of interests, of occupations. Nothing is to be denied him if he will work for it—whether it is the enjoyment of serious music, or learning mathematics at home. The great wide wonderful world, in all its fascination, is being unveiled before eyes that once rarely strayed above the ground. Out into space with our greatest telescopes, and down into the brain with our electron microscopes. Here in full color is the art of all mankind; and there, the gripping and tragic story of war. Here, the range of things a man can do in the new do-it-yourself field; and there, contact with all the major thinkers and personalities of the world's leadership. To tell you what an informed, intelligent citizen can find in broadcasting today in New York, where I am more familiar with the radio and television schedules, calls not for a speech but for a rhapsody.

—Sylvester L. Weaver, president, NBC, 1955

Most mass-entertainments are in the end what D. H. Lawrence described as "anti-life." They are full of a corrupt brightness, of improper appeals and moral evasions. . . . They tend towards a view of the world in which progress is conceived as a seeking of material possessions, equality as moral levelling, and freedom as the ground for endless irresponsible pleasure. These productions belong to a vicarious, spectators' world; they offer nothing which can really grip the brain or heart. . . . They have intolerable pretensions; and pander to the wish to have things both ways, to do as we want and accept no consequences. A handful of such productions reaches daily the great majority of the population: their effect is both widespread and uniform.

—Richard Hoggart, 1957

The more I see of television, the more I dislike and defend it. Television is not for me but for many others who do like it, but who have no time for many things that I like. It seems to me that television is: the literature of the illiterate, the culture of the lowbrow, the wealth of the poor, the privilege of the underprivileged, and the exclusive club of the excluded masses.

—LEE LOEVINGER, FCC Commissioner, 1966

Instances of people harmed by television will not be found in averages or statistics but in hospitals and prisons.

—HARRY J. SKORNIA, past president,
National Association of Educational
Broadcasters, 1965

1

The forces felt by broadcasting do not push toward more varied and ambitious programming for less-than-largest audiences; only law can compel the networks to accept a duty to be adventurous. Unfortunately, even in this age of affirmative action, law is much more effective at keeping people and businesses from doing things thought undesirable than in making them do what the government wants them to do. In the United States the federal regulatory agencies, established by Congress to channel the energies of enterprise and greed into socially productive courses, have worked well only in their earliest years, when they were pursuing well-understood targets with the hot blood of youth. Today, few people have a good word for any of them; but even against this competition, the Federal Communications Commission has a bad name. It "has drifted, vacillated and stalled in almost every major area," Dean James M. Landis wrote in a report to President-elect John F. Kennedy in 1960. "It seems incapable of policy planning, of disposing within a reasonable period of time the business before it, of fashioning procedures that are effective to deal with its problems."

But the fault lies less with the agency than with Congress, which wrote a hopelessly bad law in the Communications Act of 1934, and then refused to improve it despite mounting evidence that it could not be made to work. What is wrong with the law is what is usually wrong with a law, which is that the people who wrote it had not thought through what they hoped to accomplish with it. The immediate and obvious need was for somebody to allocate space in the frequency spectrum, so stations would not interfere with each other. Once the government established a license for the use of the "airwaves," then the possessors of that license were clearly in some manner responsible to the public for their use of it. But what their responsibility is, and how it is to be enforced, Congress never said. The FCC is instructed to award licenses according to "public convenience, interest or necessity."

Judge Henry Friendly has observed that "the standard of public convenience and necessity, introduced into the federal statute book by Transportation Act, 1920, conveyed a fair degree of meaning when the issue was whether new or duplicating railroad construction should be authorized or an existing line abandoned. . . . The standard was almost drained of meaning under section 307 of the Communications Act, where the issue was almost never the need for broadcasting service but rather who should render it." It was assumed by the Commission from early on that if there was a frequency available for allocation and somebody wanted it, that was proof enough of a public interest in the use of the spectrum space. A Circuit Court once instructed the Commission to consider whether in fact a market was sufficiently large to support another station before deciding to let one be built, but the Commission does not usually pay much attention to what courts say. ("These guys," a Commissioner once said of his colleagues, "won't even read Supreme Court opinions. I circulate them, but nobody reads them.")

When there are competing applications for a license, the FCC must somehow decide who wins. The obvious criteria involve the comparative financial capability, experience and access to engineering savvy of the competing groups, but when properties as lucrative as VHF television channels are up for grabs, it is no great trick for all competitors to present a group clearly qualified as plausible operators of a station. "These things," said a man who was there in the fifties, "are like jousts in the Middle Ages." So the Commission inquires as to the programming the applicant intends to offer the public. The Commission in 1960 offered applicants some guidance about what was expected:

The major elements usually necessary to meet the public interest, needs and desires of the community in which the station is located as developed by the industry, and recognized by the Commission, have included: 1) Opportunity for Local Self-Expression, 2) The Development and Use of Local Talent, 3) Programs for Children, 4) Religious Programs, 5) Educational Programs, 6) Public Affairs Programs, 7) Editorializing by Licensees, 8) Political Broadcasts, 9) Agricultural Programs, 10) News Programs, 11) Weather and Market Reports, 12) Sports Programs, 13) Service to Minority Groups, 14) Entertainment Programs.

Promises, of course, are cheap; and the applicant can make them knowing that the Commission has recognized that "we would be deluding ourselves and the public, if we concluded that the program proposals will be produced exactly as represented." It is useful for broadcasters to know this, because the Commission does have the power to revoke licenses, and though the law specifically denies the Commission "the power of censorship," one can reasonably assume that a failure to live up to programming promises would be legal grounds for revocation. A more real source of danger than revocation, however, is the need to get a license renewed every three years, at which time somebody can file a competing application for the channel. The renewal application must be accompanied by the logs of a week of sample days chosen at the last minute by the Commission. Presumably anyone whose logs show that he failed to live up to the promises made when he was awarded the license would be vulnerable to a challenge, and this vulnerability was asserted in the late 1940s.

Paul Porter, a large, amiable Southerner who later became a partner of Thurman Arnold and Abe Fortas in Washington's most glamorous law firm, remembered his days as Chairman of the Commission: "My old friend Cliff Durr was playing left end, and when renewals came up Cliff would make this comparison between promise and performance. If there was a raving disparity, as there often was, Cliff would say we should have a hearing, and I would go along. It got so we had half the industry on temporary licenses; we didn't have enough hearing examiners to handle the business. We discovered that of 26 clear channel [radio] stations, only one or two had farm directors; we made a lot of jobs for county agents. Then we issued a blue book—*Public Service Responsibilities of Broadcasters*—and said, Go and sin no more. It had a short-term salutary effect."

The problem was that the FCC could not—really could not—take away a license even at renewal time without exceedingly strong reason. To keep licenses continually up for grabs, with the chance that renewals could be denied as punishments or granted as rewards, would give the FCC much too much power—secret power, secretly exercised—to censor what the stations do. As Commissioner Lee Loevinger said (after he had retired from the Commission and become a lawyer for broadcasters), "The establishment of renewal roulette in the broadcasting field will certainly destroy any semblance of free press and free speech in the broadcasting media. . . . What degree of freedom or independence can exist for an enterprise which is wholly dependent upon the favor and whim of a government agency for its existence at intervals of not more than three years?"

Following this self-denying rationale, however, the FCC made it almost impossible for anybody to lose a license, no matter how heinous his behavior. The matter came to a head in 1964 with the renewal application of WLBT-TV of Jackson, Mississippi, which had consistently presented only negative, nasty stories about that city's large Negro community. The application was opposed by the Office of Communications of the United Church of Christ, which has an affiliation with Tougaloo College in Jackson. The FCC found merit in the opposition and granted only a one-year temporary license renewal, telling the station to mend its ways during that year. But temporary renewals had invariably produced permanent renewals in the end. The church went to the courts, and in 1969, in his last opinion before his nomination as Chief Justice, Judge Warren Burger of the Circuit Court for the District of Columbia told the FCC to deny renewal to WLBT and award the license elsewhere.

Also in 1969, the FCC, by a vote of 3-1 on a seven-man Commission, denied the renewal application of WHDH in Boston, partly no doubt to right an old wrong (because the original issuance of this license had been tainted by corruption back in the 1950s), but publicly on the grounds that WHDH was owned by the Boston *Herald Traveler*, and where possible newspapers ought not to be awarded television broadcast licenses. Commissioner Nicholas Johnson in a concurring opinion argued that "the standards at renewal time ought to be the same standards that would prevail if all applicants were new applicants." As Commissioner Johnson had said elsewhere that the Commission's actions in judging between competing applicants were like drawing from a hat, his position left the door open for a majority of FCC Commissioners (or, indeed, any pushy member of the staff of any one Commissioner) to coerce a station owner into doing something or not doing something by the threat that his name would not come out of the hat at renewal time.

Broadcasters got very upset about the WHDH decision, and with their help a group of 22 Senators and 118 Representatives was gathered to introduce a bill that would prohibit the FCC from denying a license renewal to a station that was giving good service. The FCC then reconsidered its position, and in January 1970 issued a "Policy Statement" to the effect that it would not require comparative hearings between a challenger and a licensee whose "program service during the preceding license term has been substantially attuned to meeting the needs and interests of its area." Commissioner Johnson dissented. "There is no question," he wrote, "but that the American people have been deprived of substantial rights by our action today." But he was forgiving about it. "They have tried," he said of his fellow Commissioners. "They really have. . . . Thus it is, with no feelings save understanding, frustration and sorrow, that I dissent." In 1971, Judge J. Skelly Wright of the D.C. Circuit Court overthrew the "Policy Statement" and required hearings for all contested renewals, writing one of those I-know-I-can't-affect-what-happens-but-I'm-here-to-tell-you-you-stink opinions in which Judge Wright has specialized (his masterpiece of the genre being his order to desegregate the 90 percent black Washington public schools). In the middle of all this ardent controversy, it is easy to forget—because nobody ever mentions the fact—that in most instances network programs will account for something like 85 percent of all the hours viewers devote to the station, whoever gets the license.

No court, of course—not even Judge Wright's—would ever uphold the FCC if it decided to lift a man's license simply because it preferred another man's set of promises or political affiliations. ("There is no property right in a license," Rosel Hyde said in 1967, when he was Chairman of the Commission, "but there are enormous *procedural* rights.") Still, the costs of preparing a contested application and going through hearing procedures and appearances before the Commission will run a station owner tens, maybe hundreds of thousands of dollars. There is thus a margin created within which any group threatening to challenge a license can operate. It doesn't cost much to get such a challenge started (though following through on it can be very expensive: Forum Communications in New York has spent over $200,000 in a serious fight to get the license for Channel 11 away from the New York *Daily News*). And any expenses a community group incurs in preparing such challenges may be picked up by a grant from the Ford Foundation or the Stern Family Fund or some other innovative charitable institution (Commissioner Johnson has dedicated some of the royalties on his book *How to Talk Back to Your Television Set* to help defray the costs of such challenges).

Often enough, what is being dealt with here is simply extortion: as payment for being relieved of the challenge to its renewal, the station agrees to prepare some programs on the plight of the oppressed black or Spanish-speaking minorities, using the members of the challenging group as well-paid consultants. Even at its worst, though, the result gets some jobs for members of minority groups and makes broadcasters conscious of the need for sensitivity to the hopes and attitudes of a section of their constituency they have usually been ignoring. Some of the clauses negotiated in these "contracts" merely force the station to behave like a broadcaster—for example, to carry network public-affairs shows when offered rather than bumping them for the sake of the few bucks to be made out of a movie. In the great majority of cases where stations have bought out a challenging group, clean or corrupt, the costs were easily within what the station could absorb from its profits.

The problem with the continuing hoohah about renewals is that it requires for its sustenance an immense phony issue of "access to media" as a problem in American life. This is the largest and, one hopes, the last flowering of a strange attitude broadcasting has long inspired—the attitude that though what I find on radio or television never persuades *me*, it does persuade many otherwise wonderful people to make the mistake of disagreeing with me. The only remedy for this condition is that I, too, must be put on radio or television to reverse the effects of misguided persuasion, or to "tell it like it is."

But in truth the only remedy for the distorted signal is professionalism behind the transmitter. Certainly, access to *media* means nothing at all, as the London, Ontario, experience and the history of public television indicate. Access to *audience* might have some value, though in fact it is not true that the Middle American does not know what the radical is saying or would agree if he knew the argument better. But access to audience must be earned, with talent. There is something bittersweet funny about the sight of all these groups of ardent young lawyers and graduate students and junior executives at the foundations, none of whom can write a song anyone would sing or a book anyone would read or a play anyone would act, none of whom holds a position which gives his thought significance in the lives of others or could gather twenty-five people to hear him speak at a meeting—"demanding" access to the great audience of an entertainment medium. Mayor Lindsay himself rarely got a rating higher than 4 on his Sunday night television show in New York, and he's a charismatic pro. The catharsis of the angry young is no doubt a consummation devoutly to be wished, but to make it the major issue of public policy in broadcasting is to abdicate responsibility on all the important questions. At an agency like the FCC—which is charged with all sorts of regulatory responsibilities other than television, burdens its staff of fewer than two thousand with the processing of more than three thousand applications a day, and on its senior levels can just barely manage to consider one important issue at a time—emphasis on access means that the hard questions about broadcasting will never be considered at all.

2

These hard questions, of course, are those of the social impact of the medium, and the possibility of meliorating that impact by government action in a democratic society. For television will not go away; it is embedded in the culture now, like frozen lasagna, golf carts and sociology departments. Those who would deny that it has been a boon to individuals in their private lives can be brushed aside: there is simply no question that television has answered the most desperate of human needs, the need for escape from boredom, escape from self. Traditionally, heaven has always been seen as a place of pretty continuous entertainment. For those multitudes who cannot escape through their work or their reading or the experience of art, television has been about as close as they could hope to come to a heaven on earth.

But men do not live as individuals: they are sustained by each other in a society. Television has been so pervasive a presence in American society that one cannot imagine what American life would be like without it. Still, some influences can be claimed for it on no better authority than obviousness and observation:

1. People go out less at night. The diminishing need for places to congregate has been a contributing factor in the flight to the suburbs and the decay of the city. The fact that the home has become the prime locus for entertainment has changed the nature of home and family in ways nobody has yet been bright enough to explore. Among the real differences between today's young and the young of previous generations is the fact that as children today's young shared more of their parents' entertainment—and less of every other aspect of their parents' lives—than the young of any previous generation in any social class except the very rich.

2. People have acquired a new kind of relationship with large

numbers of total strangers who come into their homes on a picture tube. Every television entertainer (including newsmen) has had the experience of being greeted on the street by people they do not know at all, who then suddenly withdraw on the realization that this person who was indeed in their home cannot know he has been there. In America the fact that many of these visitors have been Negro is a social event of prime importance but sometimes ambiguous meaning. Several surveys have strongly indicated that Negroes themselves believe television to be the American institution that cares most about what happens to them and is most on their side; certainly, Negroes have been more prominent on television—in sports, entertainment and public-affairs programs—than anywhere else in the large society. But the effects of the entertainment programs have probably been far more positive than the effects of the most well-meaning (*especially* the most well-meaning) public-affairs shows. The inescapable bias of news qua news has too often impaled black Americans as a class in the butterfly case of trouble, *interesting* trouble, on the short-time horizon. The experience of color on color television has been most important, and sometimes most disturbing, for Negroes themselves. "If you live in a black community," Albert Murray wrote recently, commenting on what he called the Minority Psyche Fallacy, "the world looks black." True once; but no longer.

Politically, the common statement that the constant presence of an electronic specter has made "image" substitute for reality is as simple-minded as the earlier insistence that television somehow revealed "the truth" about people. No political figure today has the "image" that a Warren G. Harding or Andrew Jackson or Caesar Augustus commanded in times prior to television. But the feeling of familiarity is new.

3. The work of establishing a unified culture in a country the size of a continent has been accomplished (apparently in the Soviet Union as well as in the United States), completing a job the national magazines began three generations ago (and thereby, though I write as one with strong personal reasons to wish 'tweren't so, making the mass magazine obsolete in terms of social function).

This final Americanization of the community has greatly weakened in fact the particular institutions of a heterogeneous society (the Sokol, the Knights of Columbus, the Negro church, the trade-union meeting hall, the DAR, the neighborhood political club, the KKK). As a prime mover in the downgrading of local phenomena and the elevation of national phenomena in the consciousness of ordinary people, television has contributed to the feeling of "powerlessness" that does afflict fair numbers of people.

4. The speed and ease of introduction of novelty have biased both consumption and production toward new—or arguably new—products; the nature of television advertising has biased industrial research toward the creation of products that yield a demonstrable, surface improvement. But the idea that television advertising is itself a major cultural influence (apart from the pressure for maximum audience that advertising creates) cannot be seriously supported in the 1970s. All the stigmata of Americanization, from snack bars on superhighways to dishwashers to supermarkets to snotty kids, have come rapidly to Europe despite the much slighter presence of television advertising there. The triumphant ad campaigns of the 1950s, which built new markets for detergents and headache remedies and life insurance and hair sprays and air travel and other estimable economic goods, were not to be found in the latter 1960s; advertising on television like advertising in print had become part of the wallpaper for most people most of the time, proving, probably, that one can become conditioned to anything.

No doubt television advertising continues to sell merchandise, probably at unit costs lower than those attributable to other general-audience media, and its pervasiveness makes it an indispensable tool for forcing new products onto already crowded shelves in the stores. Moreover, because the arrival of quality inexpensive video-tape equipment enabled local stations to make professional-looking advertising for local retailers at about the same time that the 1970 recession pushed down the price of minutes, food chains and department stores have begun to do price-oriented broadcast advertising, taking the money out of their newspaper budgets. This advertising has been effective in drawing customers into the stores, and as a class it grew rapidly even in the recession years 1970 and 1971. During this decade, local television will probably cripple the big metropolitan newspapers as network television in the 1950s and 1960s crippled *Collier's,* the *Saturday Evening Post* and *Look.* Painfully little attention has been paid to this erosion of support for the newspapers, which are in fact the only possible medium for the expression of diversity to the entire community. Typically, the FCC picked precisely the wrong moment to move against newspaper ownership of television facilities: for the 1970s, it would have been much wiser public policy to encourage the ownership of local stations by local papers.

While it remains true that a man who advertised a cancer cure on television could sell a lot of snake oil (which means that some regulation is always going to be necessary), ordinary advertising for ordinary products ought not to be taken so seriously as most academic critics seem to take it. At present, it seems more significant in shifting market shares from one brand to another than in encouraging increases in total consumption of any product. In 1971 cigarette advertising was ruled off the air completely—and sales of cigarettes in the United States *increased.* Continuing to argue that the tube makes people buy buy buy what they would otherwise shun shun shun, as J. K. Galbraith does, is like spinning prayer wheels: it may get you good marks Somewhere, but it doesn't much help you understand what's going on.

5. The universal instant availability of entertainment to a national professional standard has severely reduced the demand for entertainment to regional or local standards. "The trouble with show business today," Jack Benny told Tom Sloan of the BBC, "is that there is nowhere to go to find out how bad you can be." In sports, television killed off the minor leagues; in the cities, it killed off the night clubs. It has seriously diminished the demand for touring companies of all Broadway shows other than those that offer a look at the pubic hair of actors and actresses. (The increasing nudity in films is also a by-product of television, because that's what television can't supply.) Here, of course, television merely continues and accelerates a trend begun by the talking picture and the phonograph rec-

ord. Certainly in proportion to the population and maybe in absolute numbers, there are fewer people making a living in America today as entertainers and artists—though those few who do make a living probably live a good deal better than their ancestors.

Television itself, in America, has been extremely inhospitable to all artistic effort that is designed to remain in the memory. It is more than possible—though far from certain—that television will end up diminishing the pool of trained talent from which significant artists can be drawn, and that any reduction in this pool produces a reduction in the number of artists. Setting out on an artistic career is a bad gamble at best; if there are to be rewards only for big winners, some who could have made important contributions may be rational enough to decide that the risks are too great. Whatever the social values of amateurism, the fact is that significant contributions to an art form can be made only by those who dedicate to it full time and energy. A diminution of their numbers would endanger the history of mankind.

6. Increasing proportions of people have received increasing proportions of what they think they know from the vicarious experiences of television. This, too, extends and accelerates a trend, which John Dewey was the first to remark more than seventy years ago. Civilization is a coin with two sides; people who live in cities know a great deal less about the natural world than people who live on farms; thus, Dewey argued, education in the cities should be careful to provide as much experience as possible, even at some sacrifice of abstract reasoning. The growth of electronic media, especially television, has vastly expanded the extent to which people learn (or think they learn) at second hand, without employing the trial-and-error, reward-and-punishment, successive approximation processes which are the basic human equipment for learning. Moreover, the apparent data base is shared by young and old, neither of whom have experienced much of what they think they know.

When Spiro Agnew was riding around on his charger denouncing the young, the *New York Times* reacted angrily in an editorial acclaiming the new generation as "the best-informed in history"—but all that was really meant by the praise was that the young talked about the same currently fashionable ideas and stories that bemused their elders. To the extent that the conflict between generations in the 1960s was exacerbated by differences in perception, the cause was not a great difference in experience between the two groups (which has always been the case and never makes the real trouble: people honor each other's different experiences) but a great similarity in the vicarious experience which had become the base of knowledge for both. Of knowledge, but not of wisdom; for the consciousness of ignorance is the beginning of wisdom, and the media mask the consciousness of ignorance.

We touch here, daintily, on the McLuhan problem. Much of what McLuhan has written is simply ignorant and wrong (especially the widely accepted argument that sequence is obsolete: the heart of the television experience is remote control of the viewer's time, and the fundament of the computer, McLuhan's other example, is the rigorously sequential flow chart). The urgent statement that the medium is the message means no more (probably much less) than the old saw that the style is the man. The hopelessness of the "hot" and "cool" stuff as tools for analysis becomes obvious after about two sentences. Page after dreary page the reader is forced to observe the antics of a popular college lecturer of real but limited scholarly attainment who keeps the class hopping by saying the first thing that comes to the surface in his ragbag of a mind. But McLuhan's instinct that something new has happened with the introduction of television—a widely shared instinct, accounting for his sales and reputation—cannot be dismissed so easily.

The viewing experience *does* seem more of-a-piece than the reading experience—that is, the differences between reading a newspaper and reading a novel seem greater than the differences between looking at a televised movie and looking at a documentary. This homogenization of what ought by rights to be different experiences is the strangeness of television. The prattle about media and messages hides the truth, because it reduces complex experience to simple statement and because it falsely proclaims that other media have similar characteristics. They do not. Content changed the nature of the radio experience drastically: the Philharmonic, Jack

Benny, The Shadow and Elmer Davis offered very different experiences indeed. And the content of television is nearly as varied as that of radio in the 1940s (there is much less good music)—but everybody feels that somehow it's "all the same."

A possible explanation of this almost universal attitude is that different radio programs demanded very different levels of attention. A few were really absorbing; most could be heard while doing the dishes or school homework or while daydreaming; some could be satisfactory background for reading a book. But watching television is an activity that excludes doing anything else except eating and knitting. The requisite minimum level of attention is fairly high. At the same time, unlike films or plays in a properly designed theatre, televised pictures do not absorb the peripheral vision; and it may be that the attainable maximum level of attention is fairly low. At best, the spread between minimum and maximum is much reduced from that experienced in the use of other media.

In such an atmosphere individuality must carry greater burdens than it can manage. Thus people and issues burn themselves out with unprecedented speed. Worst of all, perhaps, television becomes ineffective at performing what has always been seen as the most important social and political role of any medium: powerful at creating celebrity, it cannot legitimate leadership or attitude. There is a spurious equality of stimuli. It should not be forgotten, of course, that radio gave legitimacy to some queer and dangerous characters, among them Adolf Hitler, Father Coughlin and Huey Long. ("I'm not going to have anything very important to say for the first few minutes, so you can call up your friends and neighbors and tell them that Huey P. Long is talking at you—United States Senator from Louisiana.") The normative quality of television—the tendency of initially impressive personalities or ideas to wear out quickly— probably limits the damage as well as the good that can be done through the use of the medium. But the subject is worth much greater attention than it has yet received.

7. Men, women and children have all been given the notion that life can be entertaining all the time. As Daniel Boorstin pointed out in *The Image*, "There was a time when the reader of an unexciting

newspaper would remark, 'How dull is the world today!' Nowadays he says, 'What a dull newspaper!' " A great deal of current societal misfortune that is investigated under political and psychological headings more probably traces to this pervasive attitude. In England, where people pay for their television service with a set tax, the matter may be stated directly: "What I want is a funny programme at 6 P.M. each day while I am eating my evening meal," a man wrote to the BBC. "I pay you six pounds a year." Similarly, John Leonard of the *New York Times Book Review* demands a news service that will make him scream (though he doesn't want a news service that might make him chuckle). . . . Most comment about the contents of the medium is suffused with a fear of being thrown back on other resources, by which one can achieve only with effort, or maybe not at all, the pleasures gained from television. A very high fraction of the world's population—probably as much as a quarter of it—has become addicted to the box. It is a phenomenon of unmeasured but clearly major importance in the conduct of all the world's business.

3

None of this seems easily controlled by a government agency. Yet for all the folly of a standard of "public convenience, interest or necessity," something must be done about an area of enterprise where competition left to itself will tend to standardize product in an unfortunate way. To overcome the homogenizing qualities of the medium requires a content so strongly different from normal programming that it probably can appeal only to minorities too small to carry the costs of production under an advertiser-supported system, or any other democratically controlled system. A degree of aristocratic intervention is clearly indicated.

At bottom, the societal problem is also the problem that directly faces the private entrepreneurs of the broadcasting system: how to allocate the revenues. The question is made much more difficult by the fact that, in the words of BBC's Huw Wheldon, "It is pro-

grammes that make policy and not policy that makes programmes."
Nobody seriously wants the government involved in making programs, and nobody who has even a nodding acquaintance with television programming in Eastern Europe will want programs made to
express policy. Still, there are some attractive ideas for government
intervention in the determination of what goes on the air. The following would seem to be the most interesting:

1. Restrict the time that can be given to commercials, and the
number of commercial interruptions. E. William Henry, while Chairman of the FCC, tried to write into regulations the National Association of Broadcasters' voluntary code on maximum commercial
minutes, but Congressional friends of the broadcasters made strong
objection. Henry's authority in Washington derived mostly from his
personal friendship with the Kennedys, and after Johnson's accession
he no longer had the clout to do anything about this quite modest
suggestion. If restriction on the number of minutes would reduce
income too far, results as valuable might be achieved by restricting
the number of interruptions. The European state broadcasting systems require the grouping of commercials in "pods" at specified
hours. These pods in fact get heavy viewing (RAI in Italy helps assure the viewing by supplying a running gag of ten-second cartoon
bursts—comic characters watching a tennis match or dodging traffic
or missing a golf ball—which interrupt the string of commercials; a
perfect way to deal with the problem). In any event, four breaks an
hour for commercial messages (the British ITA rules call for three)
would seem a reasonable maximum to impose on broadcasters who
are using scarce frequency spectrum for a token annual fee.

2. License *networks* for fairly long periods (the British model for
the ITA "contractors" is seven years), and renew licenses only on
proof of minimum performance. There is no rhyme nor reason to
the present system whereby stations are licensed and held responsible for programs disseminated by networks which are not licensed.
It is true that the FCC acts as though it had regulatory power over
the networks, because the networks make their money on their
owned stations; challenges to what the network does can easily be
directed to these stations. But the exercise of demanding program

proposals from stations which will get all their significant programming from a network has been a demeaning experience for everyone involved. Worse, it has allowed the networks to be increasingly irresponsible in their search for maximum audience.

One must tread here with great care. The charge that television is intellectually inconsequent comes with ill grace from a literary and academic community that regards Norman Mailer as a genius and Marshall McLuhan as an insightful guide and R. D. Laing as a profound philosopher—a community that made *The Greening of America* the most talked-about book of its season. When Nicholas Johnson says that "television is the candy the child molester gives your child," one is dealing with paranoia, not with policy. Much of the hostility to the networks derives from deeply seated needs of critics who *must* believe in a great conspiracy that keeps other people from being just like themselves; and some of it is motivated by the critics' desire to keep lesser breeds from having too good a time. (This is not an unusual attitude on the left, where entertainment has replaced religion as the perceived opiate of the masses; as long ago as the 1920s, Thomas Mann's Settembrini cynically warned that "Music is politically suspect.") But if there is to be any audience at all for serious programming, the nation needs the networks: "The beautiful and to many people obscene thing about the networks," Richard Jencks said while president of the CBS Broadcast Group, "is that we bring our own audience. We go to the carny tent with our crowd. Our show on the health crisis draws twenty million people, while the same show on educational television would draw half a million."

The risks that a network runs when it programs a more ambitious hour are much greater than most critics realize. Television viewing is a little like cigarette purchasing: people have brands and brand loyalties. A viewer driven off a channel of a Wednesday night because what appears there is too heavy for him after a day's work (or, worse yet, too poorly done: those who aim their sights high are more likely to miss completely) may permanently switch his Wednesday brand. "The audience for a special," Frank Stanton said in the early 1960s, when his network had rather more of them, "is in large part people who came to see something else. Typically, it's a smaller audience than you would get for your usual programming." Confidential studies at CBS indicate that advertising for a forthcoming public-affairs show will *reduce* its audience. "But," Stanton added, "if you threw out the bad specials, I'd bet my bottom dollar the audience would be higher."

To the extent that the government demands a greater quantity of more ambitious programming from a network, it guarantees a higher proportion of bad specials, and an average audience for nonstandard fare even lower than what such programs receive today. Nevertheless, some such constraints must be imposed, on the network level, where the resources exist to make successful effort possible. The sort of arrangement John Doerfer pushed on the networks when he was Chairman of the FCC could certainly be revived. Each network should be required to provide an hour a week of nonfiction programming in prime time, for the benefit of the small minority who would watch—and also, more important, for the benefit of the men producing the program, to enable them to hone skills of reportage and assembly that will in time of crisis be vital to the society.

Even more necessary is an hour—maybe ninety minutes—a week from each network for what Gary Steiner called "heavy entertainment." Now that there is a functioning National Endowment for the Arts, there is no longer any difficulty in creating an operating definition of heavy entertainment: it is a program put on by a group subsidized by NEA. In return for their access to the channel, the networks can be ordered to devote their facilities periodically to extending the audience for the publicly supported performing arts. In return for their government subsidy, the performing groups can be ordered to make programs available to the networks at minimal charge (and to permit a limited number of rebroadcasts at a small additional charge). Here again, the benefits flow two ways—to an audience most of whom would never otherwise have the chance to see the serious artistic work being financed by their tax dollars, and to the performers occasionally pushed out of their little holes in the ground and forced to think about how to please what will be even at low ratings much larger audiences than they have ever known before. Moreover, the networks remain really quite free from government control, because the choice among groups will be extremely large. This is not a matter of court- or commission-imposed "access": the network programmers' professional skills of selection are still employed—but on a more exalted level.

Commercial minutes could be sold during the natural breaks in such programs, but the networks would doubtless incur losses. Some of these could be passed on to the stronger affiliated stations by requiring those in larger markets to accept such programs without payment for their time; much of the rest could be absorbed by abolishing the recent "Westinghouse rule" limiting network feed, which has produced no programming of importance and has meant financial loss at nearly all stations in markets below the top thirty. And what net costs might be incurred would at least be related to the social responsibilities of broadcasting.

Though everybody in the industry has the willies and the foundations keep pretending they know something not perceptible to the naked eye, it is unlikely in the extreme that the television audience is going to be "fragmented" by new technology. The notion that video cassettes are going to sell like phonograph records has to be wrong: the overwhelming majority of phonograph records sold are used by their purchasers to give an aural background to other activities, while television absorbs the time. Except for local sports, there is no source of viable programming to lure people to use the multiple channels of "the wired city." Two-thirds of the nation's television sets can receive six channels off the air right now, and the networks command more than 90 percent of the prime-time audience: in fact, their proportion of the total audience for the hours they were all on the air rose slightly in fall 1971, despite notable advertising for the Public Broadcasting System and a steady growth in the number of independent stations.

Pay-TV would make a major difference, and eventually would produce programming for minority tastes, though small-audience material would stand at the end of a long, long queue. But unless the advertiser-supported system comes to collapse in the toils of increasing costs or the Puritan war against cheerful consumption, the political obstacles to a pay system are likely to be impassable. "Radio and television," says Sol Taishoff, founder and still spiritual father of *Broadcasting* magazine, "are the only things the American people get for nothing." Economists can quibble about this, but that's certainly the way most people perceive it; and few of them are going to believe the pay-TV promoters' claims of all sorts of new goodies waiting behind the cash box—for the excellent reason that the claims are mostly false. The Ford Foundation can live in a hothouse—as indeed it does, in the most stunning piece of architectural symbolism in America—but everyone else must live in a cold, hard world.

4

Is there nothing more?

"It often seems to me," Richard Hoggart wrote in *The Uses of Literacy,* "that many of the people who do know something of the process described here have too easy a tolerance towards it. There are many who feel that they 'know all the arguments about cultural debasement', and yet can take it all remarkably easily. Sometimes they confess to a rather pleasant ability to go culturally slumming, to 'enjoy looking at the ——— now and again.' I wonder how often this ease arises from the fact that, though they may know all the arguments, they do not really know the material, are not closely and consistently acquainted with the mass-produced entertainment which daily visits most people. In this way it is possible to live in a sort of clever man's paradise, without any real notion of the force of the assault outside."

When our boys were small, we did not have television in our apartment. If I had to review something, I would set up a screening or go across the hall to our neighbors, who had lots of television sets. The year when I reviewed for *Harper's Magazine,* we bought a tiny portable for my desk and told the children it was there but not for them: "Daddy watches television only when he's paid for it." Now the boys are bigger and sensible, and we have two sets, one of them color. I have seen just about every series now on the air,

once. The boys have a few things they see more or less regularly—
Flip Wilson, Ironside, Room 222, Mary Tyler Moore—and sometimes I join them. We watch Jacques Cousteau, football games and
the basketball and hockey play-offs and the World Series; I try to
catch a fraction of the documentaries, and *NET Opera,* and anything
else that looks especially interesting on public television; and the
family gets together sometimes for reruns of *Get Smart* and *The
Honeymooners.* But there are millions of people in America who see
as much television in a week as I see in a year. "And you're writing
a book about it!" says my neighbor's wife scornfully. "You fraud!"

All right. It seems to me that television has only minimal obligations to an author and music critic living in New York. We go to
opera or a concert perhaps twice a week; and if we feel like finding
out whether the theatre has improved since last year (it hasn't), we
can go to the theatre; and there is always a luxurious choice of new
and old movies. New records I have to review. Books . . . Two
evenings of many weeks are given to driving to and from a country
house. And I am by metabolism a night worker. There are occasional evenings when I might like to relax at a television set and can
find nothing of any possible interest, but by and large there is much
more on television that I would rather like to see than there is time
for me to see it. I was shocked and distressed by the CBS decision
to deprive *Sixty Minutes* and *CBS Reports* of their regular Tuesday
night time slot, but personally, Lord knows, I had other things to
do Tuesday night. And it seems to me that while people of my
tastes elsewhere in the country do not have the glutton's feast of
entertainment options found in New York, they too have substantial
resources at the university (and indeed—dare one breathe it?—at
the bookstores) to fill those evenings when television is all blah. The
decision to kill *Hee Haw* and *Family Affair,* however—both of them
shows I was physically unable to watch for more than about five
minutes—seemed to me truly inexcusable. Here without economic
rationale for the network (because both shows were profitable to
the network, though not to its owned stations in the big cities), CBS
took cherished entertainment away from country people who had
few other available pleasures.

Though I can think of exceptions—especially in the synthetic and manipulated counterculture—I usually find when I take a look at popular entertainment that what seems to me the highest order of talent in the field has risen to the top. Most of the entertainers who have lasted any length of time in television—Lucille Ball, Carol Burnett, Jackie Gleason, Dick van Dyke, Bob Hope, even (gulp!) Milton Berle—have possessed skills greater than those of most of their less successful competitors. Barbra Streisand is by every criterion I know one hell of a singer. In England, where all the episodes in a comedy series are written by the same hands, popularity comes over and over again to the work of certain individuals, and it can't be an accident. If in fact the structure of "mass-produced entertainment" rewards the higher orders of popular talent—if there is no Gresham's Law in entertainment—what remains of Hoggart's indictment?

Hoggart is concerned, as most people are, about what will happen in the future to people like himself. He worries about the loss of what he is willing to call a "saving remnant" in the working class—that portion of the young who are looking for a less instinctive, more civilized life than their parents have known. They can be seduced from these goals, Hoggart feels, by the spurious attractions of mass-produced entertainment—and in this context the high quality of talent involved in the entertainment is at best irrelevant, at worst an even more persuasive snake in the grass.

Unlike the Nick Johnsons and Harry Skornias and Thomas P. F. Hovings, who are in the afflatus business, Hoggart cares deeply about the audience; it gives his work a power that the others cannot generate. But despite the immense public attention paid to the young who have turned out as Hoggart predicted fifteen years ago (victims of "the wish to have things both ways, to do as we want and accept no consequences"), the saving remnant seems as large as ever. The adolescent whose desire for a comprehended life survives the distaste and distemper of his peer group will not be turned aside by the much less powerful forces of mass-produced entertainment. And there is no evidence whatever to back the belief—hope would be a better word—that mass entertainment can stimulate such ambitions in those who have not acquired them from genetic inheritance or family nurture. School can do it, through the example of a teacher or a friend; but television cannot. Clearly there must be an obligation in television as elsewhere in the society for people to do the best they can—retaining the courage to remember that it won't matter much.

Except for convinced Christians, who feel they offer a balm for suffering, only the cruel and unthinking will actively seek to deprive their adult neighbor of his right to live an unexamined life. "Human kind," the bird told T. S. Eliot, "cannot stand very much reality"; and the poet went on to prefigure television as

> . . . a place of disaffection
> Time before and time after
> In a dim light: neither daylight
> Investing form with lucid stillness
> Turning shadow into transient beauty
> With slow rotation suggesting permanence
> Nor darkness to purify the soul
> Emptying the sensual with deprivation
> Cleansing affection from the temporal.
> Neither plenitude nor vacancy. Only a flicker
> Over the strained time-ridden faces
> Distracted from distraction by distraction
> Filled with fancies and empty of meaning
> Tumid apathy with no concentration. . . .

That's life, of course. It applies to television just as Bing Crosby's comment about television—"Well, I'd say it's pretty good, considering it's for nothing"—applies to life. Earnestness harms more often than it helps.

All the English-language theatre worth preserving since *Gammer Gurton's Needle* would not fill 5 percent of the time the networks must program each year. Fretting about the average level of television is like complaining about old age, an activity satisfying only to the speaker at the moment of speaking. Over time, a society cannot rely on life, on moon landings and assassinations, to provide triumphs and tragedies. The important criticism of television is that its leaders have not sought for tragedy or triumph in invention and artifice. They don't do the best they can.

Notes

Page
15 FCC quote: *Statement to Senate Interstate and Foreign Commerce Committee*, June 21, 1956, CBS, New York, p. 21.
Stanton quote: *Ibid.*
16 Seldes on Zworykin: Seldes, *op. cit.*, p. 161.
21 Lyons quote: Lyons, *op. cit.*, p. 275.
22 Seldes quote: Gilbert A. Seldes, *The Public Arts*, Simon & Schuster, New York, 1956, pp. 256–257.
24 Borden quote: Borden, *op. cit.*, p. 725.
25 Weaver's assistant quote: Martin Mayer, "Television's Lords of Creation," *Harper's Magazine*, November 1956, p. 27.
Siegel quote: Martin Mayer, "ABC," *Show* magazine, October 1961, p. 59.
29 Influence of Nielsen on programming: Figures from *The Television Audience 1970*, A. C. Nielsen Co.

Chapter 2. I Call Hello Out There But Nobody Answer

30 Reith quote: William P. Dizard, *Television: A World View*, University of Syracuse Press, 1966, p. 319.
Nielsen lady's dog: *Broadcast Ratings, Hearings Before a Subcommittee of the Committee on Interstate and Foreign Commerce, House of Representatives*, 1963, Pt. 3, p. 1448.
Nielsen quote: Nielsen, *The Responsibilities of Marketing Research: An Address Delivered on the Occasion of the Dedication of Nielsen House at Buchrain, Lucerne, Switzerland, September 20, 1966*, p. 20.
33 BBC pamphlet: *Audience Research, Methods and Services*, 1966, p. 24.
35 Fred Allen quote: In Dizard, *op. cit.*, p. 9.
Sloan quote: In Tom Sloan, *Television Light Entertainment, BBC Lunch-Time Lectures*, 8th Series, London, 1969, p. 5.
38 ARF study: *Recommended Standards for Radio and Television Program Audience Measurement*, ARF, New York, 1954, pp. 2–3.
41 Vital weaknesses: *Broadcast Ratings*, Pt. 3, p. 1270.
42 Beville quote: *Broadcast Ratings*, Pt. 4, p. 1837.
43 Barnathan quote: *Ibid.*, p. 1871.
Last CONTAM study: "An Experiment in Rating Research Methodology," *Advertising Research Foundation Proceedings, 11th Annual Conference*, October 14, 1969, p. 24.
50 Schramm and associates quote: Wilbur Schramm, Jack Lyle and Edwin B. Parker, *Television in the Lives of Our Children*, Stanford University Press, 1961, pp. 66–67 (paper edition).

Chapter 3. The Mystical Business of Selling Time

58 Share estimates: "Mad. Ave.'s TV Morning Line," *Variety*, September 16, 1970, p. 29.

Chapter 4. Flip Wilson and Other Prime-Time Phenomena

73 Sloan quote: Tom Sloan, *op. cit.*, p. 18.
Hoggart quote: "How Do You Assess the Quality of Life Within a Society? We Have Not Even Asked the Question, Let Alone Answered

Page

It. Richard Hoggart talks to Barry Turner in Paris," *New Academic*, No. 2, May 13, 1971, London, p. 8.

87 Sloan quote: *Op. cit.*, pp. 8–9, 10.

90 Merle Miller and Evan Rhodes, *Only You, Dick Daring!*, Bantam ed., New York, 1965, p. 13.

97 H. Hotelling: "Stability in Competition," *Economic Journal*, Vol. 41, 1929, pp. 52–53. See discussion in Edward Hastings Chamberlin, *The Theory of Monopolistic Competition*, Harvard University Press, Cambridge, Mass., 1933, 1942 ed., pp. 208–213.

98 Johnson quote: "Dissenting Opinion of Nicholas Johnson: The ITT-ABC Merger Case ['In the Matter of Applications by American Broadcasting Company for Assignment of Licenses of Stations . . .'']," Washington, December 21, 1966, mimeo, pp. xi, xii.
Klein quote: Paul Klein, "The Men Who Run TV Aren't That Stupid. . . . They Know Us Better Than You Think," *New York* magazine, January 21, 1971, p. 21.

101 Barnouw and Rice quotes: Erik Barnouw, *The Image Empire*, Oxford University Press, New York, 1970, p. 33.

102 Attenborough quote: *BBC-2*, BBC, London, 1966, pp. 4–5.

103 Anonymous *Harper* reviewer quote: Martin Mayer, "How Good Is Television at Its Best? Part II: More Than Plenty of Drama," *Harper's Magazine*, September 1960, p. 86.

Chapter 5. Other Dayparts, Other Customs

118 Arthur Shulman and Roger Youman: *How Sweet It Was*, Bonanza Books, New York, 1966, p. 317.
Strike It Rich: *Ibid.*, p. 339.

119 Does no demonstrable harm: Seldes, *The Public Arts*, p. 102.
Transmit the truth: *Ibid.*, p. 101.
Sinister as well as hateful: *Ibid.*, p. 107.
W. Lloyd Warner and William E. Henry: *The Radio Day Time Serial: A Symbol Analysis*, Genetic Psychology Monographs, XXXVII, pp. 3–71, cited in Seldes, *The Great Audience*, p. 234.

120 Harris poll: *Life*, September 10, 1971, p. 42.
Sales figures: *Broadcasting*, January 25, 1971, p. 45.

126 William J. Baumol and William G. Bowen: *Performing Arts: The Economic Dilemma*, Twentieth Century Fund, New York, 1966.

Chapter 6. *Sesame Street* and Saturday Morning

128 Halloran quote: J. D. Halloran, *The Effects of Mass Communication*, Leicester University Press, 1964, p. 46.
Bettelheim quote: Bruno Bettelheim: *The Informed Heart*, Paladin ed., London, 1965, p. 53.
Schramm quote: Wilbur Schramm, Jack Lyle and Edwin B. Parker, *op. cit.*, p. 95.

129 On children's rankings of *Prix Jeunesse* winners: Publications of the Internationales Zentralinstitut für das Jugend-und Bildungsfernsehen, No. 4, *The Scarecrow*, Munich, 1970, p. 6.

130 My favorite: *Ibid.*, No. 3, *Patrik and Putrik*, Munich, 1969, p. 18.

Page
130 Photograph sequencing tasks: No. 3, *Clown Ferdl*, Munich, 1969, p. 43.
131 James Q. Wilson: "Violence, Pornography and Social Science," *The Public Interest*, Winter 1971, p. 49.
      J. D. Halloran, R. L. Brown and D. C. Chaney: *Television and Delinquency*, Leicester University Press, 1970, pp. 27–29.
      Himmelweit quote: H. T. Himmelweit, A. N. Oppenheim and Pamela Vince, *Television and the Child*, Oxford University Press, London, 1958.
132 Fritz Redl: In Schramm *et al.*, p. 143.
      Chesterton quote: In Halloran *et al.*, pp. 29–30.
138 Joan Cooney quotes: In Cooney, "The Potential Use of Television in Preschool Education," mimeographed, undated, pp. 10, 27, 33.
146 Cooney quote: *Ibid.*, pp. 6–7.
149 Gibbon and Palmer quote: "Pre-Reading in *Sesame Street*," mimeographed, undated, unpaginated.
152 Gibbon and Palmer quote: *Ibid.*
156 Schramm quote: Schramm *et al.*, *op. cit.*, p. 76.
      ETS researchers: Samuel Ball and Gerry Ann Bogatz, *A Summary of the Major Findings of "The First Year of Sesame Street: An Evaluation,"* Educational Testing Service, Princeton, N. J., October 1970, pp. 4–5.
159 Blackwell quote: Frank Blackwell and Mary Jackman, "Sesame Street: A Report on a Monitoring Study," National Council for Educational Technology, London, 1971, mimeo, p. 16.
160 Sociological study: "Mass Media," *The Center Forum*, October 20, 1968, Center for Urban Education, New York, p. 1.

Chapter 7. Sports: The Highest and Best Use

165 Wheldon quote: *BBC Handbook*, London, 1971, p. 15.
179 Johnson quote: William O. Johnson, Jr., *Super Spectator and the Electric Lilliputians*, Little, Brown, Boston, 1971, p. 89.
189 Johnson quote: *Ibid.*, pp. 198–199.
190 Ogilvie and Tutko quote: Bruce C. Ogilvie and Thomas A. Tutko, "Sport: If You Want to Build Character, Try Something Else," *Psychology Today*, October 1971, p. 61.

Chapter 8. The Nightly Network News

192 Curran quote: Charles Curran, *Broadcasting and Society*, BBC pamphlet, London, 1971, p. 15.
      Kolade quote: "How Influential Is TV News?," *Columbia Journalism Review*, New York, Summer 1970, p. 22.
199 Whale quote: John Whale, *The Half-Shut Eye*, St. Martin's Press, London, 1969, p. 29.
205 Johnson quote: Barry G. Cole, *Television*, The Free Press, New York, 1970, p. 324.
      Salant reply: *Ibid.*, p. 332.
      Khrushchev data: Charles Winick, *Taste and the Censor in Television*, Fund for the Republic, 1959, p. 31.
207 Goldstein quote: Richard N. Goldstein, "The Union Role in TV News," *Industrial and Labor Relations Report Card*, New York State School of

Page

Industrial and Labor Relations, Cornell University, Ithaca, N. Y., Vol. XVII, No. 1, Summer 1971, pp. 3, 4.
208 Ben H. Bagdikian: *The Information Machines,* Harper & Row, New York, 1971, p. 90.
214 Pool quote: *Columbia Journalism Review,* Summer 1970, p. 21.
215 Agnew quote: *Alfred I. Du Pont–Columbia University Survey of Broadcast Journalism,* 1969–1970, pp. 133–135, 138.
216 Honolulu: *Ibid.:* p. 35.
217 British researcher: Malcolm Warner, "TV Coverage of International Affairs," *Television Quarterly,* Vol. VII, No. 2, p. 74.
219 Stanton quote: In Friendly, *Due to Circumstances Beyond Our Control,* p. 212.
220 CBS Guidelines: In William Small, *To Kill a Messenger,* Hastings House, New York, 1970, p. 73.
Carpenter quote: Edmund Carpenter, *They Became What They Beheld,* Outerbridge & Dienstfrey, New York, 1970, unpaginated, section "Media as Codifiers."
Schaeffer quote: Pierre Schaeffer, *Machines à Communiquer,* Editions du Seuil, Paris, 1970, p. 22. Translation by author.
Pilkington Commission quote: *Report of the Commission on Broadcasting,* HMSO, London, 1960, par. 102.

Chapter 9. Right Before Your Eyes: The Political Nexus

223 Annual Report to the Shareholders of CBS, 1968, p. 13.
Riesman quote: David Riesman, "Political Crusades and the University," address at University of New Mexico, June 1970; in *Dialogue, USIA,* Vol. 4, 1971, p. 86.
Wiggin quote: Maurice Wiggin, "Birds of a Feather," *Sunday Times* of London, June 13, 1971.
224 White quote: William S. White, "Television's Role in Politics," in Cole, ed., *Television,* p. 83.
239 Roosevelt speeches: Edward W. Chester, *Radio, Television and American Politics,* Sheed & Ward, New York, 1969, p. 33.
Boake Carter story: *Ibid.,* p. 174.
240 FCC prohibition of opinion: Mayflower Broadcasting Corp., 8 FCC 333.
FCC approval of opinion: "Report on Editorializing by Broadcast Licensees," in Part II, *Federal Register,* Vol. 29, No. 145, Washington, July 25, 1964, p. 10424.
242 White quotes: *Red Lion Broadcasting Corp.* v. *FCC,* 375 US 367, at 388, 389–90.
245 John Whale: *The Half-Shut Eye,* p. 199.
Robert MacNeil: *The People Machine,* Harper & Row, New York, 1968, pp. 137, 159.
248 Fingerhut quote: Vic Fingerhut, "A Limit on Campaign Spending—Who Will Benefit?," *The Public Interest,* Fall 1971, p. 4.
$60 million figure: *Broadcasting,* June 21, 1971, p. 82.

Chapter 10. Five Thousand Words with Pictures

256 MacNeil quotes and pages: from *Proceeding Against Frank Stanton and Columbia Broadcasting System, Inc., Report of the Committee on*

Page

Interstate and Foreign Commerce, House of Representatives, 92nd Congress, 1st Session, Washington, p. 135.

257   Minority report: Ibid., p. 217.

Henkin interview and script: See both in their entirety, Subpoenaed Material Re Certain TV News Documentary Programs, Hearings, Special Subcommittee on Investigations, interview on pp. 251–259, script on pp. 234–245.

259   Henkin quotes at hearings: Ibid., pp. 56–57.

260   WTTG case: Ibid., p. 306.

Laurent quotes: Ibid., p. 143.

261   Polar bear script: Ibid., p. 171.

263   Friendly quote: Friendly, op. cit., pp. 133–134.

Note: Mass Media, Report of the Special Senate Committee on the Mass Media, Vol. 1, The Uncertain Mirror, Ottawa, 1970, p. 106.

265   New CBS rules: Congressional Record, July 12, 1971, p. H6953.

266   CBS newsmen's letters: Ibid., pp. H6577–6579.

267   Minority report: In Proceeding, op. cit., p. 237.

269   Burger opinion: Office of Communications of the United Church of Christ v. FCC, 16 P&F Radio Reg. 2nd, p. 2095.

Stanton quote: Hearings, p. 101.

Salant quote: Hearings, p. 46.

Seldes quotes: From Seldes, The Public Arts; first quote, p. 223; second quote, p. 222.

272   Holmes quote: Hanson v. Globe Newspaper Co., 159 Mass. 299.

274   Stern quote: Variety, November 10, 1971, p. 28.

275   Six cents: Friendly, op. cit., p. 170.

Chapter 11. Local Television and the Meaning of Diversity

277   FCC Report: In TV Network Program Procurement, p. 205.

282   Brown quote: Les Brown, Television, Harcourt Brace Jovanovich, New York, 1971, p. 178.

299   Westinghouse quote: Group W Policy Manual, p. 50.

306   Friendly quote: Friendly, Due to Circumstances Beyond Our Control, p. 193.

307   Du Pont–Columbia quote: Survey of Broadcast Journalism, 1969–70, p. 11.

Chapter 12. Public Television and the Meaning of Diversity

313   Hoggart quote: Richard Hoggart, The Uses of Literacy, Pelican ed., London, 1958, p. 338.

Corporation for Public Broadcasting Annual Report 1970, p. 14.

315   Ford grants figures: Ford Foundation Annual Report 1960, p. 128.

316   Hart quote: Educational Broadcasting, December 1971, p. 21.

321   Footnote: Public Television: A Program for Action, the Report of the Carnegie Commission on Educational Television, Bantam ed., New York, 1967, p. 55.

323   Carnegie Report: Ibid., p. 17.

324   Johnson quote: Johnson, How to Talk Back to Your Television Set, Bantam ed., New York, 1970, p. 162.

800,000 viewers: Seldes, The Public Arts, p. 271.

Page
325 Leonard quote: *Look* magazine, November 17, 1970.
Opportunity Line figures: H. S. Dordick, L. G. Chester, S. I. Firstman and R. Birtz, *Telecommunications in Urban Development*, RAND Corporation 6069-RC, Santa Monica, 1969, pp. 20–21.
331 Karayn quote: *New York Times*, August 25, 1971, p. 75.
Carnegie Report: *op. cit.*, p. 59.
335 Jonathan Miller quote: Jonathan Miller, *Marshall McLuhan*, Viking Press, New York, 1971, p. 124.
344 *L'Express* quote: *"La Bande à Schaeffer,"* February 22, 1971, p. 82.

Chapter 13. Cable Television: Hick, Hook, Hoke, Hooey Us

346 Singer quote: *Issues for Study in Cable Communications: An Occasional Paper from the Alfred P. Sloan Foundation*, New York, 1970, pp. 7–8.
Switzer quote: *Mass Media*, Vol. I, pp. 214–215.
348 Loevinger quote: Lee Loevinger, "Program Regulation," *Federal Bar Journal*, Vol. XX, 1966, p. 11.
Television of abundance: *On the Cable: The Television of Abundance*, Report of the Sloan Commission, McGraw-Hill, New York, 1971, p. 43.
350 Gould quote: Jack Gould, "Triumph in Pay TV," *New York Times*, March 19, 1961, Sec. II, p. 13.
354 Stewart quote: *Fortnightly Corp.* v. *United Artists Television, Inc.*, 392 US 398–9.
369 FCC letter: FCC 71–787 63303, August 5, 1971, pp. 1–2.
370 Number of channels: *Ibid.*, p. 27.
Five minutes in duration: *Ibid.*, p. 29.
80% of the time: *Ibid.*, p. 30.
Two-way communication: *Ibid.*, p. 31.
Flick across the dial: *Ibid.*, p. 32.
372 Unable to find any viewing: "The Effect of CATV on Television Viewing," Television Bureau of Canada, Toronto, p. 14.
Mass Media, Vol. 1, p. 218.
373 Feldman quote: Nathaniel E. Feldman, *Cable Television: Opportunities and Problems in Local Program Origination*, RAND Corporation, Santa Monica, R-570-FF, p. 19.
374 Union contract: *Senza Sordino*, ICSOM, December 1971, p. 1.
375 Leland Johnson quote: Leland L. Johnson, *Cable Television and the Question of Protecting Local Broadcasting*, RAND Corporation, Santa Monica, 1970, 5-595-MF, p. 21.
376 Stewart quote: *Fortnightly*, pp. 401–402.
Friendly quote: *The Federal Administrative Agencies*, p. 171.
Sloan quote: *On the Cable*, p. 57.
377 Johnson quote: Johnson, *op. cit.*, p. 10.

Chapter 14. If There Is No Answer, What Is the Message?

381 Weaver quote: Sylvester L. Weaver, *Television and the Intellectual*, NBC, New York, 1955, pp. 5–6.
Hoggart quote: Richard Hoggart, *The Uses of Literacy*, Penguin ed., p. 340.

Page
382   Loevinger quote: Lee Loevinger, "The Limits of Technology in Broadcasting," in *Journal of Broadcasting*, University of Southern California, Los Angeles, Vol. 10, No. 4, pp. 295–296.
Skornia quote: Harry J. Skornia, *Television and Society*, McGraw-Hill paperbacks, 1965, New York, p. 145.
Landis quote: James M. Landis, *Report on Regulatory Agencies to the President-Elect*, printed for the Senate Committee on the Judiciary, 86th Congress, 1960, p. 54.
383   Friendly quote: Henry J. Friendly, *The Federal Regulatory Agencies*, p. 55.
384   Categories of program: *FCC Broadcast Primer*, FCC, Washington, p. 5.
Program proposals: Tribune Co., 9 Radio Reg. 719 (FCC 1954).
Power of censorship: Federal Communications Act, Section 326.
385   Loevinger quote: Loevinger, *Statement to the Subcommittee on Communications of the United States Senate Committee on Commerce Regarding S. 2004*, p. 16.
386   Johnson quote: WHDH, Inc. 16 FCC 2nd, p. 27.
Like drawing from a hat: Flower City Television Corp., 9 FCC 2nd, 249.
Policy Statement: FCC Public Notice FCC 70–62 40869, January 15, 1970, p. 2.
Johnson dissent: *Ibid.*, dissenting opinion, p. 1, p. 8.
390   Murray quote: Albert Murray, *South to a Very Old Place*, McGraw-Hill, New York, 1971, p. 65.
392   Benny quote: Tom Sloan, *Television Light Entertainment*, BBC Lunch-Time Lectures, 8th Series, p. 7.
395   Long quote: In *Broadcasting*, October 18, 1971, p. 30.
Boorstin quote: Daniel Boorstin, *The Image*, Atheneum, New York, 1962, p. 7.
396   Letter to BBC: Kenneth Lamb, *The BBC and Its Public*, BBC Lunch-Time Lectures, 8th Series, BBC, London, 1969, p. 4.
Leonard comment: John Leonard, "A Chuckle Instead of a Scream," *New York Times News of the Week in Review*, October 10, 1971, p. 7.
Wheldon quote: *In The Public Interest: A Six-Part Explanation of BBC Policy*, BBC, London, January 1971, p. 17.
398   Johnson quote: *New York Times*, October 19, 1971, p. 50, col. 4.
401   Hoggart quote: *Op. cit.*, p. 344.
404   Eliot quote: T. S. Eliot, "Burnt Norton," first published in *Collected Poems, 1909–1935*, Harcourt, Brace & Co., New York, 1936, p. 217; later the first of the *Four Quartets*.
Crosby quote: in Shulman and Youman, *How Sweet It Was*, p. 8.

Index

ABC (American Broadcasting Company), 6, 25–26, 94, 111, 155; as advertising medium, 120; children's programs, 162; conventions, political, 229–230, 233–234, 236; documentaries, 253; and local stations, 280, 281, 286, 295, 297, 304; news programs, 196, 197, 200–202, 205–207, 209; political broadcasts, 243; program scheduling, 98, 99, 104–105; ratings, 46, 49, 280; sports broadcasting, 165–178, 180, 182–183, 185–187, 189; talk shows, 114; time sales, 54, 56–58, 62, 64

ABC (English theatre chain), 289

Academy Awards, 87

Access, public, to television, 360, 370, 388–389

Ace, Goodman, 77, 78

Action for Children's Television, 161–163

Adam-12, 51

Adler, Peter Herman, 329

Advertising: in cable television, 366; CBS as largest medium, 120; of cigarettes, discontinued, 60–62, 392; cultural effects of, 391–392; data services, 67–68; Fairness Doctrine in, 242; on local stations, 4, 64–70; in magazines, 70–71; networks and, 13–16; 100,000 homes as market size, 4; political, 246–249; products as sponsors, 14; program choice influenced by, 101; program control and, 27–28; radio, 10–11, 14; ratings and, 50–52; restriction of, 397; *see also* Commercials; Sponsorship; Time, selling

Advertising Research Foundation, 38

Advocates, The, 325

AFTRA, (American Federation of Television and Radio Artists), 198, 199

Agnew, Spiro, 221, 393; on control of television news, 215–216; Des Moines speech, 215–218

Air Force Journal, 255

Air Power, 255

Alcoa Corporation, 28

Alcoa Playhouse, 24

Alderney, 288, 290

All in the Family, 35, 55, 80, 88, 94, 329

Allen, Fred, 14, 25, 35, 118

Allen, Gracie, 77

Allen, Steve, 78, 110, 114

Alsop, Joseph, 252

Altoona, Pa., 357

American Association of Advertising Agencies, 45

American Association of University Women, 315

American Basketball Association, 183

American Broadcasting Company, *see* ABC

American Civil Liberties Union, 252

American Conservatory Theatre, 323

American Football Conference, 64

American Football League, 187

American Research Bureau, 36, 37, 49, 324–325; *ARB Network Television Program Analysis,* 66; in time sales, 66

American Telephone and Telegraph Company, *see* AT&T

Amos 'n' Andy, 31

Ampex, 206

Anchor men in news programs, 195–196

Andersonville Trial, The, 330

413